T0136873

Studies in Fuzziness and Soft Computing

Volume 355

About this Series

The series "Studies in Fuzziness and Soft Computing" contains publications on various topics in the area of soft computing, which include fuzzy sets, rough sets, neural networks, evolutionary computation, probabilistic and evidential reasoning, multi-valued logic, and related fields. The publications within "Studies in Fuzziness and Soft Computing" are primarily monographs and edited volumes. They cover significant recent developments in the field, both of a foundational and applicable character. An important feature of the series is its short publication time and world-wide distribution. This permits a rapid and broad dissemination of research results.

More information about this series at http://www.springer.com/series/2941

Andrés Jiménez-Losada

Models for Cooperative Games with Fuzzy Relations among the Agents

Fuzzy Communication, Proximity Relation and Fuzzy Permission

Andrés Jiménez-Losada
Departamento de Matemática Aplicada II
Universidad de Sevilla
Seville
Spain

ISSN 1434-9922 ISSN 1860-0808 (electronic)
Studies in Fuzziness and Soft Computing
ISBN 978-3-319-85919-4 ISBN 978-3-319-56472-2 (eBook)
DOI 10.1007/978-3-319-56472-2

Softcover reprint of the hardcover 1st edition 2017

Printed on acid-free paper

This Springer imprint is published by Springer Nature
The registered company is Springer International Publishing AG
The registered company address is: Gewerbestrasse 11, 6330 Cham, Switzerland

A book, as a journey, begins with concern
and it ends with melancholy

José Vasconcelos

To my wife and my daughters, thanks
for being capable of putting up with me.

Foreword

Cooperative game theory can be seen as an axiomatic approach to analyze situations in which the players can make coalitions with worth. In a cooperative game with transferable utility, the worth of a coalition is the maximal profit or the minimal cost for the players in their own coalition. A solution concept is a distribution of the profit (cost) of the great coalition among the players. Stable sets, core, kernel, and several types of values are solution concepts, but I believe that the Shapley value is adapted to very different situations and it enables a great variety of extensions through their combinatorial properties.

This book analyzes values for cooperative games with fuzzy information among the players. Each model uses a different relation: a particular coalition, a coalition structure, a communication structure, a priori union system, a permission structure, or a coercive structure, among other combinatorial structures. It is very remarkable that the work of the author on the value of Shapley has extended the classic models of Aumann and Dreze, Myerson, Owen, Gilles, and van den Brink to the situations of fuzzy information between the players.

Mathematics is learned by doing, exploring new models with new tools, and applying the knowledge obtained. Still remains much work to make, especially in the field of applications of the models to propose new methods of conflict resolution in economic, social, cultural, and political situations that can be analyzed from the new perspective of fuzzy approximation.

When I started my research on the Shapley value in games constrained or defined on combinatorial structures, I could not imagine the level of development achieved by this field in which the Prof. Jiménez-Losada has obtained new and interesting characterizations of the Shapley value in the context of fuzzy information.

I hope that this book will be found useful as a text to start working in a new and exciting field of research defined in its title "Models for Cooperative Games with Fuzzy Relations among the Agents."

Seville, Spain
January 2017

J. Mario Bilbao

Preface

Game theory is a mathematical discipline which studies situations of competition and cooperation among several agents or players. This is a consistent definition with its large number of applications. These applications come from economy, sociology, engineering, policy, computation, psychology, or biology. I focus this book in the cooperative branch. This branch analyzes only the outcomes that result in situations of cooperation, those cases where players are grouped in coalitions. In the classical model, there are several basic suppositions. Players are symmetric, and in each coalition, any player is as important as the rest and they cooperate at the same level. Coalitions are all feasible, in the sense that the worth of each coalition does not depend on any particular relation among the players. The use of techniques from determined mathematical structures and fuzzy sets has allowed us to describe better real problems by new models in cooperative games. Numerous studies have been introduced describing certain additional information about the players or the feasibility of the coalitions which modify the cooperation behavior. Fuzzy coalitions defined different levels of participation in a continuous model of cooperation. My work for several years closely with my colleagues in the research group have been focussed in the analysis of games with restricted cooperation, first by certain classical mathematical structures and currently fuzzy cooperation structures.

This book has a double vocation, one to be a treatise and one to be a practical manual. It is treatise on games with a bilateral fuzzy relation among the players. The idea is presenting the difference in the models, and then, I focus on one particular classical solution for games, the Shapley value. It is self-contained, in the sense that all the mean contents about the topic are included. I present several fuzzy models that we have studied particularly in their respective papers but, in the same context, given certain degree of generality. Each model is analyzed first in crisp case and later in the fuzzy option. All the models contain certain nuances which allow the reader to see the results as newness about the subject. But this book is also a good manual for students and researchers, in the sense that all the proofs are included showing different ways of analysis in these situations. So, I opt in each model for an axiomatization in different way. Usually, I present axioms and properties in the most feasible general way. But also I study the last one in particular cases because the refinement of the model permits to use specific properties. Any person, only with a common knowledge base about maths, can follow the book. There are also a lot of examples which show the application of the different proposed formulas and concepts, in numbers: 104 definitions, 39 theorems, and 96 propositions with their proofs, 110 examples, 37 tables, and 53 figures. I hope that the reader will find useful this work.

Seville, Spain
January 2017

Andrés Jiménez-Losada

Acknowledgements

I would like to be thankful to all those people that made possible this book with their contributions and their support: my colleges in the research group, my Ph.D. students, my parents, my wife, and my daughters. Also I thank the support of the following organizations: Department of Applied Mathematics II and Institute of Mathematics Research, both at the University of Seville.

Thank you very much.

Contents

Symbols

N	Players
i	Player
n	Number of players
2^N	Coalitions
S	Coalition
$\mathcal{G}^N$	Games
$\wedge, \vee$	Minimum, maximum
u_T	Unanimity game
Δ_T^v	Dividend
v^{svg}	Saving game
v^{dual}	Dual game
$\mathcal{G}_{sa}^N$	Superadditive games
$\mathcal{G}_c^N$	Convex games
$\mathcal{G}_m^N$	Monotone games
$\mathcal{G}_s^N$	Simple games
$v\|_T$	Subgame
f	Value
Θ^N	Permutations
$<_T$	T-ordering
$<_i$	i-ordering
$m^\theta(v)$	Marginal vector
$\phi(v)$	Shapley value
$\int_c$	Choquet integral
$[0,1]^N$	Fuzzy coalitions
τ	Fuzzy coalition
e^S	Canonical vector
$[\tau]_t$	t-cut
$im(\tau)$	Images of fuzzy set
$supp(\tau)$	Support of fuzzy set

v^{cr}	Crisp version of game
$\phi^{cr}(v)$	Crisp Shapley value
$\phi^{d}(v)$	Diagonal value
pl	Partition function
v^{pl}	Fuzziness of game
ml	Multilinear extension
pr	Proportional extension
ch	Choquet extension
$R(N)$	Binary relations
r	Binary relation
N^r, N^ρ	Domain
$L(r), L(\rho)$	Links
$N/r, N/\rho$	Components
$[0,1]^{N\times N}$	Fuzzy binary relations
ρ	Fuzzy binary relation
τ^ρ	Fuzzy domain
r^S, ρ^τ	Relations of coalitions
v_T	Restricted game
$\phi(v,T)$	Extended Shapley value
$v^{T\,dual}$	*T*-dual game
v_τ^{pl}	*pl*-restricted game
τ^{pl}	*pl*-game
$\phi^{pl}(v,\tau)$	*pl*-extended Shapley value
P_0^N	Coalition structures
$\mathscr{B}$	Coalition structure
$v_{\mathscr{B}}$	Coalitional game
$\mu(v,\mathscr{B})$	Coalitional value
$v^{\mathscr{B}\,dual}$	$\mathscr{B}$-dual game
G^N	Communication structures
$g(v,r)$	Myerson graph worth
v/r	Vertex game
r_{-ij}	Deleting a link in a graph
r_{-i}	Isolating a vertex in a graph
$\mu(v,r)$	Myerson value
v^{rdual}	*r*-dual game
FG^N	Fuzzy communication structure
ρ^t_{-ij}	Decreasing the level of a link
$\gamma(v,\rho)$	Fuzzy graph worth
pg	Proportional by graphs extension
cg	Choquet by graphs extension
$\gamma^{pl}(v,\rho)$	*pl*-worth
CV	Choquet by vertices extension
$(v/\rho)^{pl}$	*pl*-vertex game

$(v/\rho)^{\gamma}$	γ-vertex game
pl^{*}	Induced partition
pb	Probabilistic extension
P^{N}	A priori union systems
$w^{v,\mathscr{B}}$	Quotient game
$v_{p}^{\mathscr{B}}$	p-union game
$\omega(v,\mathscr{B})$	Owen value
C^{N}	Cooperation structures
$\omega(v,r)$	Myerson-Owen value
$null^{v}(r)$	Null components
FC^{N}	Proximity relations
pp	Prox-proportional extension
$[N/\rho]_{pl}$	pl-groups
ρ_{pl}^{K}	K-scaling
$\omega^{pl}(v,\rho)$	pl-Myerson-Owen value, pl-Owen value
FP^{N}	Similarity relations
A^{N}	Permission structures
σ^{r}	Sovereign part
r°	Quasi-reflexive interior
$\hat{r}$	Transitive closure
v^{r}	Permission game
$\bar{\sigma}^{r}$	Coercive part
$\bar{v}^{r}$	Coercive game
$\delta(v,r)$	Local permission value
$\hat{\delta}(v,r)$	Permission value
$\bar{\delta}(v,r)$	Coercive value
σ^{ρ}	Fuzzy sovereign part
ρ°	Weakly reflexive interior
v_{pl}^{ρ}	pl-permission game
$\bar{\sigma}^{\rho}$	Fuzzy coercive part
$\bar{v}_{pl}^{\rho}$	pl-coercive game
$\delta^{pl}(v,r)$	Local pl-permission value
$\hat{\rho}$	Fuzzy transitive closure
$\hat{\delta}^{pl}(v,r)$	pl-permission value
$\bar{\delta}^{pl}(v,r)$	pl-coercive value
η^{ρ}	Probabilistic part

Chapter 1
Cooperative Games

1.1 Introduction

This chapter corresponds to an introduction about cooperative games with transferable utility, namely the framework within which the book is developed. It does not seek to be fully exhaustive because this book is not a handbook about cooperative games but it sets out to be consistent and comprehensive in order to be useful throughout the rest of the chapters.

The study of the theory of games started in von Neumann and Morgenstern [29] in 1944 with the publication of *Theory of Games and Economic Behavior,* although there is some earlier research in von Neumann [28] in 1928. Game theory is a mathematical discipline which studies situations of competition and cooperation among several agents (players). This is a consistent definition with the large number of applications. These applications come from economy, sociology, engineering, policy, computation, psychology or biology. Game theory is divided into two branches, called the non-cooperative and cooperative branches. They differ in how they formalize interdependence among players. In the non-cooperative theory, a game is a detailed model of all the moves available to the players. By contrast, the cooperative theory abstracts away from this level of detail, and describes only the outcomes that result when players are grouped together (coalitions). This research is focused on cooperative models. In von Neumann and Morgenstern [29] the authors described coalitional games in characteristic function form, also known as transferable utility games (games for short). The characteristic function of a game is a real-valued function on the family of coalitions. The real number assigned to each coalition is interpreted as the utility of the cooperation among this group of players. In these cases the worth of a coalition can be allocated among its players in any way. The adjective transferable refers to the assumption that a player can transfer any part of his utility to another player.

Solving a game means determining which coalition or coalitions are formed and obtaining a vector at the end of the game with the corresponding individual payoffs for

A. Jiménez-Losada, *Models for Cooperative Games with Fuzzy Relations among the Agents*, Studies in Fuzziness and Soft Computing 355,
DOI 10.1007/978-3-319-56472-2_1

the cooperation of the players (payoff vector). The classic model of game considers that the grand coalition (the coalition of all the players) will be formed and assumes that there are no restrictions in cooperation, therefore every subset of players can form a different coalition. A value for games is a function assigning a payoff vector for each game. The most known value was introduced by Shapley [24] in 1953. The Shapley value determines the payoff vector for a game by an explicit formula using the worths of all the coalitions in the game, and it is sustained in a set of reasonable properties called axioms which identify uniquely the value. A large list of values have been defined since then: nucleolus [23], Banzhaf value [2, 18, 21], compromise value [27], etc. This book is focused on the Shapley value.

This chapter introduces cooperative games, the Shapley value and those results about them that we will use in the next chapters. For further learning the reader can use books about games in general as Driessen [6], Owen [19], Bilbao [3], Peters [20] or González-Díaz [11]. More particularly, the book "The Shapley value", ed. Roth [22] is a broad vision about this value.

1.2 Games

Cooperative games analyze situations among a finite set of agents where they cooperate to get a determined benefit (profits, debts, costs,...). In this book we consider games in coalitional form with transferable utility. In this kind of games the benefit obtained by the cooperation can be allocated among the agents in any way, so the individual utilities of them have a transfer system permitting that an agent can lose utility to other one in order to keep the cooperation. Hence the benefit is a payoff number for the set of agents. The coalitional form implies that it is known a mapping determining the benefit for any subset of agents. Given a finite set N we denote as 2^N its set of power, namely the family of subsets of N, and with the tiny letter n the cardinality of N.

Definition 1.1 Let N be a finite set of elements named players. A cooperative game with transferable utility over N is a mapping $v : 2^N \to \mathbb{R}$ which assigns a worth to each subset of players (coalition), satisfying that $v(\emptyset) = 0$. The family of cooperative games with transferable utility over N is denoted as $\mathcal{G}^N$.

From now on we use game instead cooperative game with transferable utility. To define a game we need then two elements: the set of players N and the mapping v (named usually characteristic function). In the next examples we see how the games can modeled different situations.

Example 1.1 We consider a production economy in which there are several peasants and one landowner. This model has been studied in Shapley and Shubik [26] and Chetty et al. [4]. We present it here from [6]. The peasants contribute only with their work and they are of the same type. The landowner hires the peasants to cultivate his land. If t peasants are hired by the landowner, then the monetary value of the harvest obtained is denoted by $h(t) \in \mathbb{R}$. The mapping $h : \{0, 1, \ldots, m\} \to \mathbb{R}$ is named production function where m is the total number of peasants. In what follows, it is required that h satisfies these two conditions:

1. the landowner by himself does not produce anything, i.e., $h(0) = 0$.
2. mapping h is nondecreasing, i.e., $h(t+1) \geq h(t)$ for each $h \in \{0, 1, \ldots, m-1\}$.

Both conditions imply that h is a nonnegative mapping. We consider the landowner as player 1 and the peasants as players $2, \ldots, m+1$. Then this situation can be modeled as a game with $m+1$ players with characteristic function v given by

$$v(S) = \begin{cases} 0, & \textit{if } 1 \notin S \\ h(|S|-1), & \textit{if } 1 \in S. \end{cases}$$

The value of any coalition that contains only peasants is 0 because they do not have any land. Even more, the worth of each coalition that contains the landowner is equal to the monetary value of the harvest that is obtained by the peasants that are in that coalition. Obviously $v(\emptyset) = 0$.

Example 1.2 Control games were proposed by Feltkamp [9]. A control situation is defined by a set of agents N who decide about a good. They, as group, have the control of the good but they want to determine the power of each one. The control game is $v \in \mathcal{G}^N$ with $v(N) = 1$ and $v(S) \in \{0, 1\}$. If $v(S) = 1$ then S is able to control the good.

Example 1.3 Littlechild and Owen [14] considered a game to solve a cost allocation problem of setting landing strips for different kinds of aircraft in an airport. Suppose N the set of planes at this airport and consider m different sizes of planes ordered from smallest to largest. For each $j = 1, \ldots, m$ we denote as N_j the set of aircrafts with size j and $c_j \geq 0$ the cost of a appropriate strip for this size. Of course $c_j < c_{j+1}$ for every $j = 1, \ldots, m-1$, and the strip constructed for a size $j = 2, \ldots, m$ is also appropriated for $j-1$. The authors used the game over N defined for all non-empty coalition S as

$$v(S) = \bigvee \{c_j : j \in \{1, \ldots, m\}, N_j \cap S \neq \emptyset\},$$

to determine what part of cost of the biggest strip is attributable to each plane.

Example 1.4 A bankruptcy problem arises in how to share the capital of a firm among its creditors when quantity of the debts is greater than the capital. O'Neill [16] proposed a game to solve this situation. The players are the set of creditors N. If we denote Q the available capital and $q_i > 0$ the debt of the firm with creditor $i \in N$ then the game is

$$v(S) = \left[Q - \sum_{i \in N \setminus S} q_i \right] \vee 0, \forall S \subseteq N.$$

Of course we suppose that $\sum_{i \in N} q_i > Q$.

Example 1.5 Games have been used in decision problems since [29]. Particularly they proposed to introduce a game in order to analyze the power of the members in a decision committee. The players in N are the members of the committee. A decision system is defined to determined when a proposal is approved, for instance a fixed quota $q > n/2$ of the members supporting it. Observe in this situation that the most important goal is to get a coalition which be able to approve the proposal, regardless of other considerations about it. So, the game is defined for each coalition S as $v(S) = 1$ if $|S| \geq q$ and $v(S) = 0$ otherwise.

Example 1.6 Owen [18] defined linear production games. Suppose a set of agents N in m production processes with l resources. It is known: for each agent i an initial endowment $b^i \in \mathbb{R}^l_+$, the profits for each unit of each product is the vector $c \in \mathbb{R}^m_+$ and the quantity of resource j to get a unit of product k as a positive matrix A. The *linear production game* is given by $v \in \mathcal{G}^N$ with

$$v(S) = \bigvee \left\{ c \cdot x : x \in \mathbb{R}^m_+,\ Ax \leq \sum_{i \in S} b^i \right\}$$

for each non-empty $S \subseteq N$.

Example 1.7 Baeyens et al. [1] used games to propose a solution to allocate the benefits of the cooperation among a group of wind power producers. By an stochastic process and using statistical tools the authors determine how to calculate the maximal expected profit for the aggregation of the power output of the producers as single entity into a forward energy market. In this case N is the set of producers, and for each coalition S they determine the worth $v(S)$ as the expected benefit of the coalition in the aggregation of their systems. This game is used to allocate the benefit among the producers.

From now on we usually consider fixed the set of players N, and we study the family of games $\mathcal{G}^N$.

Following Example 1.5, the decision system is not always based in the size of the group supporting the proposal but also in the nature of the group. Unanimity games are examples of these ones. In a unanimity game over N there exists a particular group of members T whose support is necessary and sufficient to approve the proposal. We will see later that these games play a very important role into the family of games $\mathcal{G}^N$.

Definition 1.2 Let T be a non-empty coalition in N. The unanimity game $u_T \in \mathcal{G}^N$ is defined for all $S \subseteq N$ as

$$u_T(S) = \begin{cases} 1, \text{ if } T \subseteq S \\ 0, \text{ otherwise.} \end{cases}$$

There are two internal operations defined over $\mathcal{G}^N$, product by an scalar and sum.

- If $c \in \mathbb{R}$ and $v \in \mathcal{G}^N$ then $cv \in \mathcal{G}^N$ with $cv(S) = c(v(S))$ for all $S \subseteq N$.
- If $v, w \in \mathcal{G}^N$ then $v + w \in \mathcal{G}^N$ with $(v + w)(S) = v(S) + w(S)$ for all $S \subseteq N$.

The family of games over N with these operations has structure of vectorial space, and the set of unanimity games $\{u_T : T \subseteq N, T \neq \emptyset\}$ is a basis of the space. Furthermore the dimension of this vectorial space is 2^{n-1}.

Proposition 1.1 *If $v \in \mathcal{G}^N$ then there exist a unique set of numbers $\{\Delta_T^v : T \subseteq N, T \neq \emptyset\}$ satisfying*

$$v = \sum_{\{T \subseteq N, T \neq \emptyset\}} \Delta_T^v u_T \quad \text{with} \quad \Delta_T^v = \sum_{S \subseteq T} (-1)^{|T|-|S|} v(S).$$

These numbers are called the (Harsanyi) dividends of the game.

Proof Consider $S \subseteq N$,

$$\begin{aligned} \sum_{\{T \subseteq N, T \neq \emptyset\}} \Delta_T^v u_T(S) &= \sum_{\{T \subseteq S, T \neq \emptyset\}} \sum_{R \subseteq T} (-1)^{|T|-|R|} v(R) \\ &= \sum_{R \subseteq S} \left[\sum_{\{R \subseteq T \subseteq S\}} (-1)^{|T|-|R|} \right] v(R) = v(S), \end{aligned}$$

because

$$\sum_{\{R \subseteq T \subseteq S, T \neq \emptyset\}} (-1)^{|T|-|R|} = \sum_{t=0}^{s-r} \binom{s-r}{t} (-1)^t = (1-1)^{s-r} = \begin{cases} 1, \text{ if } s = r \\ 0, \text{ otherwise.} \end{cases}$$

We prove now that the unanimity games are linearly independent. Let 0 be the null game $0(S) = 0$ for all coalition S and the linear combination

$$\sum_{\{T\subseteq N, T\neq\emptyset\}} a_T u_T = 0.$$

Suppose there exists $T' \subseteq N$, $T' \neq \emptyset$ with $a_{T'} \neq 0$. We can take T' minimal by inclusion satisfying the above condition. Using Definition 1.2 we have

$$\sum_{\{T\subseteq N, T\neq\emptyset\}} a_T u_T(T') = a_{T'} = 0.$$

□

The Harsanyi dividends [12] of a game $v \in \mathcal{G}^N$ can be also obtained by a recurrence formula, using that for each coalition S,

$$v(S) = \sum_{\{T\subseteq S, T\neq\emptyset\}} \Delta_T^v. \tag{1.1}$$

So, we have for each player $i \in N$ that the dividend of his individual coalition is $\Delta_{\{i\}}^v = v(\{i\})$ and for any other coalition S

$$\Delta_S^v = v(S) - \sum_{\{T\subsetneq S, T\neq\emptyset\}} \Delta_T^v. \tag{1.2}$$

This formula allows us to interpret the dividend of a coalition S as the marginal profit obtained by it, the increase of benefit due to the formation of this coalition.

Example 1.8 Suppose $N = \{1, 2, 3\}$ and $v \in \mathcal{G}^N$ with $v(S) = |S| - 1$ for every non-emptyset coalition. Next table includes the dividends of the coalitions calculated using (1.2). So, $v = u_{\{1,2\}} + u_{\{1,3\}} + u_{\{2,3\}} - u_N$ (Table 1.1).

Dividends are the coefficients of the games with regard to the unanimity games and then they work well with the operations of the vectorial space.

Table 1.1 Dividends of the game v

S	$\{1\}$	$\{2\}$	$\{3\}$	$\{1,2\}$	$\{1,3\}$	$\{2,3\}$	N
Δ_S^v	0	0	0	1	1	1	−1

Proposition 1.2 *Let $v, w \in \mathcal{G}^N$ and $a, b \in \mathbb{R}$. For each non-empty coalition $S \subseteq N$ the dividend satisfies*

$$\Delta_S^{av+bw} = a\Delta_S^v + b\Delta_S^w.$$

Proof Proposition 1.1 implies that dividends are unique. Hence, as

$$\begin{aligned} av + bw &= a \sum_{\{S \subseteq N : S \neq \emptyset\}} \Delta_S^v u_S + b \sum_{\{S \subseteq N : S \neq \emptyset\}} \Delta_S^w u_S \\ &= \sum_{\{S \subseteq N : S \neq \emptyset\}} [a\Delta_S^v u_S + b\Delta_S^w] u_S. \end{aligned}$$

□

Now we present several properties which are interesting for a game. After the definition we will explain each of them.

Definition 1.3 Let $v \in \mathcal{G}^N$.

1. Game v is additive if $v(S) + v(T) = v(S \cup T)$ for all $S, T \subseteq N$ with $S \cap T = \emptyset$.
2. Game v is superadditive (subadditive) if $v(S) + v(T) \leq (\geq) v(S \cup T)$ for all $S, T \subseteq N$ with $S \cap T = \emptyset$.
3. Game v is convex (concave) if $v(S) + v(T) \leq (\geq) v(S \cup T) + v(S \cap T)$ for all coalitions S, T.
4. Game v is 0-normalized if $v(\{i\}) = 0$ for each player $i \in N$.
5. Game v is a $\{0, 1\}$-game if $v(S) = 0$ or 1 for all coalition S.
6. Game v is anonymous if $v(S) = v(T)$ when $|S| = |T|$.
7. Game v is monotone if $v(S) \leq v(T)$ when $S \subseteq T$.
8. Game v is simple if v is a monotone $\{0, 1\}$-game.

An additive game v is identified with a vector denoted with the same letter $v \in \mathbb{R}^N$ verifying for any non-empty coalition S,

$$v(S) = \sum_{i \in S} v_i.$$

Therefore, additive games represent situations of ineffectual cooperation.

Suppose that the worths in a game v are considered as profits. Superadditivity means that players have the incentive to cooperate getting bigger coalitions. From the individual point of view we get, using several times superadditivity, that players are interested in forming coalitions,

$$v(S) \geq \sum_{i \in S} v(\{i\}).$$

But also for each coalition $S \subset N$, there is a possibility to bargain with the rest of the players being benefited,

$$v(N) - v(N \setminus S) \geq v(S).$$

A *profit allocation problem* is a game $v \in \mathcal{G}^N$ which is superadditive and the worth of each coalition is interpreted as the profit that this coalition can guarantee whatever strategy of the other players. There are games which are not exactly profit games (see Example 1.4) where this property has the same interpretation. But superadditivity condition is not a stimulus to cooperate for the players in other situations, rather the opposite. A *cost allocation problem* is a subadditive game $v \in \mathcal{G}^N$ studying situations where the agents cooperate in order to allocate the costs of a common project, so the worth of a coalition is interpreted as the cost of a similar project for the coalition. The subadditivity implies, in a similar way than the superadditivity, an interest of the agents in cooperating. If $v \in \mathcal{G}^N$ is superadditive (subadditive) then $-v$ is subadditive (superadditive).

Example 1.9 Game v in Example 1.1 is a profit allocation problem. We test the superadditivity. If S, T are disjoint coalitions in N, $S \cap T = \emptyset$, then the landowner is at most in one of them. Suppose $1 \in S$ (the same with T, otherwise is trivial). Hence $v(S) = h(|S| - 1)$ and $v(T) = 0$. As h is nondecreasing and nonnegative we get $h(|S| - 1) \leq h(|S| + |T| - 1)$.

The superadditivity (subadditivity) condition can be extended to all pair of coalitions by convexity (concavity). Convex games are perhaps the games with the best properties. Obviously convex games are superadditive too, but the opposite is not always true. Baeyens et al. [1] showed that the game of the wind energy producers commented in Example 1.7 is not convex but it is superadditive. Example 1.1 is convex if and only if h is a convex function: $h(t + 1) - f(t) \geq h(r + 1) - h(r)$ when $t > r$, but it is always superadditive.

Example 1.10 The airport game, Example 1.4, is concave. Let S, T coalitions. We take players $j_S \in S$, $j_T \in T$ satisfying $c_{j_S} = v(S)$ and $c_{j_T} = v(T)$. Given $j \in S \cap T$ and $j' \in S \cup T$ we get

$$c_{j_S} + c_{j_T} = c_{j_S} \wedge c_{j_T} + c_{j_S} \vee c_{j_T} \geq c_j + c_{j'}.$$

Hence this game is also a cost allocation problem.

There exist several equivalent conditions for a game to be convex. We show one of them that we will use in the book.[1]

Proposition 1.3 *Game v is convex if and only if for all player $i \in N$ and coalitions $S \subset T \subseteq N \setminus \{i\}$ it holds*

$$v(S \cup \{i\}) - v(S) \leq v(T \cup \{i\}) - v(T).$$

Proof Suppose a convex game v. Let $S \subset T \subseteq N \setminus \{i\}$. By definition of convexity

$$v(S \cup \{i\}) + v(T) \leq v(T \cup \{i\}) + v(S).$$

On the other hand we consider any S, T coalitions. If $S \subseteq T$ then the proof is trivial. Now set $S \setminus T = \{i_1, \ldots, i_m\}$. The equivalent condition for player i_1 and coalitions $S \cap T$ and T implies $v((S \cap T) \cup \{i_1\}) - v(S \cap T) \leq v(T \cup \{i_1\}) - v(T)$. Hence

$$v(T) - v(S \cap T) \leq v(T \cup \{i_1\}) - v((S \cap T) \cup \{i_1\}).$$

Sequentially,

$$\begin{aligned} v(T) - v(S \cap T) &\leq v(T \cup \{i_1\}) - v((S \cap T) \cup \{i_1\}) \\ &\leq v(T \cup \{i_1, i_2\}) - v((S \cap T) \cup \{i_1 i_2\}) \\ &\leq \cdots \leq v(T \cup S) - v(S). \end{aligned}$$

□

Next example uses the above proposition as in Driessen [6] to show that the bankruptcy game is convex.[2]

Example 1.11 We consider the bankruptcy game (Example 1.4). Let $i \in N$ and $S \subset T \subseteq N \setminus \{i\}$. Foremost observe that $Q - \sum_{j \in N \setminus S} q_j + q_i \leq Q - \sum_{j \in N \setminus T} q_j + q_i$. We have

$$v(S \cup \{i\}) + v(T) = \left(\left[Q - \sum_{j \in N \setminus S} q_j + q_i \right] \vee 0 \right) + \left(\left[Q - \sum_{j \in N \setminus T} q_j \right] \vee 0 \right)$$

$$= \left[Q - \sum_{j \in N \setminus S} q_j + q_i \right] \vee \left[Q - \sum_{j \in N \setminus T} q_j \right] \vee \left[2Q - \sum_{j \in (N \setminus T) \cup (N \setminus S)} q_j + q_i \right] \vee 0$$

[1]similar condition can be stated for concave games using the other inequality.

[2]it is possible to prove it directly from the definition.

$$\leq \left[Q - \sum_{j \in N \setminus T} q_j + q_i\right] \vee \left[2Q - \sum_{j \in (N \setminus T) \cup (N \setminus S)} q_j + q_i\right] \vee 0$$

$$= \left(\left[Q - \sum_{j \in N \setminus T} q_j + q_i\right] \vee 0\right) + \left(\left[Q - \sum_{j \in N \setminus S} q_j\right] \vee 0\right) = v(T \cup \{i\}) + v(S).$$

A 0-normalized game (Definition 1.3) supposes that it is only possible to get profits by cooperation. The 0-*normalization* of a game $v \in \mathcal{G}^N$ is a new game v_0 defined for each coalition S as

$$v_0(S) = v(S) - \sum_{i \in S} v(\{i\}). \tag{1.3}$$

Each cost allocation problem is usually identified with a profit allocation problem $v^{svg} \in \mathcal{G}^N$ over N, called the *saving game*, defined for any coalition S as

$$v^{svg}(S) = \sum_{i \in S} v(\{i\}) - v(S), \tag{1.4}$$

the opposite of a 0-normalization. It is an easy exercise to test the superadditivity of the saving game for cost allocation problems, let S, T coalitions with $S \cap T = \emptyset$

$$v^{svg}(S) + v^{svg}(T) = \sum_{i \in S \cup T} v(\{i\}) - (v(S) + v(T)) \leq v^{svg}(S \cup T).$$

Moreover, in a similar way, the reader can test that if a cost allocation problem v is concave then the corresponding saving game v^{svg} is convex.

There is another classification from interpretation of the games. A *optimistic game* is a game where the worth of a coalition represents the best benefit that this coalition can obtain for the cooperation (what we could get). On the other hand, an *pessimistic game* considers that the worth of a coalition is the insured benefit that this coalition can obtain in the best situation for the players outside. There is also a way to transform one case in the other. If $v \in \mathcal{G}^N$ is an optimistic game then the *dual game* $v^{dual} \in \mathcal{G}^N$, defined for each coalition S as

$$v^{dual}(S) = v(N) - v(N \setminus S), \tag{1.5}$$

is a pessimistic game and vice versa.

Example 1.12 Bankruptcy games (Example 1.4) can be defined just another form, if S is a coalition

$$w(S) = Q \wedge \sum_{i \in S} q_i.$$

Game w is optimistic while v in Example 1.4 is pessimistic. Moreover $w = v^{dual}$.

If $v \in \mathcal{G}^N$ is interpreted as profits then v^{dual} can be seen as costs but the dual of a profit allocation problem is not always a cost allocation problem.

Example 1.13 Take the game over $N = \{1, 2, 3\}$ given by $v(N) = 1$, $v(S) = 5/6$ if $|S| = 2$ and $v(S) = 0$ if $|S| = 1$. It is easy to test that this game is superadditive. Now the dual game is $v^{dual}(N) = 1$, $v^{dual}(S) = 1$ if $|S| = 2$ and $v^{dual}(S) = 1/6$ if $|S| = 1$. We obtain

$$v^{dual}(\{1, 2\}) = 1 \geq 1/3 = v^{dual}(\{1\}) + v^{dual}(\{2\}).$$

If $v \in \mathcal{G}^N$ is convex then v^{dual} is concave and vice versa. Suppose v convex, and S, T coalitions. We have

$$\begin{aligned} v^{dual}(S \cup T) + v^{dual}(S \cap T) &= 2v(N) - [v(S \cup T) + v(S \cap T)] \\ &\leq 2v(N) - [v(S) + v(T)] = v^{dual}(S) + v^{dual}(T) \end{aligned}$$

It is possible to define also equivalence classes in $\mathcal{G}^N$.

Definition 1.4 Let $v, w \in \mathcal{G}^N$. Game w is strategically equivalent to v if there is a positive number $a > 0$ and an additive game $b \in \mathbb{R}^N$ satisfying

$$v = aw + b.$$

Suppose $w = av + b$ strategically equivalent to v. Game av is interpreted as an scale of the worths, and b is seen as an individual premium or tax. The 0-normalization of a game v is strategically equivalent to v.

Proposition 1.4 *Being strategically equivalent is an equivalence relation in* $\mathcal{G}^N$.

Proof REFLEXIVITY. Each game v is strategically equivalent to itself taking $a = 1$ and $b = 0 \in \mathbb{R}^N$.
SYMMETRY. If w is strategically equivalent to v, $w = av + b$ with $a > 0$ and $b \in \mathbb{R}^N$, then v is strategically equivalent to w with

$$v = \frac{1}{a}w + \frac{-1}{a}b.$$

TRANSITIVITY. Let v_2 be a game strategically equivalent to v_1 and v_3 strategically equivalent to v_2. Since $v_2 = a_1 v_1 + b_1$ and $v_3 = a_2 v_2 + b_2$, with $a_1, a_2 > 0$ and $b_1, b_2 \in \mathbb{R}^N$, then

$$v_3 = a_1 a_2 v_1 + (a_2 b_1 + b_2).$$

□

Simple games (Definition 1.3) represent decision situations in committees. Coalitions S satisfying $v(S) = 1$ are named *winning coalitions* and those verifying $v(S) = 0$ are *losing coalitions*. If v is a simple game then $W(v)$ represents the set of winning coalitions. Unanimity games (Definition 1.2) and Example 1.5 are simple games, but Example 1.2 is a $\{0, 1\}$-game which is not simple. Interesting examples of simple games are the voting games.

Example 1.14 A voting situation in a committee or parlament formed by a set N of individual agents or groups is represented by a pair $[q; w]$ where $w \in \mathbb{R}^N$ and $q \in R_{++}$, defining a simple situation of decision about a motion. Each weight w_i is the worth of the vote of a man or group (for instance the number of seats of the group or the number of persons represented by her). Number q is named quota and it means the worth using the above weights which must be reached to adoption the motion by the committee. A voting situation $[q; w]$ can be represented by a simple game $v \in \mathcal{G}_s^N$ where

$$v(S) = \begin{cases} 1, & \text{if } w(S) \geq q \\ 0, & \text{otherwise.} \end{cases}$$

The analysis of this particular family of games differs from usual study of games because the worths of the coalitions are not profits or costs, they only present a qualitative discrimination. Two internal operations for simple games are *meet* and *join*, if v, w are simple games then $v \vee w$, $v \wedge w$ are new simple games defined for each coalition S as

$$(v \vee w)(S) = v(S) \vee w(S)$$
$$(v \wedge w)(S) = v(S) \wedge w(S).$$

Although these operations are usually used for simple games, they can apply to any pair of games.

Proposition 1.5 *The family of simple games is a distributive lattice with operations meet and join.*

Proof Obviously $v \vee w$ and $v \wedge w$ are $\{0, 1\}$-games and N is a winning coalition for both of them. Hence we test that they are also monotone. Let $S \subseteq T$. As $v(S) \leq v(T)$ and $w(S) \leq w(T)$ then

Table 1.2 Distributive property of the lattice of simple games

v	w	u	$w \vee u$	$v \wedge w$	$v \wedge u$	$v \wedge (w \vee u)$	$(v \wedge w) \vee (v \wedge u)$
0	0	0	0	0	0	0	0
0	1	0	1	0	0	0	0
0	1	1	1	0	0	0	0
1	0	0	0	0	0	0	0
1	1	0	1	1	0	1	1
1	1	1	1	1	1	1	1

$$(v \vee w)(S) \leq (v \vee w)(T), (v \wedge w)(S) \leq (v \wedge w)(T).$$

The top of the lattice is game $\hat{1}$ with $\hat{1}(S) = 1$ for all non-empty coalition S, and the bottom is game $\hat{0}$ which verifies $\hat{0}(S) = 0$ for all coalition S. Next table shows that the lattice is distributive, namely $v \wedge (w \vee u) = (v \wedge w) \vee (v \wedge u)$. Supposed S any coalition, we have seen in the table the different cases depending on the worths of coalition S (Table 1.2). □

Unanimity games also allow describe the simple games using the lattice operations. Let v be a simple game. A coalition $S \subseteq N$ is *minimal winning* in v if $v(S) = 1$ and $v(T) = 0$ for all $T \subsetneq S$. Each simple game has at least one minimal winning coalition because it always has winning coalitions. The set of minimal coalitions v is denoted as $W_m(v)$. Minimal coalitions are *mutually exclusive* in the sense that if $S \in W_m(v)$ then there is not $T \subset S$ with $T \in W_m(v)$.

Proposition 1.6 *Let v be a simple game. $W_m(v)$ is the only non-empty family U of mutually exclusive coalitions satisfying*

$$v = \bigvee_{S \in U} u_S.$$

Proof Consider $T \subseteq N$ a winning coalition, $v(T) = 1$. There exists $S \subseteq T$ with $S \in W_m(v)$ and then $u_S(T) = 1$. Therefore

$$\bigvee_{S \in W_m(v)} u_S(T) = 1.$$

Now suppose $T \subseteq N$ a losing coalition, $v(T) = 0$. From the monotonicity of v we have $u_S(T) = 0$ for all $S \in W_m(v)$. Thus

$$\bigvee_{S \in W_m(v)} u_S(T) = 0.$$

Hence

$$v = \bigvee_{S \in W_m(v)} u_S.$$

Suppose U, U' two non-empty sets of mutually exclusive coalitions verifying

$$v = \bigvee_{T \in U} u_T = \bigvee_{S \in U'} u_S.$$

If $T \in U$ then $v(T) = 1$. There is $S \in U'$ with $u_S(T) = 1$ and then $S \subseteq T$. For this S there exists $T' \in U$ with $u_{T'}(S) = 1$ and then $T' \subseteq S$. As U is mutually exclusive we obtain $S = T$. □

Remark 1.1 Each property of games described in this section defines a particular subfamily of games which will be used throughout the book. So,

$$\begin{aligned}
\mathcal{G}_{sa}^N &\equiv \text{superadditive games,} \\
\mathcal{G}_c^N &\equiv \text{convex games,} \\
\mathcal{G}_0^N &\equiv \text{0-normalized games,} \\
\mathcal{G}_a^N &\equiv \text{anonymous games,} \\
\mathcal{G}_m^N &\equiv \text{monotone games.} \\
\mathcal{G}_s^N &\equiv \text{simple games,}
\end{aligned}$$

Remember also that additive games are identified with N-dimensional vectors, $\mathbb{R}^N$.

Games can be defined in any finite set of players, therefore it is possible to restrict the activity in the game $v \in \mathcal{G}^N$ to a particular set of its players $T \subset N$ by a new game over T reducing the function v to 2^T.

Definition 1.5 Let $v \in \mathcal{G}^N$. For each coalition T, the subgame over T is $v|_T \in \mathcal{G}^T$ with $v|_T(S) = v(S)$ for all $S \subseteq T$.

Given a game v, when it is not confuse we will use the same notation v for its subgames. So, if $v \in \mathcal{G}^N$ then $v \in \mathcal{G}^T$, with $T \in 2^N$, is the subgame of v over T.

1.3 Values for Games

The classical coalitional game theory assumes a determined behavior for the players. They always are interested in cooperation, then they look for the great coalition. Hence, we suppose in a game $v \in \mathcal{G}^N$ that players are interested in forming N. In the above section we explained that there may have been conditions in the game to motivate this behavior of the players (superadditivity or subadditivity). Taking into account the several ways to transform a subadditive game in a superadditive one, many authors consider games as superadditive games (Shapley [24] or Owen [19]). But there exist situations where this condition is not true (Example 1.2). A game $v \in \mathcal{G}^N$ is *flexible* if players can change the mapping looking for the best option independently of the above behavior. A *partition* of a coalition $S \subseteq N$ is a set of different non-empty subcoalitions $\{T_k\}_{k=1}^m$ with $T_p \cap T_q = \emptyset$ for all $p, q = 1, \ldots, m$ and $\bigcup_{k=1}^m T_k = S$. So, they can construct a partition in coalitions of the set of players instead of N and consider the subgames for these coalitions, if the great coalition is not attractive. Other option is using a new mapping getting the superadditivity instead of v. The superadditive extension of v is another game v^{sa} given by

$$v^{sa}(S) = \bigvee \left\{ \sum_{k=1}^{m} v(T_k) : \{T_k\}_{k=1}^m \text{ partition of } S \right\}. \tag{1.6}$$

Obviously, $v^{sa} \in \mathcal{G}_{sa}^N$. We consider in this book non flexible games, following the classical theory, then players in a coalition S form S and their worth is $v(S)$. They cannot change this fact.

Given a game $v \in \mathcal{G}^N$, a *payoff vector* for v is a vector $x \in \mathbb{R}^N$ whose component x_i for each $i \in N$ is interpreted as the individual utility obtained for player i from her cooperation in N to get $v(N)$. To solve a game consists in finding a "reasonable" payoff vector. Following Remark 1.1, this vector can be also seen as an additive game, namely we turn the game into an additive one. So, if we have a payoff vector $x \in \mathbb{R}^N$ then we can use for each coalition S,

$$x(S) = \sum_{i \in S} x_i.$$

Demanding conditions to the payoff vectors the set of reasonable solutions was restricted in several ways (stable set [29], core [10], kernel [5],...) named set solution concepts. Shapley [24] introduced the concept of value as a mapping obtaining a unique payoff vector for each game.

Definition 1.6 A value for games over N is a mapping $f : \mathcal{G}^N \to \mathbb{R}^N$. For each $v \in \mathcal{G}^N$ the image $f(v)$ is interpreted as a payoff vector of the game.

Nowadays there are a lot of values for games (the Shapley value [24], the nucleolus [23], the Banzhaf value [18], the compromise value [27],...). Some of them have an explicit formula and others are defined by an algorithm. Shapley also provided his value with certain conditions (axioms) such that it is the unique value satisfying them (an axiomatization). This fact gains importance now in order to choose the adequate value for each situation. Hence defining a new value implies to find out an axiomatization of it in order to distinguish between the others. Next we describe several of these properties for a value. From now on let f be a value for games over N.

Supposed that the great coalition N is formed in a game $v \in \mathcal{G}^N$ obtaining $v(N)$. Assuming that they use the same currency as measure of the payoff, the problem to solve in the game is how to allocate $v(N)$ among the players in N. If the payoff vector is an answer of the above question then it is named efficient. Thus an interesting axiom for f is the following one.

Efficiency. For each game $v \in \mathcal{G}^N$ it holds

$$f(v)(N) = v(N).$$

If we consider a change of scale in the worths of the game then the payoffs should be proportional to this change. Individual premiums or taxes should be assumed only for the player involved. This condition expresses a determine relationship between the value f and the equivalence of games.

Covariance. Given two strategically equivalent games, $w = av + b$ with $a > 0$ and $b \in \mathbb{R}^N$, it holds

$$f(w) = af(v) + b.$$

One the most controversial axioms is the additivity. Considering that it is not possible to transfer utility between two different games with the same set of players, then the payoffs of the sum of games should be the sum of the payoffs in each game. Although this fact is not always feasible in certain situations this condition is also a good tool for calculating.

Additivity. If $v, w \in \mathcal{G}^N$ then

$$f(v + w) = f(v) + f(w).$$

A broader condition than additivity is linearity.

Linearity. For all $v, w \in \mathcal{G}^N$ and $a, b \in \mathbb{R}$,

$$f(av + bw) = af(v) + bf(w).$$

While the only known information about the players are the worths of the coalitions, the payoff vectors should not distinguish among equivalent players for the characteristic function. Let $v \in \mathcal{G}^N$. Two players $i, j \in N$ are *symmetric* in v if for all $S \subseteq N \setminus \{i, j\}$ it holds

$$v(S \cup \{i\}) = v(S \cup \{j\}).$$

Symmetry.[3] For every game v and two different symmetric players $i, j \in N$ in v, it happens

$$f_i(v) = f_j(v).$$

When we take N as the set of players we identify each player to a specific label, one player to 1, another one to 2 and go on. But these labels must be harmless to the game. Let $v \in \mathcal{G}^N$ be a game and $\theta \in \Theta^N$ be a permutation over N. In this case we interpret θ as an interchange of labels. If S is a coalition then the same coalition with the new labels is $\theta(S) = \{\theta(i) : i \in S\}$ and the permutation game is $\theta v \in \mathcal{G}^N$ with $\theta v(\theta S) = v(S)$.

Anonymity.[4] Given a game $v \in \mathcal{G}^N$ and $\theta \in \Theta^N$ it holds for every player $i \in N$

$$f_{\theta(i)}(\theta v) = f_i(v).$$

If a value verifies anonymity then it also satisfies symmetry. Suppose $i, j \in N$ symmetric players, and the order $\theta_{ij} \in \Theta^N$ consisting in changing player i with player j in the natural order, namely $\theta_{ij}(i) = j$, $\theta_{ij}(j) = i$ and $\theta_{ij}(k) = k$ if $k \neq i, j$. We can test that $\theta_{ij} v = v$ since i, j are symmetric. So, using anonymity we get $f_j(v) = f_{\theta_{ij}(i)}(\theta v) = f_i(v)$. But the opposite is not always true.

A player $i \in N$ is *necessary* for a game v if $v(S) = 0$ when $S \subseteq N \setminus \{i\}$. This condition is really important in the game if v is monotone. In that case the necessary players must receive the highest payoffs.

Necessary player. If $i \in N$ is a necessary player for $v \in \mathcal{G}_m^N$ then for all $j \in N \setminus \{i\}$

$$f_j(v) \leq f_i(v).$$

Necessary player axiom seems independent of the symmetry and anonymity, and so are. But the utility of the axiom in the proofs is similar in many cases, in the sense that it implies equal treatment for all the necessary players, i.e. in a monotone game v two necessary players i, j satisfy $f_i(v) = f_j(v)$.

---

[3] Also named equal treatment.

[4] Also named symmetry.

The *marginal contribution* of a player $i \in N$ to a coalition $S \subseteq N \setminus \{i\}$ is measured as $v(S \cup \{i\}) - v(S)$. This marginal contribution can be described in other way, consider in this case $S \subseteq N$ with $i \in S$, and then the contribution into S is taken as $v(S) - v(S \setminus \{i\})$. Marginal contributions are used in the description of several axioms.

A player $i \in N$ is a *dummy player* in a game $v \in \mathcal{G}^N$ if $v(S \cup \{i\}) - v(S) = v(\{i\})$ for all $S \subseteq N \setminus \{i\}$, namely the marginal contributions are constant. It is reasonable to assume that if a player always contributes to any coalition with the same quantity then her payoff is this number.

Dummy player. If $i \in N$ is a dummy player in $v \in \mathcal{G}^N$ then

$$f_i(v) = v(\{i\}).$$

Next one is a particular case. A player $i \in N$ is a *null player* in a game $v \in \mathcal{G}^N$ if $v(S \cup \{i\}) = v(S)$ for all $S \subseteq N \setminus \{i\}$. It is reasonable to assume that if a player does not contribute to any coalition then her payoff is zero.

Null player. If $i \in N$ is a null player in $v \in \mathcal{G}^N$ then

$$f_i(v) = 0.$$

Now we comment several axioms related with ordering in games. Supposing fixed a non-empty coalition T, we say that two games verify $v <_T w$ if $v(T) < w(T)$ and $v(S) = w(S)$ for all $S \neq T$. Players in T should be beneficiaries of the extra payoff in w in front of v. This circumstance can be described in two ways.

Coalitional monotonicity. If two games verify $v <_T w$ for a non-empty coalition T then for all $i \in T$,

$$f_i(v) \leq f_i(w).$$

Weakly coalitional monotonicity. If two games verify $v <_T w$ for a non-empty coalition T then

$$f(v)(T) \leq f(w)(T).$$

Marginal contributions determine a partial order among the games for each player. Let $i \in N$. Given two games $v, w \in \mathcal{G}^N$ we note $v <_i w$ if for all $S \subseteq N \setminus \{i\}$

$$v(S \cup \{i\}) - v(S) \leq w(S \cup \{i\}) - w(S).$$

If a player contributes in a game w more than in other v then this player should have more payoff in w than v. Observe that $v <_{\{i\}} w$ is not the same that $v <_i w$.

Marginality. If two games verify $v <_i w$ for a player i then

$$f_i(v) \leq f_i(w).$$

Suppose that a game represents a situation where players proceed in an independent and rationality way. Each of them hopes to get better her payoff from the cooperation with regard to her individual action. Otherwise she can decide not to cooperate.

Individual stability. For each game $v \in \mathcal{G}^N$ and each player $i \in N$ it holds

$$f_i(v) \geq v(\{i\}).$$

If we think of this condition extended for any coalition, we suppose that the sum of the payoff obtained for the players in the coalition should be greater than the worth of the coalition.

Coalitional stability. For each game $v \in \mathcal{G}^N$ and each non-empty coalition $S \subset N$ it holds

$$f(v)(S) \geq v(S).$$

Both conditions are reasonable in a profit context. That is why this property is usually required only for superadditive games.

Using only three of these properties we can determine in a unique way the value of the unanimity games.

Proposition 1.7 *If f is a value over $\mathcal{G}^N$ satisfying efficiency, null player and symmetry then for all $T \subseteq N$, $T \neq \emptyset$,*

$$f_i(u_T) = \begin{cases} \dfrac{1}{|T|}, & \text{if } i \in T \\ 0, & \text{otherwise.} \end{cases}$$

Proof Observe that if $i \notin T$ then i is a null player for game u_T. Null player property implies that $f_i(u_T) = 0$. Suppose now $i, j \in T$ two different players. They are symmetric in u_T because

$$u_T(S \cup \{i\}) - u_T(S) = 0 = u_T(S \cup \{j\}) - u_T(S)$$

for all $S \subseteq N \setminus \{i, j\}$. Thus $f_i(u_T) = f_j(u_T)$. So, using efficiency we have, fixed any player $i \in T$,

$$\sum_{i \in N} f_i(u_T) = |T| f_i(u_T) = u_T(N) = 1.$$

□

Until now we have considered the set of players fixed but actually it can be any finite set. More generally the concept of value is given for all the games.

Definition 1.7 A value for games f assigns to each non-empty finite set N of players a value f^N for games over N.

From this point of view it is possible to consider properties modifying the set of players. Next one is an example of them. Let f be a value for games. The losses of profits produced for a player by the desertion of another one in the game are the same that if we exchange their roles.

Balanced contributions. Let N be a set of players with $|N| > 1$ and $v \in \mathcal{G}^N$. It holds for all $i, j \in N, i \neq j$,

$$f_i^N(v) - f_i^{N\setminus\{j\}}(v) = f_j^N(v) - f_j^{N\setminus\{i\}}(v).$$

Observe that we have used subgames (Definition 1.5) of v in the above formulation. Usually, we will only use the notation f^N when there is likelihood of confusion with the set of players N, otherwise f will represent the value for games over N.

We are not interested in payoffs of benefits for the players in a simple game over N. For these situations a payoff vector is interpreted as a *power vector*, namely a vector $x \in \mathbb{R}^N$ such that x_i is interpreted as the power or influence of player i in the decision situation defined by the simple game.

Definition 1.8 A power index for simple games over N is $f : \mathcal{G}_s^N \to \mathbb{R}^N$ such that $f(v)$ is interpreted as a power vector for each $v \in \mathcal{G}_s^N$.

Additivity and linearity are not internal conditions for power indices. We introduce another axiom using the lattice operations in this context. Let f be a power index. If the players decide to transfer the best options for two games in another game, the combination with the game using the worst options implies the same power allocation than the initial games.

Transference. If $v, w \in \mathcal{G}_s^N$ then

$$f(v \vee w) + f(v \wedge w) = f(v) + f(w).$$

This axiom is possible to apply also to all the games.

1.4 The Shapley Value

We focus this book on the analysis of the Shapley value in different ways for fuzzy relations among the players. Perhaps because this value is the most known and used of all but also because it is the most powerful in properties and versatile for different contexts. The existence of an explicit formula also permit to work with it in a not difficult way. But this value has also inconveniences, that is why in some situations others are used.

We denote again as Θ^N the set of permutations of N. Each $\theta \in \Theta^N$ defines a total order $<_\theta$ over N. Shapley [24] supposed that if players become members of the coalition following the order defined by a particular permutation θ, then each of them obtains her marginal contribution when she is incorporated to the coalition. So we get a payoff vector for each permutation θ.

Definition 1.9 Let $v \in \mathcal{G}^N$. For every $\theta \in \Theta^N$ the marginal vector $m^\theta(v) \in \mathbb{R}^N$ is defined as

$$m_i^\theta(v) = v(S_\theta^i \cup \{i\}) - v(S_\theta^i) \forall i \in N,$$

where $S_\theta^i = \{j \in N : j <_\theta i\}$.

If we consider the chosen permutation of N in a random way then we can take the expected value. This fact is the thrust of the definition of the Shapley value.

Definition 1.10 The Shapley value ϕ is the mapping $\phi^N : \mathcal{G}^N \to \mathbb{R}^N$ for games over every finite set N satisfying for each $v \in \mathcal{G}^N$

$$\phi^N(v) = \frac{1}{n!} \sum_{\theta \in \Theta^N} m^\theta(v).$$

Example 1.15 Suppose the bankruptcy problem (see Example 1.4) with three creditors, $N = \{1, 2, 3\}$. The capital of the unfortunate firm is $Q = 50000$ €. The demands of the creditors are $q_1 = 25000$ €, $q_2 = 20000$ €, $q_3 = 40000$ € respectibely. So, the (pessimistic) game from Example 1.4 is calculated in the next table (Table 1.3). For each permutation we determine the marginal vector of the game (see Table 1.4). For instance if we take $\theta = (1, 2, 3)$ then

$$m^\theta(v) = (v(\{1\}), v(\{1, 2\}) - v(\{1\}), v(\{1, 2, 3\}) - v(\{1, 2\})) = (0, 10000, 40000).$$

Table 1.3 Game v

S	{1}	{2}	{3}	{1, 2}	{1, 3}	{2, 3}	N
$v(S)$	0	0	5000	10000	30000	25000	50000

Table 1.4 Marginal vectors of game v

θ	$m^\theta(v)$	θ	$m^\theta(v)$
(1, 2, 3)	(0, 10000, 40000)	(1, 3, 2)	(0, 30000, 20000)
(2, 1, 3)	(10000, 0, 40000)	(2, 3, 1)	(25000, 0, 25000)
(3, 1, 2)	(25000, 20000, 5000)	(3, 2, 1)	(25000, 20000, 5000)

Finally we calculate the average of the marginal vectors,

$$\phi(v) = (14166.6, 13333.3, 22500).$$

Example 1.16 Let $v \in \mathbb{R}^N$ be an additive game. In this case all the marginal contributions of a player $i \in N$ are the same, if $S \subseteq N \setminus \{i\}$ then

$$v(S \cup \{i\}) - v(S) = \sum_{j \in S \cup \{j\}} v_j - \sum_{j \in S} v_j = v_i.$$

Hence $m_i^\theta(v) = v_i$ for each $\theta \in \Theta^N$ and $i \in N$, then $\phi(v) = v$.

Now we see several interesting properties of the Shapley value. The Shapley value is a linear function.

Proposition 1.8 *The Shapley value satisfies linearity, additivity and covariance.*

Proof Let $i \in N$. For each $T \subseteq N$ we have that $(av + bw)(T) = av(T) + bw(T)$. So

$$\begin{aligned}\phi_i(av + bw) &= \frac{1}{n!} \sum_{\theta \in \Theta^N} \left[(av + bw)(S_\theta^i \cup \{i\}) - (av + bw)(S_\theta^i)\right] \\ &= a \sum_{\theta \in \Theta^N} \left[v(S_\theta^i \cup \{i\}) - v(S_\theta^i)\right] + b \sum_{\theta \in \Theta^N} \left[w(S_\theta^i \cup \{i\}) - w(S_\theta^i)\right] \\ &= a\phi_i(v) + b\phi_i(w).\end{aligned}$$

Linearity implies directly additivity. Moreover, linearity and Example 1.16 obtain covariance. □

The Shapley value is an allocation of the worth of the great coalition N.

Proposition 1.9 *The Shapley value satisfies efficiency.*

Proof Suppose at least two players, $n \geq 2$, because the proof is trivial if there is only one player. We do the sum of all the payoffs $\phi_i(v)$ of the players in N for a game v,

$$\phi(v)(N) = \sum_{i \in N} \phi_i(v) = \frac{1}{n!} \sum_{\theta \in \Theta^N} \sum_{i \in N} m_i^\theta(v).$$

Let $\theta \in N$. Observe that if $\theta(i) = \theta(j) + 1$ then $S_\theta^i = S_\theta^j \cup \{j\}$ thus

$$v(S_\theta^i \cup \{i\}) - v(S_\theta^i) + v(S_\theta^j \cup \{j\}) - v(S_\theta^j) = v(S_\theta^i \cup \{i\}) - v(S_\theta^j).$$

Hence, as $S_\theta^{i_1} = \emptyset$ and $S_\theta^{i_N} \cup \{i_N\} = N$ for the first player i_1 and the last player i_N in order θ,

$$\sum_{i \in N} m_i^\theta(v) = v(N).$$

We get then $\phi(v)(N) = v(N)$. □

Remark 1.2 In the proof of the above proposition we have proved that all the marginal vectors are also efficient, namely if v is a game and $\theta \in \Theta^N$ then $m^\theta(v)(N) = v(N)$.

Proposition 1.10 *The Shapley value satisfies anonymity and symmetry.*

Proof Let $\theta \in \Theta^N$. For each $\theta' \in \Theta^{\theta(N)}$ there exists only one order $\theta'' \in \Theta^N$ such that for all $i \in N$ we have $\theta'(\theta(i)) = \theta(\theta''(i))$. Thus, for each player i we have $S_{\theta'}^{\theta(i)} = \theta(S_{\theta''}^i)$. Moreover, $\Theta^N = \Theta^{\theta(N)}$. We obtain

$$\begin{aligned}
\phi_{\theta(i)}(\theta v) &= \frac{1}{n!} \sum_{\theta' \in \Theta^{\theta(N)}} m_{\theta(i)}^{\theta'}(\theta v) = \frac{1}{n!} \sum_{\theta' \in \Theta^{\theta(N)}} \left[\theta v \left(S_{\theta'}^{\theta(i)} \cup \theta(i)\right) - \theta v \left(S_{\theta'}^{\theta(i)}\right)\right] \\
&= \frac{1}{n!} \sum_{\theta'' \in \Theta^N} \left[\theta v \left(\theta(S_{\theta''}^i \cup \{i\}\right)\right) - \theta v \left(\theta(S_{\theta''}^i)\right)\right] = \phi_i(v).
\end{aligned}$$

We said in the above section that anonymity implies symmetry. □

Also it verifies those properties for special players.

Proposition 1.11 *The Shapley value satisfies necessary player, dummy player and null player.*

Proof Let $i \in N$ be a dummy player in a game $v \in \mathcal{G}^N$. For every coalition S with $i \notin S$ we have $v(S \cup \{i\}) - v(S) = v(\{i\})$. Thus, Definition 1.10 implies

$$\phi_i(v) = v(\{i\})\frac{1}{n!}\sum_{\theta \in \Theta^N} 1 = v(\{i\}).$$

Remember that null player is a particular case of dummy player.

Suppose $v \in \mathcal{G}_m^N$, therefore all the marginal vectors are non negative. Let $i, j \in N$ with i a necessary player. For each $\theta \in \Theta^N$ with $j <_\theta i$ we have $m_j^\theta(v) = 0$. Otherwise $i \in S_\theta^j$ and taking θ' the order swapping i, j in θ we have $S_\theta^j \cup \{j\} = S_{\theta'}^i \cup \{i\}$. Using that v is monotone,

$$m_j^\theta(v) = v(S_\theta^j \cup \{j\}) - v(S_\theta^j) \leq v(S_{\theta'}^i \cup \{i\}) = m_i^{\theta'}(v).$$

So, $\phi_j(v) \leq \phi_i(v)$. □

We test now the ordering properties and show that the Shapley value verifies all of them.

Proposition 1.12 *The Shapley value satisfies coalitional monotonicity, weakly coalitional monotonocity and marginality.*

Proof Suppose $v <_T w$. For each player $i \in T$ and $\theta \in \Theta^N$ we obtain $v(S_\theta^i) = w(S_\theta^i)$. Thus for all permutation θ,

$$v(S_\theta^i \cup \{i\}) - v(S_\theta^i) \leq w(S_\theta^i \cup \{i\}) - w(S_\theta^i).$$

We get $\phi_i(v) \leq \phi_i(w)$ from the definition of the value, and then the coalitional monotonicity. Colitional monotonicity implies weakly coalitional monotonicity. The Shapley value is an average of marginal contributions, so marginality follows directly from definition. □

Now we study the stability conditions for the Shapley value. We consider only superadditive games.

Proposition 1.13 *The Shapley value satisfies individual stability over $\mathcal{G}_{sa}^N$.*

Proof Let $v \in \mathcal{G}_{sa}^N$. For each player $i \in N$ and $\theta \in \Theta^N$ we have

$$v(S_\theta^i \cup \{i\}) \geq v(S_\theta^i) + v(\{i\}),$$

thus $v(S_\theta^i \cup \{i\}) - v(S_\theta^i) \geq v(\{i\})$. So,

$$\phi_i(v) = \frac{1}{n!} \sum_{\theta \in \Theta^N} v(S_\theta^i \cup \{i\}) - v(S_\theta^i) \geq v(\{i\}) \frac{1}{n!} \sum_{\theta \in \Theta^N} 1 = v(\{i\}).$$

□

But this value failes to extend the stability for superadditive games to all coalition.

Example 1.17 Suppose the following superadditive, anonymous and 0-normalized game v over $N = \{1, 2, 3\}$ given by

$$v(S) = \begin{cases} 0, & \text{if } |S| \leq 1 \\ 25, & \text{if } |S| = 2 \\ 30, & \text{if } S = N. \end{cases}$$

The Shapley value of an anonymous game attaches the same payoff for all the players because they are symmetric and this value verifies symmetry. Since the Shapley value is also efficient then

$$\phi(v) = (10, 10, 10).$$

But this payoff vector is not stable for any coalition $S = \{i, j\}$,

$$\phi(v)(S) = 20 \leq v(S) = 25.$$

If we restrict the domain of the value to the convex games then we reach the stability for all the coalitions.

Proposition 1.14 *The Shapley value satisfies coalitional stability over $\mathcal{G}_c^N$.*

Proof Let $v \in \mathcal{G}_c^N$. We will prove that all the marginal vectors are stable for all the coalitions. Consider $\theta \in \Theta^N$ and $T \subseteq N$ non-empty coalition. For each player $i \in T$, besides S_θ^i (Definition 1.9), we also take $T_\theta^i = \{j \in T : \theta(j) < \theta(i)\}$, verifying $T_\theta^i \subseteq S_\theta^i$. We apply Proposition 1.3 to S_θ^i, T_θ^i obtaining, by convexity,

$$m_i^\theta(v) = v(S_\theta^i \cup \{i\}) - v(S_\theta^i) \geq v(T_\theta^i \cup \{i\}) - v(T_\theta^i).$$

Observe again that if $T = \{i_1, \ldots, i_t\}$ with $i_1 <_\theta \cdots <_\theta i_t$ then $T_\theta^{i_k} = T_\theta^{i_{k-1}} \cup \{i_k\}$, $T_\theta^{i_t} = T$ and $T_\theta^{i_1} = \emptyset$. So,

$$m^\theta(v)(T) \geq \sum_{i \in T} v(T_\theta^i \cup \{i\}) - v(T_\theta^i) = v(T).$$

Finally,

$$\phi_i(v)(T) = \frac{1}{n!} \sum_{\theta \in \Theta^N} m^\theta(v)(T) \geq v(T).$$

□

If the game v is subadditive or concave we obtain stability (individual and coalitional) with the other inequality. In the second section we introduced two ways to study a subadditive game by a superadditive one, and also a way to change from a pessimistic game to an optimistic one. Shapley value satisfies the following relationships with regard to these transformations.

Proposition 1.15 *The Shapley value satisfies the following equalities for a game $v \in \mathcal{G}^N$,*

1) $\phi(v) = -\phi(-v)$.
2) $\phi_i(v) = v(\{i\}) - \phi_i(v^{svg})$ for all $i \in N$.
3) $\phi(v) = \phi(v^{dual})$.

Proof The first two equalities follow from the linearity of the Shapley value (Proposition 1.8) and Example 1.16. We will prove the last one. For each permutation $\theta \in \Theta^N$ we take the dual permutation $\theta^d \in \Theta^N$ given by $\theta^d(i) = n - \theta(i) + 1$. Observe that this relation of duality is a bijective mapping over Θ^N. Let $i \in N$. By definition of dual permutation we have $S_{\theta^d}^i = (N \setminus S_\theta^i) \setminus \{i\}$. Calculating the payoff of the dual game (1.5),

$$\begin{aligned}
\phi_i(v^{dual}) &= \frac{1}{n!} \sum_{\theta \in \Theta^N} v^{dual}(S_\theta^i \cup \{i\}) - v^{dual}(S_\theta^i) \\
&= \frac{1}{n!} \sum_{\theta \in \Theta^N} v(N \setminus S_\theta^i) - v((N \setminus S_\theta^i) \setminus \{i\}) \\
&= \frac{1}{n!} \sum_{\theta^d \in \Theta^N} v\left(S_{\theta^d}^i \cup \{i\}\right) - v\left(S_{\theta^d}^i\right) = \phi_i(v).
\end{aligned}$$

□

There are several formulations of the Shapley value, we present here three. The first one, in Definition 1.10, describes the value by the marginal vectors. The second one uses once each marginal contribution of a player.

Theorem 1.1 *The Shapley value of a game $v \in \mathcal{G}^N$ for each player $i \in N$ can be calculated as*

$$\phi_i(v) = \sum_{S \subseteq N \setminus \{i\}} c_s^n \left[v(S \cup \{i\}) - v(S)\right] = \sum_{\{S \subseteq N : i \in S\}} c_{s-1}^n \left[v(S) - v(S \setminus \{i\})\right],$$

where $|S| = s$ and

$$c_s^n = \frac{s!(n-s-1)!}{n!}.$$

Proof Let $v \in \mathcal{G}^N$ and $i \in N$. From Definitions 1.9 and 1.10 we get

$$\phi_i(v) = \frac{1}{n!} \sum_{\theta \in \Theta^N} v(S_\theta^i \cup \{i\}) - v(S_\theta^i).$$

Of course for each coalition $S \subseteq N \setminus \{i\}$ there exists at least one permutation θ verifying $S_\theta^i = S$. Moreover, $\theta \in \Theta^N$ verifies $S_\theta^i = S$ when $j \in S$ if and only if $j <_\theta i$. The number of this kind of permutations is $s!(n-s-1)!$ with $|S| = s$. We obtain grouping the family of all the permutations related to a each $S \subseteq N \setminus \{i\}$,

$$\phi_i(v) = \frac{1}{n!} \sum_{S \subseteq N \setminus \{i\}} s!(n-s-1)! \left[v(S \cup \{i\}) - v(S)\right].$$

The other equality follows from doing $T = S \cup \{i\}$. □

Example 1.18 We turn to Example 1.15. Now, we use the optimistic version of the bankruptcy game in Example 1.12. Following Proposition 1.15 the results are the same in both cases. Table 1.5 represents the worths of the coalitions in the dual game. The calculation of the Shapley value is done by the formula in Theorem 1.1. The marginal contributions are determined in Table 1.6. The coefficients only depend on the cardinality of the coalitions and they are: $c_0^3 = 1/3$, $c_1^3 = 1/6$ and $c_2^3 = 1/3$. So, $\phi(v^{dual}) = (14166.6, 13333.3, 22500)$.

Table 1.5 Game v^{dual}

S	$\{1\}$	$\{2\}$	$\{3\}$	$\{1, 2\}$	$\{1, 3\}$	$\{2, 3\}$	N
$v^{dual}(S)$	25000	20000	40000	45000	50000	50000	50000

Table 1.6 Marginal contributions v^{dual} for each player

$v^{dual}(\{1\})$	$v^{dual}(\{1,2\}) - v^{dual}(\{2\})$	$v^{dual}(\{1,3\}) - v^{dual}(\{3\})$	$v^{dual}(N) - v^{dual}(\{2,3\})$
25000	25000	10000	0
$v^{dual}(\{2\})$	$v^{dual}(\{1,2\}) - v^{dual}(\{1\})$	$v^{dual}(\{2,3\}) - v^{dual}(\{3\})$	$v^{dual}(N) - v^{dual}(\{1,3\})$
20000	20000	10000	0
$v^{dual}(\{3\})$	$v^{dual}(\{1,3\}) - v^{dual}(\{1\})$	$v^{dual}(\{2,3\}) - v^{dual}(\{2\})$	$v^{dual}(N) - v^{dual}(\{1,2\})$
40000	25000	30000	5000

We can obtain a third formulation of the Shapley value using the dividends (Proposition 1.1) of the game. If the dividend of a coalition is the benefit margin attributed only to the formation of this coalition, then the Shapley value allocates these margins among the players concerned in an egalitarian way.

Theorem 1.2 *Let $v \in \mathcal{G}^N$. The Shapley value satisfies for each player $i \in N$ that*

$$\phi_i(v) = \sum_{\{T \subseteq N : i \in T, T \neq \emptyset\}} \frac{\Delta_T^v}{|T|}.$$

Proof Given a game v and a player i, from Propositions 1.1 and 1.8 we have

$$\phi_i(v) = \sum_{\{T \subseteq N, T \neq \emptyset\}} \Delta_T^v \phi_i(u_T).$$

As Shapley value verifies null player (Proposition 1.11), symmetry (Proposition 1.10) and efficiency (Proposition 1.9) then Proposition 1.7 implies the result. □

Example 1.19 We get the Shapley value of the bankruptcy game in Example 1.15 (the pessimistic version) again but using dividends. The worths of the coalitions are in Table 1.3. Next table obtains the dividends of the coalitions, using formula (1.2) (Table 1.7). So, we have

Table 1.7 Dividends of the game v

S	$\{1\}$	$\{2\}$	$\{3\}$	$\{1,2\}$	$\{1,3\}$	$\{2,3\}$	N
Δ_S^v	0	0	5000	10000	25000	20000	−10000

$$\phi_1(v) = \frac{1}{2}10000 + \frac{1}{2}25000 - \frac{1}{3}10000 = 14166.6.$$

Finally $\phi(v) = (14166.6, 13333.3, 22500)$.

We present now two different axiomatizations of the Shapley value using the above properties. The first one is the most classical axiomatization of this value, similar to the original one given by Shapley [24].

Theorem 1.3 *The Shapley value is the only value for games over N satisfying linearity, efficiency, null player and symmetry.*

Proof Shapley value verifies the four axioms from Propositions 1.8, 1.9, 1.10 and 1.11.

Proposition 1.7 and linearity imply only one option for a value f satisfying the four axioms. □

Remark 1.3 It is enough additivity instead linearity because it is an easy exercise to repeat the proof of Proposition 1.7 for game cu_T with $T \neq \emptyset$ and $c \in \mathbb{R}$, obtaining a unique payoff vector for this game. So, from additivity,

$$f(v) = \sum_{\{T \subseteq N : T \neq \emptyset\}} f\left(\Delta_T^v u_T\right).$$

When an axiomatization is introduced for a value it is advisable to show the logical independence of the axioms, namely we test the necessity of all of them.

Remark 1.4 We find values different to the Shapley value verifying all the axioms except one of them.

- Consider value f^1 defined for each $v \in \mathcal{G}^N$ as $f^1(v) = a\phi(v)$ with $a \in \mathbb{R} \setminus \{1\}$. As ϕ satisfies null player, symmetry and linearity then f^1 does too.
- Let $\theta \in \Theta^N$. The marginal vector for each game using this permutation can be considered as a payoff vector. Hence we can define $f^2(v) = m^\theta(v)$ for all $v \in \mathcal{G}^N$. Suppose $n \geq 2$, $f^2 \neq \phi$ from Definition 1.10. Remark 1.2 showed that f^2 verifies efficiency. If i is a null player then all her marginal contributions are zero, also the marginal contribution in the order θ. So, f^2 satisfies null player. Let $v, w \in \mathcal{G}^N$, and $a, b \in \mathbb{R}$ we get

$$f_i^2(av + bw) = m_i^\theta(av + bw) = (av + bw)(S_\theta^i \cup \{i\}) - (av + bw)(S_\theta^i)$$
$$= af_i^2(v) + bf_i^2(w).$$

We get also linearity. f^2 satisfies efficiency, null player and linearity.

- The egalitarian value is defined as

$$f_i^3(v) = \frac{v(N)}{n} \forall v \in \mathcal{G}^N, \forall i \in N.$$

Obviously $f^3 \neq \phi$ verifies efficiency, symmetry and linearity.
- The following value tries to modify the egalitarian value to get the null player property. For each $v \in \mathcal{G}^N$ we denote as $Null(v) = \{i \in N : i \text{ null player in } v\}$ and $null(v) = |Null(v)|$. Consider the value defined for v as

$$f_i^4(v) = \begin{cases} \dfrac{v(N)}{n - null(v)}, & \text{if } i \notin Null(v) \\ 0, & \text{if } i \in Null(v). \end{cases}$$

Obviously the value satisfies null player and also efficiency and symmetry (if two players are symmetric then or both are null players or neither is null). $f^4 \neq \phi$ if $null(v) \geq 1$. If $null(v) = n$ then $v = 0$ and we take $f^4(0) = 0$.

Other axiomatizations can be obtained in a similar way of the above one swapping symmetry axiom by anonymity one or necessary player one.

Young [30] used marginality to axiomatize the Shapley value jointly anonymity and another axiom.[5] Here we propose to use marginality with efficiency and symmetry.

Theorem 1.4 *The Shapley value is the only value for games over N satisfying efficiency, symmetry and marginality.*

Proof Shapley value satisfies the three axioms from Propositions 1.9, 1.10 and 1.12.

Suppose now $f^1 \neq f^2$ verifying the three axioms. Each game $v \in \mathcal{G}^N$ is written as

$$v = \sum_{\{T \subseteq N : T \neq \emptyset\}} \Delta_T^v u_T,$$

using Theorem 1.1. Let $p(v) = |\{T \subseteq N : T \neq \emptyset, \Delta_T^v \neq 0\}|$. If $p(v) = 0$ then $v = 0$ and every player is a null player, therefore they are symmetric. Using efficiency and symmetry we have $f^1(v) = 0 = f^2(v)$. Hence we can look for any game with the smallest number $p(v) \geq 1$ such that $f^1(v) \neq f^2(v)$. We denote as

$$P(v) = \bigcap_{\{T \subseteq N : T \neq \emptyset, \Delta_T^v \neq 0\}} T.$$

[5] Strongly monotonicity.

All the players in $P(v)$ are symmetric in v, namely if $i, j \in P(v)$ and $S \subseteq N \setminus \{i, j\}$ then using (1.1) we have $v(S \cup \{i\}) = 0 = v(S \cup \{j\})$. So, $f_i^1(v) = f_j^1(v)$ and $f_i^2(v) = f_j^2(v)$. If $i \notin P(v)$ then we can define

$$w = \sum_{\{T \subseteq N : i \in T\}} \Delta_T^v u_T,$$

with $p(w) < p(v)$. The marginal contributions of player i in this new game are the same as in v. Let $S \subseteq N \setminus \{i\}$. We obtain from (1.1),

$$w(S \cup \{i\}) - w(S) = \sum_{\{T \setminus \{i\} \subseteq S : i \in T, \Delta_T^v \neq 0\}} \Delta_T^v = v(S \cup \{i\}) - v(S).$$

Marginality (in both sides) implies that

$$f_i^1(v) = f_i^1(w) = f_i^2(w) = f_i^2(v).$$

Finally, we take any $i \in P(v)$ and then by efficiency

$$0 = \sum_{j \in N} f^1(v) - f^2(v) = p(v)[f_i^1(v) - f_i^2(v)].$$

We get $f^1(v) = f^2(v)$, and this is not possible. □

Remark 1.5 We test the logical independence of the axioms.

- f^1 in Remark 1.4 satisfies symmetry and marginality.
- f^2 in Remark 1.4 satisfies efficiency and marginality.
- f^3 in Remark 1.4 satisfies efficiency and symmetry.

We will see another axiomatization given by Myerson [15] using properties as value in the general sense.

Proposition 1.16 *The Shapley value satisfies balanced contributions.*

Proof Let $N, n > 1$, and $v \in \mathscr{G}^N$. Theorem 1.1 implies for all $i, j \in S, i \neq j$,

$$\begin{aligned}\phi_i^N(v) - \phi_j^N(v) &= \sum_{\{S \subseteq N : i \in S\}} c_{s-1}^n [v(S) - v(S \setminus \{i\})] \\ &- \sum_{\{R \subseteq N : j \in R\}} c_{r-1}^n [v(R) - v(R \setminus \{j\})] \\ &= \sum_{\{T \subseteq N : i, j \in T\}} c_{t-1}^n [v(T \setminus \{j\}) - v(T \setminus \{i\})]\end{aligned}$$

$$+ \sum_{\{S \subseteq N\setminus\{j\}: i \in S\}} c^n_{s-1}[v(S) - v(S \setminus \{i\})]$$
$$- \sum_{\{R \subseteq N\setminus\{i\}: j \in R\}} c^n_{r-1}[v(R) - v(R \setminus \{j\})]$$

$$\begin{aligned}
\phi_i^N(v) - \phi_j^N(v) = & \sum_{\{T \subseteq N: i,j \in T\}} c^n_{t-1}[v(T \setminus \{j\}) - v(T \setminus \{i\})] \\
& + \sum_{\{T \subseteq N: i,j \in T\}} c^n_{t-2}[v(T \setminus \{j\}) - v(T \setminus \{i, j\})] \\
& - \sum_{\{T \subseteq N: i,j \in T\}} c^n_{t-2}[v(T \setminus \{i\}) - v(T \setminus \{i, j\})] \\
= & \sum_{\{T \subseteq N: i,j \in T\}} (c^n_{t-1} + c^n_{t-2})[v(T \setminus \{j\}) - v(T \setminus \{i\})] \\
= & \ \phi_i^{N\setminus\{j\}}(v) - \phi_j^{N\setminus\{i\}}(v).
\end{aligned}$$

The last equality is true because $c^n_{t-1} + c^n_{t-2} = c^{n-1}_{t-1} + c^{n-1}_{t-2}$. □

Myerson [15] presented his axiomatization in the context of games with conference structure, we explain it on the classical family of games.

Theorem 1.5 *The Shapley value is the only value for games satisfying efficiency and balanced contributions.*

Proof We know that the Shapley value satisfies efficiency (Proposition 1.9) and balanced contributions (Proposition 1.16).

Let f, g be two values for games verifying both axioms and $v \in \mathcal{G}^N$. We work by induction in n. Efficiency implies $f^N(v) = g^N(v)$ for any game with $n = 1$. We suppose true the uniqueness when $n < k$, $k \geq 2$. Consider any $i_0 \in N$. For each $j \in N \setminus \{i_0\}$ balanced contributions says that

$$\begin{aligned}
f_j^N(v) - f_{i_0}^N(v) &= f_j^{N\setminus\{i_0\}}(v) - f_{i_0}^{N\setminus\{j\}}(v) \\
&= g_j^{N\setminus\{i_0\}}(v) - g_{i_0}^{N\setminus\{j\}}(v) = g_j^N(v) - g_{i_0}^N(v).
\end{aligned}$$

Therefore we obtain $f_j^N(v) - g_j^N(v) = f_{i_0}^N(v) - g_{i_0}^N(v)$. Using efficiency

$$0 = v(N) - v(N) = \sum_{i \in N} f_j^N(v) - g_j^N(v) = |N|[f_{i_0}^N(v) - g_{i_0}^N(v)].$$

Thus $f_{i_0}^N(v) = g_{i_0}^N(v)$. □

Shapley and Shubik [25] studied the Shapley value as a power index when it is applied to simple games. In a simple game the marginal contributions of a player i are 1 or 0, then only those coalitions where i changes the situation with her incorporation are important.

Definition 1.11 Let $v \in \mathcal{G}_s^N$. A coalition $S \subseteq N \setminus \{i\}$ is a swing for player i if $v(S \cup \{i\}) = 1$ and $v(S) = 0$. The set of swings for player i in v is denoted as $SW_i(v)$.

A swing for a player is a coalition where the incorporation of this player is critical. Observe that if $S \in W_m(v)$ then $S \setminus \{i\}$ is a swing for all player $i \in S$. We get from Theorem 1.1 a special formula of the Shapley–Shubik index using the swings of the players besides the definition.

Definition 1.12 The Shapley–Shubik index is a power index over $\mathcal{G}_s^N$ defined for each simple game v as $\phi(v)$. For all $i \in N$ and $v \in \mathcal{G}_s^N$ it holds

$$\phi_i(v) = \frac{|\{\theta \in \Theta^N : S_\theta^i \in SW_i(v)\}|}{n!} = \sum_{S \in SW_i(v)} c_s^n.$$

Example 1.20 We consider a voting situation [26; 20, 15, 6, 5], see Example 1.14, for a parliament with four parties $\{1, 2, 3, 4\}$. The simple game is defined by the minimal winning coalitions,

$$W_m(v) = \{\{1, 2\}, \{1, 3\}, \{2, 3, 4\}\}.$$

Table 1.8 contains the swings for each party. Observe that all the swings have cardinality 2 or 3 and the Shapley coefficients for these cardinalities with four players are the same $c_1^4 = c_2^4 = 1/12$. So, the Shapley–Shubik index is

$$\phi(v) = (5/12, 3/12, 3/12, 1/12).$$

Table 1.8 Swings of v

i	$SW_i(v)$
$\{1\}$	$\{\{2\}, \{3\}, \{2, 3\}, \{2, 4\}, \{3, 4\}\}$
$\{2\}$	$\{\{1\}, \{1, 4\}, \{3, 4\}\}$
$\{3\}$	$\{\{1\}, \{1, 4\}, \{2, 4\}\}$
$\{4\}$	$\{\{2, 3\}\}$

Despite the difference of seats between parties 2 and 3 their power is the same. The weights of parties 3 and 4 are very closed but the power does not.

The Shapley–Shubik index for a player determines the probability of finding an order of the players where the incorporation of the player is critical for the decision. But certain properties of the Shapley value are not feasible in simple games, as additivity, linearity or covariance because they use non-internal operations.

Dubey [7] showed an axiomatization of the Shapley–Shubik index using transference instead linearity or additivity.

Theorem 1.6 *The Shapley–Shubik index is the only power index over $\mathcal{G}_s^N$ satisfying efficiency, null player, symmetry and transference.*

Proof Since Propositions 1.9, 1.10 and 1.11 we know that Shapley value satisfies efficiency, symmetry and null player for all the games and then also for simple games. We will see that the Shapley–Subik index verifies transference. Let $v, w \in \mathcal{G}_s^N$ be two simple games. We can consider v, w as elements in $\mathcal{G}^N$. Working in $\mathcal{G}^N$, we prove the claim $(v \vee w) + (v \wedge w) = v + w$ by Table 1.9. As the Shapley value verifies additivity (Proposition 1.8) we get transference.

Suppose now an index f satisfying the axioms. We get the uniqueness by induction on the number of minimal coalitions. Let $v \in \mathcal{G}_s^N$ with $|W_m(v)| = 1$ (at least there is one). So, $v = u_S$ with $W_m(v) = \{S\}$, and we know the uniqueness from Proposition 1.7. Taking to be true the uniqueness if $|W_m(v)| = k - 1$ we will prove it for $|W_m(v)| = k$. Let $T \in W_m(v)$. We define another simple game,

$$w = \bigvee_{S \in W_m(v) \setminus \{T\}} u_S.$$

This another game satisfies that $W_m(w) = W_m(v) \setminus \{T\}$ from Proposition 1.6 and the fact that $W_m(v) \setminus \{T\}$ is mutually exclusive. Proposition 1.6 again implies

$$v = \bigvee_{S \in W_m(v)} u_S = w \vee u_T.$$

Table 1.9 Transfer property of the Shapley–Shubik index

v	w	$v+w$	$v \vee w$	$v \wedge w$	$(v \vee w) + (v \wedge w)$
0	0	0	0	0	0
1	0	1	1	0	1
1	1	2	1	1	2

We compute now $w \wedge u_T$ using that $\mathcal{G}_s^N$ is a distributive lattice (Proposition 1.5)

$$w \wedge u_T = \left(\bigvee_{S \in W_m(v) \setminus \{T\}} u_S \right) \wedge u_T = \bigvee_{S \in W_m(v) \setminus \{T\}} (u_S \wedge u_T) = \bigvee_{S \in W_m(v) \setminus \{T\}} u_{S \cup T}.$$

Observe that $(u_S \wedge u_T)(R) = 1$ if and only if both $S, T \subseteq R$. Family $\{S \cup T : S \in W_m(v) \setminus \{T\}\}$ is also mutually exclusive because $W_m(v)$ is too, and then by Proposition 1.6

$$W_m(w \wedge u_T) = \{S \cup T : S \in W_m(v) \setminus \{T\}\}.$$

As $|W_m(w)| = |W_m(w \wedge u_T)| = k - 1$ and $|W_m(u_T)| = 1$ we get using transference

$$\phi(v) = \phi(w) + \phi(u_T) - \phi(w \wedge u_T).$$

□

Until now we have seen that the Shapley value satisfies a lot of interesting conditions. Furthermore we have several easy formulas to calculate it and it is a linear function. But this value is not a panacea, that is why there are other interesting values although this one is the most used. The main criticisms to the Shapley value are the following.

- Additivity is a nice mathematical property but the interpretation is in doubt. Depending on the situation players could bargain utility from one of the game to use in the other.
- The combinatorial nature of the Shapley formula implies a random determined formation of the coalitions.
- The value does not always stable.

Also transference, null player, efficiency and symmetry are called into question for simple games, see Laruelle and Valenciano [13] and Einy and Haimanko [8] for other axiomatizations without these axioms.

References

1. Baeyens, E., Bitar, E.Y., Khargonekar, P.P., Poolla, K.: Coalitional aggregation of wind power. IEEE Trans. Power Syst. **28**(4), 3774–3784 (2013)
2. Banzhaf, J.F.: Weighted voting doesn't work: a mathematical analysis. Rutgers Law Rev. **19**(2), 317–343 (1965)
3. Bilbao, J.M.: Cooperative Games on Combinatorial Structures. Theory and Decision Library, vol. 26. Springer, Heidelberg (2000)
4. Chetty, V.K., Dasgupta, D., Raghavan, T.E.S.: Power and distribution of profits. Discussion paper 139. Indian Statistical Institute, Delhi Center, New Delhi (1976)
5. Davis, M., Maschler, M.: Econometric Research Program. Memorandum, vol. 58. Princeton University, Princeton (1963)

6. Driessen, T.S.H.: Cooperative Games, Solutions and Applications. Kluwer Academic Publishers, Dordrecht (1988)
7. Dubey, P.: On the uniqueness of the Shapley value. Int. J. Game Theory **4**, 131–139 (1975)
8. Einy, E., Haimanko, O.: Characterizations of the Shapley–Shubik power index without the efficiency axiom. Games Econ. Behav. **73**(2), 615–621 (2011)
9. Feltkamp, V.: Alternative axiomatic characterizations of the Shapley and Banzhaf value. Int. J. Game Theory **24**, 179–186 (1995)
10. Gillies, D.B.: Solutions to general non-zero-sum games. In: Tucker, A.W., Luce R.D. (eds.) Contributions to the Theory Games IV. Annals of Mathematics Studies, vol. 40, pp. 47–85. Princeton University Press, Princeton (1959)
11. González-Díaz, J., García-Jurado, I., Fiestras-Janeiro, M.G.: An Introductory Course on Mathematical Game Theory. Graduate Studies in Mathematics, vol. 115. American Mathematical Society, Providence (2010)
12. Harsanyi, J.C.: A bargaining model for the cooperative n-person game. In: Tucker, A.W., Luce, D.R. (eds.) Contributions to the theory of games, vol. 4, pp. 325–355. Princeton University Press, Princeton (1959)
13. Laruelle, A., Valenciano, F.: Shapley-Shubik and Banzhaf indices revised. Math. Oper. Res. **26**(1), 89–104 (2001)
14. Littlechild, S.C., Owen, G.: A simple expression for the Shapley value in a special case. Manag. Sci. **20**(3), 370–372 (1973)
15. Myerson, R.: Conference structures and fair allocation rules. Int. J. Game Theory **9**, 169–182 (1980)
16. O'Neil, B.: A problem of rights arbitration from the Talmud. Math. Soc. Sci. **2**, 345–371 (1982)
17. Owen, G.: Multilinear extensions and the Banzhaf value. Naval Res. Logist. Q. **22**(4), 741–750 (1975)
18. Owen, G.: On the core of linear production games. Math. Program. **9**, 358–370 (1975)
19. Owen, G.: Game Theory. Academic Press, Dublin (1995)
20. Peters, H.: Game Theory, a Multi-Leveled Approach. Springer, Berlin (2008)
21. Penrose, L.: The elementary statistics of majority voting. J. R. Stat. Soc. **109**, 53–57 (1946)
22. Roth, A.E.: The Shapley value. Essays in honor of Lloyd S. Shapley. Cambridge University Press, Cambridge (1988)
23. Schmeidler, D.: The nucleolus of a characteristic function game. SIAM J. Appl. Math. **17**, 1163–1170 (1969)
24. Shapley, L.S..: A value for n-person games. In: Kuhn H.W., Tucker A.W. (eds.) Contributions to the theory of games, vol. II. Annals of Mathematical Studies v. 28, pp. 307–317 (1953)
25. Shapley, L.S., Shubik, M.: A method for evaluating the distribution of power in a committee system. Am. Polit. Sci. Rev. **48**, 787–792 (1954)
26. Shapley, L.S., Shubik, M.: On market games. J. Econ. Theory **1**, 9–25 (1969)
27. Tijs, S.H.: Bounds for the core and the τ-value. In: Moeschlin, O., Pallaschke., D. (eds.) Game Theory and Mathematical Economics. North-Holland, Amsterdam (1981)
28. von Neumann, J.: Zur Theorie der Gesellschaftsspiele. Mathematische Annalen **100**, 295–320 (1928)
29. von Neumann, J., Morgenstern, O.: Theory of Games and Economic Behavior. Princeton University Press, Princeton (1944)
30. Young, H.P.: Monotonic solutions of cooperative games. Int. J. Game Theory **14**, 65–72 (1985)

Chapter 2
Fuzzy Coalitions and Fuzziness of Games

2.1 Introduction

In a cooperative game the mean tool to analyze the situation is the knowledge of the worths of the coalitions. In an coalition it is supposed that all the players cooperate at the same level or certainty. Zadeh [20] introduced fuzzy sets to represent different degrees of membership for the elements of a set. Later Aubin [1] proposed to use fuzzy sets to define fuzzy coalitions allowing an asymmetry of the participation of the players. This fact supposes to replace the discrete scope by the continuous one. Several interesting surveys about fuzzy games (games with fuzzy coalitions) are given in Butnariu and Klement [7], Branzei et al. [5] and Borkotokey and Mesiar [3]. There is another different way in the analysis of fuzzy games, using vague payoffs. The reader can use Mares [12] to study this model. Classical games are an specific family of fuzzy games, that we name crisp games. This book works with crisp games but some information about fuzzy games helps to comprehend better the developing of our study. The mean difficulty to define a solution for a fuzzy game is how to incorporate the whole information of the game. The diagonal value, (Aumann and Shapley [2] and Aubin [1]) and the crisp Shapley value (Branzei et al. [4]) are two different approach to solve a fuzzy game in the Shapley way. But neither of them uses all the information of the fuzzy game.

A fuzziness of a crisp game is a fuzzy game determined from the crisp one. Remember that we deal with non-flexible games, therefore we look for fuzziness following the classical behavior of the players: to form the maximal feasible coalition. Fuzzy cooperation can be used as probabilistic data or real membership of the players. In the second point of view the rule of the maximal cooperation is reached by two different ways: maximal level of cooperation of maximal group of players involved. Three fuzziness following all these comments have been studied in the literature, the multilinear extension [15], the proportional extension [6] and the Choquet extension [18].

A. Jiménez-Losada, *Models for Cooperative Games with Fuzzy Relations among the Agents*, Studies in Fuzziness and Soft Computing 355,
DOI 10.1007/978-3-319-56472-2_2

2.2 Brief Overview About Fuzzy Sets

Fuzzy sets were introduced by Zadeh [20]. In this book we only deal with fuzzy sets in a finite set of points. Let K be a finite set. A *fuzzy set* A in K is given by a membership function over K, namely $\tau_A : K \to [0, 1]$. We denote the family of fuzzy sets in K as $[0, 1]^K$. Here the fuzzy set A is identified to its membership function τ_A, and then we will use $\tau \in [0, 1]^K$ to refer a fuzzy set. Each non-empty subset $Q \subseteq K$ is identified with the fuzzy set $e^Q \in [0, 1]^K$ given by $e^Q(i) = 1$ if $i \in Q$ and $e^Q(i) = 0$ otherwise. Particularly for the complete set K we will use e^K. The empty-set is also a fuzzy set that we denote as 0 with $0(i) = 0$ for all $i \in K$. The classical subsets are named *crisp sets* in order to distinguish them into the fuzzy sets.

A fuzzy set τ' is contained in another one $\tau \in [0, 1]^K$ if $\tau'(i) \leq \tau(i)$ for all $i \in K$, we say that τ' is a *fuzzy subset* of τ and type $\tau' \leq \tau$. If $\tau' \leq \tau$ then the *difference* is a new fuzzy set $\tau - \tau'$ with $(\tau - \tau')(i) = \tau(i) - \tau'(i)$. The *complement* of a fuzzy set τ is $e^K - \tau \in [0, 1]^K$.

A T-*norm* is a binary relation $T : [0, 1] \times [0, 1] \to [0, 1]$ verifying the following properties: (1) Commutativity $T(a, b) = T(b, a)$, (2) Associativity $T(a, T(b, c)) = T(T(a, b), c)$, (3) Monotonicity $T(a, b) \leq T(c, d)$ if $a \leq b$, $c \leq d$ and (4) Identity element $T(1, a) = a$. The corresponding T-*conorm* to the T-norm T is the dual binary relation

$$(1 - T)(a, b) = 1 - T(1 - a, 1 - b).$$

Using different T-norms we introduced different operations for fuzzy sets. If T is a T-norm then $T(\tau, \tau')(i) = T(\tau(i), \tau'(i))$ for all $\tau, \tau' \in [0, 1]^N$ and $i \in N$. We consider three of them.

- *Intersection and union.* $T(a, b) = a \wedge b$, $(1 - T)(a, b) = a \vee b$. If $\tau, \tau' \in [0, 1]^K$ then, the intersection and the union are respectively $\tau \wedge \tau'$ and $\tau \vee \tau'$. These operations satisfy the Morgan laws in a fuzzy sense:

$$e^K - (\tau \vee \tau') = (e^K - \tau) \wedge (e^K - \tau'),$$
$$e^K - (\tau \wedge \tau') = (e^K - \tau) \vee (e^K - \tau').$$

- *Cosum and sum.*[1] $T(a, b) = 0 \vee (a + b - 1)$, $(1 - T)(a, b) = 1 \wedge (a + b)$. If $\tau, \tau' \in [0, 1]^K$ then, the cosum and the sum are respectively $\tau \oplus \tau'$ and $\tau + \tau'$.
- *Product and coproduct.*[2] $T(a, b) = ab$ and $(1 - T)(a, b) = a + b - ab$. The product and the coproduct of τ, τ' are the fuzzy sets respectively $\tau \times \tau'$, $\tau \otimes \tau'$.

[1] This T-norm is usually named Lukasiewicz norm.

[2] Product is understood as the usual probabilistic intersection.

The *support* of a fuzzy set $\tau \in [0, 1]^K$ is the set

$$supp(\tau) = \{i \in K : \tau(i) > 0\}.$$

The *image* of τ is the set of numbers

$$im(\tau) = \{t \in (0, 1] : \exists i \in K \text{ with } \tau(i) = t\}.$$

The maximum and the minimum non-zero numbers in the image of $\tau \neq 0$ are denoted as $\vee\tau$ and $\wedge\tau$ respectively. Usually the image of a fuzzy set is given as an ordered set including zero or also one although these numbers are not in the image, we take

$$im_0(\tau) = \{0 = \lambda_0 < \lambda_1 < \cdots < \lambda_m\} = im(\tau) \cup \{0\}, \tag{2.1}$$
$$im_0^1(\tau) = \{0 = \lambda_0 < \lambda_1 < \cdots < \lambda_m = 1\} = im(\tau) \cup \{0, 1\}. \tag{2.2}$$

The *total membership* of a fuzzy set τ is something as the cardinality in a set, namely

$$|\tau| = \sum_{i \in K} \tau(i).$$

Two fuzzy sets τ_1, τ_2 are *comonotone* if for all $i, j \in K$

$$[\tau_1(i) - \tau_1(j)][\tau_2(i) - \tau_2(j)] \geq 0. \tag{2.3}$$

Let $\tau \in [0, 1]^K$ be a fuzzy set over the finite set K. For each $t \in (0, 1]$, the *t-cut* is the crisp set of all the elements in K with level greater or equal than t in τ, namely

$$[\tau]_t = \{i \in K : \tau(i) \geq t\}. \tag{2.4}$$

As K is finite there is a finite quantity of different cuts. If $im_0(\tau) = \{\lambda_0 < \cdots < \lambda_k\}$ then we use $[\tau]_k = [\tau]_{\lambda_k}$ for $k = 1, \ldots, m$. So, $[\tau]_t = [\tau]_k$ if $t \in (\lambda_{k-1}, \lambda_k]$ and $k = 1, \ldots, m$. The cuts of τ generate a decreasing sequence of sets, $[\tau]_m \subset \cdots \subset [\tau]_1$.

Choquet [9] defined a *capacity* over K as a monotone set function $v : 2^K \to \mathbb{R}$ verifying $v(\emptyset) = 0$, actually a monotone game (see Definition 1.3) over K. If we drop monotonicity, v is a game, then is named *signed capacity*. The Choquet integral [9] was first defined for capacities and later Schmeidler [16] and Waegenaere and Wakker [10] extended the concept to signed capacities. If $\tau \in [0, 1]^K$ and v is a signed capacity over K then the (signed) *Choquet integral* of τ regard to v is

$$\int_c \tau \, dv = \sum_{k=1}^{m} (\lambda_k - \lambda_{k-1}) v([\tau]_k), \tag{2.5}$$

with $im_0(\tau) = \{\lambda_0 < \lambda_1 < \cdots < \lambda_m\}$. Next properties of the integral will be used in the book.

(C1) $\int_c \tau \, d(v_1 + v_2) = \int_c \tau \, dv_1 + \int_c \tau \, dv_2$

(C2) $\int_c \tau \, dv_1 \leq \int_c \tau \, dv_2$ if $v_1 \leq v_2$

(C3) $\int_c \tau \, d(av) = a \int_c \tau \, dv$

(C4) $\int_c t\tau \, dv = t \int_c \tau \, dv$ if $t \in [0, 1]$

(C5) $\int_c (\tau_1 + \tau_2) \, dv = \int_c \tau_1 \, dv + \int_c \tau_2 \, dv$ if τ_1, τ_2 are comonotone

(C6) $\int_c \tau \, dv = a \bigvee_{i \in K} \tau(i)$ if $v([\tau]_t) = a$ for all $t \in (0, 1]$

(C7) $\int_c e^Q \, dv = v(Q)$

(C8) $\int_c \tau_1 \, dv \leq \int_c \tau_2 \, dv$ when $\tau_1 \leq \tau_2$ and v monotone

(C9) $\int_c \tau \, dv$ is a continuous function regard to τ.

(C10) $\int_c \tau \, dv = \sum_{q=1}^{p} (t_q - t_{q-1}) v([\tau]_{t_q})$, for any finite set of numbers in [0, 1] containing the image of τ (and zero), namely $\{0 = t_0 < t_1 \cdots < t_p\} \supseteq im_0(\tau)$.

2.3 Fuzzy Coalitions

Fuzzy coalitions were defined by Aubin [1]. Given the finite set of players N, a fuzzy coalition is a collective decisional entity where members may have gradual degrees of membership, as Butnariu and Klement says in [7]. So, a fuzzy coalition introduces an asymmetric relation among the players beyond the mapping of the game. A fuzzy set in the set of players is used to define a fuzzy coalition.

Definition 2.1 A fuzzy coalition of the set of players N is any fuzzy set in N. Therefore $[0, 1]^N$ denotes the family of fuzzy coalitions.

The support of a fuzzy coalition represents the set of active players and the image the different levels of participation.

Fuzzy coalitions can be interpreted in other ways. In a probabilistic sense, if $\tau \in [0, 1]^N$ then $\tau(i)$ is the probability of being in the coalition. Thus each fuzzy coalition represents a particular situation of possibility.

Aubin [1] also introduced the concept of game with fuzzy coalitions or fuzzy game.[3]

Definition 2.2 A fuzzy game (game with fuzzy coalitions) over N is a real mapping over the fuzzy coalitions $v : [0, 1]^N \to \mathbb{R}$ with $v(0) = 0$.

Branzei et al. [5] proposed the next example of fuzzy game about a public good.

Example 2.1 Suppose N a set of agents who cooperate in order to get a new facility for joint use. The cost of the facility depends on the total participation of the agents by monotone increasing function c. The profit obtained from the facility is determined for each player using an individual monotone increasing function p_i depending on her participation. This situation can be described by the fuzzy game over N

$$v(\tau) = \sum_{i \in N} p_i(\tau(i)) - c(|\tau|).$$

Example 2.2 In a university committee there two groups of delegates, professors and students. Next system to adopt a rule incorporates more options for the teachers, independently on the number of elements in each group. We have $N = \{1, 2\}$ with 1 the team of professors and 2 the student's one. In a fuzzy coalition $\tau \in [0, 1]^N$ we interpret

$$\tau(i) = \frac{|\text{delegates in team } i \text{ in favor of the rule}|}{|\text{delegates in team } i|}.$$

So, the proposition is

$$v(\tau) = \begin{cases} 1, & \text{if } \tau(1) \geq 2/3 \text{ and } \tau(2) \geq 1/2 \\ 0, & \text{otherwise.} \end{cases}$$

Example 2.3 Linear production games (Example 1.6) can be better explained as fuzzy games. In the crisp case players use their whole endowments when they cooperate. The fuzzy version permits to use the initial endowments partially. So, considering c, b, A as in Example 1.6, the *fuzzy linear production game* is defined for each $\tau \in [0, 1]^N$ as

$$v(\tau) = \bigvee \{c \cdot x : Ax \leq \tau \cdot b\}$$

Example 2.4 Butnariu and Klement [7] proposed to represent rate problems for services (electricity or water) in bulk. The individual services are the players and the bulks of services are the coalitions. Each customer (or kind of customers) is identified

[3]The notion of fuzzy game is also used for cooperative games with fuzzy payoffs, see Mares [12]. The concept of game on $[0, 1]^N$ can be studied in a different way as game with overlapping coalitions, see Chalkiadakis et al. [8].

with the particular bulk she consumes. Fuzzy coalitions represent bulks for services where the degree membership of an individual one is exactly the share of this service consumed by a particular customer. The worth of each fuzzy coalition is the cost of the particular bulk of consume.

The idea of value was extended also for fuzzy coalitions.

Definition 2.3 A value for fuzzy games over N is a mapping assigning a payoff vector in $\mathbb{R}^N$ to each fuzzy game over N. Generally a value f for fuzzy games determines a value f^N for each finite set N.

There are several intents to extend the Shapley value to fuzzy games. However it is not an easy problem in the sense that it is not possible to consider all the feasible variations (marginal contributions) of a player in the game, namely we cannot take all the information in a fuzzy game. We show two examples of these mappings. The first option is defined taking the crisp version of a fuzzy game.

Definition 2.4 If v is a fuzzy game over N then the crisp version is given by $v^{cr} \in \mathcal{G}^N$ with

$$v^{cr}(S) = v(e^S),$$

for all $S \subseteq N$.

Branzei et al. [4] introduced a Shapley value for fuzzy games.

Definition 2.5 The crisp Shapley value is defined for each v fuzzy game over N as

$$\phi^{cr}(v) = \phi(v^{cr}).$$

This concept is very limited because it only uses the crisp information of the game, thus if two fuzzy games v, v' verify $v^{cr} = v'^{cr}$ then $\phi^{cr}(v) = \phi^{cr}(v')$.

To describe axioms for the crisp Shapley value is not complicated from the usual axioms of the classical Shapley value. But the uniqueness cannot be obtained from the unanimity games (the set of fuzzy games over N is not a finite vectorial space). We use the axiomatization in Theorem 1.5. If v is a fuzzy game over N and $S \subset N$ then the *fuzzy subgame* $v|_S$ is a new fuzzy game over S that we denote usually again as v where $v(\tau) = v(\tau^0)$ for each $\tau \in [0, 1]^S$ and

$$\tau^0(i) = \begin{cases} \tau(i) & \text{if } i \in S \\ 0 & \text{if } i \in N \setminus S. \end{cases}$$

Theorem 2.1 *The crisp Shapley value is the only value for fuzzy games satisfying efficiency (over e^N) and balanced contributions*

Proof As the Shapley value is efficient (Proposition 1.9) for games over N then

$$\phi^{cr}(v)(N) = \phi(v^{cr})(N) = v^{cr}(N) = v(e^N).$$

Observe that if v is a fuzzy game over N then the crisp version of v as fuzzy subgame over S is the same that the subgame over S of the crisp version of v, namely $(v^{cr})|_S = (v|_S)^{cr}$. Balanced contributions follows since the Shapley value satisfies it (Proposition 1.16) and the above fact,

$$\begin{aligned}(\phi^{cr})_i^N(v) - (\phi^{cr})_j^N(v) &= \phi_i^N(v^{cr}) - \phi_j^N(v^{cr}) = \phi_i^{N\setminus\{j\}}(v^{cr}) - \phi_j^{N\setminus\{i\}}(v^{cr}) \\ &= (\phi^{cr})_i^{N\setminus\{j\}}(v) - (\phi^{cr})_j^{N\setminus\{i\}}(v).\end{aligned}$$

The proof of the uniqueness is exactly the same that in Theorem 1.5. □

The Shapley value was extended also to the class of fuzzy games by the own Aubin [1]. This extension is named the diagonal value. The first problem is that the diagonal value was only defined for a particular class of fuzzy games, those continuously differentiable. Later this value has been extended to another more general family of fuzzy games, see Butnariu and Klement [7] or Mertens [14].

Definition 2.6 Let v be a continuously differentiable fuzzy game. The diagonal value is defined for v and $i \in N$ as

$$\phi_i^d(v) = \int_0^1 D_i v(te^N)\,dt.$$

The diagonal value supposes that coalitions are formed by aggregation of small differences for one player but taking a symmetric membership of all the players. As we will see in Sect. 2.4 the expression of the diagonal value is based in another formulation previously known of the usual Shapley value. Aubin [1] provided the value with an axiomatization. Let f be a value for games with fuzzy coalitions. Besides efficiency[4], anonymity and linearity he introduced next axioms.

[4] Aubin named it pareto optimality.

Continuously. f is a continuous operator.

Consider $P = \{S_1, \ldots, S_m\}$ a partition of N in groups and $M = \{1, \ldots, m\}$. For each fuzzy coalition $\tau \in [0, 1]^M$ of groups we induce a fuzzy coalition $\tau^P \in [0, 1]^N$ for players as $\tau^P(i) = \tau(k)$ if $i \in S_k$. For each fuzzy game v over N a new fuzzy game over M is defined as

$$v^P(\tau) = v(\tau^P).$$

Atomicity. Let v be a fuzzy game over N and P a partition of N. It holds for all $k \in M$,

$$f_k^M(v^P) = \sum_{i \in S_k} f_i^N(v).$$

Theorem 2.2 *The diagonal value is the only value for continuously differentiate fuzzy games satisfying efficiency, linearity, anonymity, continuously and atomicity.*

Proof The diagonal value verifies continuously and linearity by construction. Adding all the payoffs of the players we get by the Barrow law for the linear integral

$$\sum_{i \in N} \int_0^1 D_i v(te^N)\, dt = \int_0^1 Dv(te^N) \cdot e^N\, dt = v(e^N).$$

Anonymity follows from $\theta(te^N) = te^N$ and $D_{\theta(i)}\theta v(te^N) = D_i v(te^N)$. We test atomicity. Let $P = \{S_1, \ldots, S_m\}$ be a partition of N in groups and $M = \{1, \ldots, m\}$. The fuzzy coalition $\tau = te^M \in [0, 1]^M$ induces $\tau^P = te^N \in [0, 1]^N$. Thus, for each $k \in M$ the chain rule applied to $y_k = e^{S_k} \cdot x$ with $x = (x_i)_{i \in N}$ for each k implies

$$D_k(v^P)(te^M) = Dv(te^N) \cdot e^{S_k} = \sum_{i \in S_k} D_i v(te^N).$$

To prove the uniqueness we consider a value f satisfying all the axioms. Polynomial functions are dense in the set of continuous differentiate fuzzy games, therefore by continuously and linearity we only have to study fuzzy games as $v(\tau) = \tau_1^{p_1} \cdots \tau_n^{p_n}$. First consider the game $v(\tau) = \tau_1 \cdots \tau_n$. Anonymity and efficiency imply $f_i(v) = \dfrac{1}{n}$ following the proof in Proposition 1.7. Now, for the general case, suppose p_k copies of τ_k for each $k \in N$. If S_k is set of copies of k and $N' = \bigcup_{k \in N} S_k$ then $\{S_1, \ldots, S_n\}$ is a partition of N'. So, atomicity says that if $i \in S_k$ then

$$f_i(v) = \frac{p_k}{p_1 + \cdots + p_n}.$$

□

These solutions, the crisp Shapley value and the diagonal one, are different as is showed in [5].

Example 2.5 Suppose v a fuzzy game over $N = \{1, 2\}$ given by

$$v(\tau) = \tau(1)\tau^2(2) \ \forall \tau \in [0, 1].$$

The crisp version is $cr(v)(\{1\}) = cr(v)(\{2\}) = 0$, $cr(v)(\{1, 2\}) = 1$. Hence

$$\phi^{cr}(v) = (1/2, 1/2).$$

Now $Dv(\tau) = (\tau^2(2), 2\tau(1)\tau(2))$ and $Dv(te^N) = (t^2, 2t^2)$. Thus,

$$\phi_1^d(v) = \int_0^1 t^2\, dt = \frac{1}{3}, \quad \phi_2^d(v) = \int_0^1 2t^2\, dt = \frac{2}{3},$$

and $\phi^d(v) = (1/3, 2/3)$.

2.4 Fuzziness of Games

Let $v \in \mathcal{G}^N$ be a game. We look for a process to evaluate a fuzzy coalition by v. In a crisp coalition e^S all the players participate in the coalition at the same level 1. Considering that the worth of a coalition is proportional to the membership of the players we know how to evaluate coalitions as te^S, $v(te^S) = tv(S)$. With this premise, Aubin [1] proposed a way to estimate the worth of a fuzzy coalition.

Definition 2.7 Let $\tau \in [0, 1]^N$ be a fuzzy coalition. A partition by levels of τ is a finite sequence $\{(S_k, s_k)\}_{k=1}^m$ satisfying:

(1) $S_k \subseteq N$ and $s_k > 0$ for each $k = 1, \ldots, m$,
(2) $\sum_{k=1}^m s_k e^{S_k} = \tau$.

A partition function is a mapping pl over $[0, 1]^N$ taking a partition by levels $pl(\tau)$ for each fuzzy coalition τ.

Observe that if $\{(S_k, s_k)\}_{k=1}^m$ is a partition by levels of $\tau \in [0, 1]^N$ then

$$\sum_{\{k: i \in S_k\}} s_k = \tau(i), \tag{2.6}$$

for every player $i \in N$. Moreover each $S_k \subseteq supp(\tau)$. If $\tau = 0$ then no partition is necessary but if we technically need to take one we will consider $\{(\emptyset, t)\}$ with any $t \in [0, 1]$. Fixed a game, we can give a partition function determining a fuzzy game.

Definition 2.8 Let $v \in \mathcal{G}^N$ and let pl be a partition function. The fuzziness of v obtained from pl is the fuzzy game v^{pl} verifying for all $\tau \in [0, 1]^N$ and $pl(\tau) = \{(S_k, s_k)\}_{k=1}^m$,

$$v^{pl}(\tau) = \sum_{k=1}^{m} s_k v(S_k).$$

Generally, these fuzziness of a game are different as we will see later but they coincide for additive games.

Example 2.6 If $v \in \mathbb{R}^N$ is an additive game over N then for all partition function and $\tau \in [0, 1]^N$ it holds $v^{pl}(\tau) = \tau \cdot v$. We get using (2.6) with $pl(\tau) = \{(S_k, s_k)\}_{k=1}^m$,

$$v^{pl}(\tau) = \sum_{k=1}^{m} s_k \sum_{i \in S_k} v_i = \sum_{i \in N} \left[\sum_{\{k: i \in S_k\}} s_k \right] v_i = \sum_{i \in N} \tau(i) v_i = \tau \cdot v.$$

The own Aubin [1] considered a fuzziness of games. If $v \in \mathcal{G}^N$ then for any coalition $\tau \in [0, 1]^N$,

$$\overline{v}(\tau) = \bigvee \left\{ \sum_{k=1}^{m} s_k v(S_k) : \{(S_k, s_k)\}_{k=1}^m \text{ partition by levels of } \tau \right\}. \quad (2.7)$$

In this case we take a partition function pl choosing an optimal partition by levels in the above supremum. But this definition is really a total flexible option, furthermore if S is a coalition then $\overline{v}(e^S) = v^{sa}(S)$, see (1.6). We are interested in non-flexible fuzziness (as far as we can control) thus it seems natural to require $v^{pl}(e^S) = v(S)$ for all coalition S. But, as we will see later in the next chapters it is better to specify more this idea.

Definition 2.9 A partition function pl in N is an extension if $pl(S) = \{(S, 1)\}$ for all non-empty coalition S. The fuzziness of a game obtained by an extension is named also an extension of the game.

The fuzziness (2.7) proposed by Aubin is not an extension. Partition functions can depend on the game, as the Aubin's one. When the partition function is independent on the game we guarantee that the fuzziness working well with the vectorial space $\mathcal{G}^N$. So, the sum and scalar product satisfy

$$(av_1 + bv_2)^{pl} = av_1^{pl} + bv_2^{pl}.$$

Suppose a partition function independent of the game pl from now on. Obviously the crisp Shapley value of an extension of a game is the Shapley value of the original game, because $(v^{pl})^{cr} = v$ for all $v \in \mathcal{G}^N$.

Theorem 2.3 *If pl is an extension then for all game $v \in \mathcal{G}^N$*

$$\phi^{cr}(v^{pl}) = \phi(v).$$

But about the diagonal value we cannot say anything in general because fuzziness are not always continuously differentiate functions.

Three interesting extensions which do not depend on the game have been studied in the literature. One of them from the probabilistic point of view and the others following the membership interpretation in games. Suppose that we interpret number $\tau(i)$ in a fuzzy coalition τ as the membership of player i. Following the classical model players should look for the maximal cooperation. But in a fuzzy situation this fact can raise in two different ways:

- They look for the biggest level of cooperation, or
- they look for the biggest crisp coalition.

We will analyze these three options in the next subsections.

The Multilinear Extension

The first one was introduced by Owen [15] outside the context of fuzzy coalitions, and later by Meng and Zhang [13]. In this case component $\tau(i)$ in a fuzzy coalition is interpreted as the probability for player i to cooperate, and then $1 - \tau(i)$ the probability of non cooperating. Following the philosophy of the classical game if S was formed then the profit of this coalition is $v(S)$ and this fact is not variable.

Definition 2.10 Let $v \in \mathcal{G}^N$ be a game. The multilinear extension of v is a fuzziness of v defined for each $\tau \in [0, 1]^N$ as

$$v^{ml}(\tau) = \sum_{S \subseteq N} \left[\prod_{i \in S} \tau(i) \prod_{i \notin S} (1 - \tau(i)) \right] v(S).$$

We consider the mapping ml given for every $\tau \in [0,1]^N$ by

$$ml(\tau) = \left(S, \prod_{i \in S} \tau(i) \prod_{i \notin S} (1 - \tau(i)) \right)_{\{S \subseteq N : [\tau]_1 \subseteq S \subseteq supp(\tau)\}}. \quad (2.8)$$

Namely we use any coalition containing player for sure, those in $[\tau]_1$, and not containing players out of the support, those impossible for cooperating. The worth $v^{ml}(\tau)$ is the expecting one of cooperation, particularly each level for a coalition is the probability of forming this coalition.

Lemma 2.1 *Let S be a coalition in N. If $\tau \in [0,1]^N$ is a fuzzy coalition then*

$$\sum_{T \subseteq S} \prod_{i \in T} \tau(i) \prod_{i \in S \setminus T} (1 - \tau(i)) = 1.$$

Proof We prove the equality by induction in the cardinality of S. If $S = \{i\}$ then there are two options, $T = \emptyset$ or $T = S$, hence $1 - \tau(i) + \tau(i) = 1$. Suppose true when $|S| < k$ and take S with $|S| = k$. Let any $j \in S$,

$$\begin{aligned}
\sum_{T \subseteq S} \prod_{i \in T} \tau(i) \prod_{i \in S \setminus T} (1 - \tau(i)) &= \sum_{\{T \subseteq S : j \in T\}} \tau(j) \prod_{i \in T \setminus \{j\}} \tau(i) \prod_{i \in S \setminus T} (1 - \tau(i)) \\
&\quad + \sum_{\{T \subseteq S : j \notin T\}} (1 - \tau(j)) \prod_{i \in T} \tau(i) \prod_{i \in S \setminus (T \cup \{j\})} (1 - \tau(i)) \\
&= \sum_{T \subseteq S \setminus \{j\}} \left[\tau(j) \prod_{i \in T} \tau(i) \prod_{i \in (S \setminus \{j\}) \setminus T} (1 - \tau(i)) \right. \\
&\quad \left. + (1 - \tau(j)) \prod_{i \in T} \tau(i) \prod_{i \in (S \setminus \{j\}) \setminus T} (1 - \tau(i)) \right] \\
&= \prod_{i \in T} \tau(i) \prod_{i \in (S \setminus \{j\}) \setminus T} (1 - \tau(i)) = 1,
\end{aligned}$$

because $|S \setminus \{j\}| < k$. □

Remark 2.1 The reader can think τ as a set of probability distributions, one for each player, thus the probability to obtain a set of them is 1. If we take the distributions of a particular coalition then the probability to obtain a subset of this coalition is also 1.

We test now that ml gets a partition by levels for each fuzzy coalition and it is an extension.

Proposition 2.1 *Mapping ml is an extension.*

Proof First we see that ml is a partition function, namely it obtains a partition by levels for each fuzzy coalition $\tau \in [0,1]^N$. Every level in $ml(\tau)$ is non-zero by the election of S. Moreover, if coalition S does not verify $[\tau]_1 \subseteq S \subseteq supp(\tau)$ then

$$\prod_{i\in S}\tau(i)\prod_{i\notin S}(1-\tau(i)) = 0.$$

Now, for each player $j \in N$ we have

$$\begin{aligned}\sum_{\{[\tau]_1\subseteq S\subseteq supp(\tau),\, j\in S\}}\prod_{i\in S}\tau(i)\prod_{i\notin S}(1-\tau(i)) &= \sum_{\{S\subseteq N:\, j\in S\}}\prod_{i\in S}\tau(i)\prod_{i\notin S}(1-\tau(i))\\ &= \tau(j)\sum_{\{S\subseteq N:\, j\in S\}}\prod_{i\in S\setminus\{j\}}\tau(i)\prod_{i\notin S}(1-\tau(i))\\ &= \tau(j)\sum_{\{S\subseteq N\setminus\{j\}\}}\prod_{i\in S}\tau(i)\prod_{i\in(N\setminus\{j\})\setminus S}(1-\tau(i))\\ &= \tau(j),\end{aligned}$$

applying the above lemma to $N \setminus \{j\}$. Finally we test that for any coalition S we get $ml(e^S) = \{(S,1)\}$, the only coalition in $ml(e^S)$ is S because $[e^S]_1 = S = supp(e^S)$ and $\prod_{i\in S} e^S(i)\prod_{i\notin S}(1-e^S(i)) = 1$. □

Example 2.7 Suppose $v \in \mathcal{G}^N$ any game over $N = \{1,2,3,4,5\}$. Consider the fuzzy coalition $\tau = (0.2, 0, 0.7, 1, 0.2)$. The information contained in τ says that for instance the probability of cooperating player 3 is 0.7 and so her probability of non cooperating is 0.3. What is certain in τ is that player 2 will never cooperate and player 4 is always willing to cooperate. The expecting worth of τ is

$$\begin{aligned}v^{ml}(\tau) = {} & 0.192v(\{4\}) + 0,048v(\{1,4\}) + 0.448v(\{3,4\}) + 0.048v(\{4,5\})\\ & + 0.112v(\{1,3,4\}) + 0.012v(\{1,4,5\}) + 0.112v(\{3,4,5\})\\ & + 0.028v(\{1,3,4,5\})\end{aligned}$$

Example 2.8 Suppose the game v over $N = \{1,2\}$ defined as $v(\{1\}) = 1$, $v(\{2\}) = 3$ and $v(\{1,2\}) = 8$. Let $\tau = (x,y) \in [0,1]\times[0,1]$ be any fuzzy coalition. The multilinear function is the quadratic polynomial (the hyperbolic paraboloid in Fig. 2.1),

$$v^{ml}(x,y) = 4xy + x + 3y,$$

with the unit square as domain.

Example 2.9 Let $T \subseteq N$ be a non-empty set. We calculate the multilinear extension of the unanimity game u_T. For each $\tau \in [0,1]^N$,

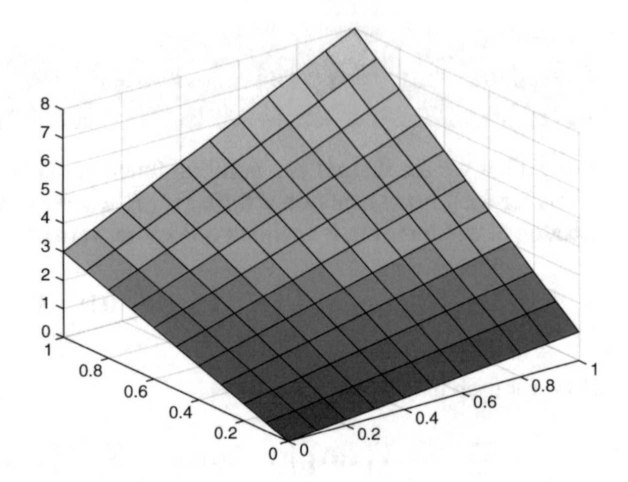

Fig. 2.1 The multilinear extension v^{ml}

$$(u_T)^{ml}(\tau) = \sum_{T\subseteq S}\left[\prod_{j\in S}\tau(j)\prod_{j\notin S}(1-\tau(j))\right] = \prod_{i\in T}\tau(i)\sum_{T\subseteq S}\left[\prod_{j\in S\setminus T}\tau(j)\prod_{j\notin S}(1-\tau(j))\right]$$

$$= \prod_{i\in T}\tau(i)\sum_{S\subseteq N\setminus T}\left[\prod_{j\in S}\tau(j)\prod_{j\in(N\setminus T)\setminus S}(1-\tau(j))\right] = \prod_{i\in T}\tau(i),$$

using Lemma 2.1 with $N\setminus T$. Thus $(u_T)^{ml}(\tau)$ is the probability of containing T.

Using the above example we get another formula of the multilinear extension using the dividends (Proposition 1.1) of the games.

Proposition 2.2 *For all game $v \in \mathcal{G}^N$ it holds*

$$v^{ml}(\tau) = \sum_{\{S\subseteq supp(\tau):S\neq\emptyset\}} \Delta_S^v \prod_{i\in S}\tau(i).$$

Proof Given two games v_1, v_2 we know that

$$(av_1 + bv_2)^{ml} = av_1^{ml} + bv_2^{ml},$$

because the extension is independent on the game. Proposition 1.1 and Example 2.9 imply the result. Observe that if a coalition S contains a player i with $\tau(i) = 0$ then $u_S^{ml}(\tau) = 0$. □

The properties of games in Definition 1.3 can be extended to fuzzy games. In Branzei et al. [5] there is an extended analysis of these properties in the fuzzy context. We focus the idea only about what happens with the fuzziness if we take a monotone, superadditive or convex game. Example 2.6 proved that the fuzziness of any additive game v is the linear function with v as coefficients.

Proposition 2.3 *Let $\tau, \tau' \in [0, 1]^N$ be two fuzzy coalitions.*

(1) If $v \in \mathcal{G}_m^N$ and $\tau \leq \tau'$ then $v^{ml}(\tau) \leq v^{ml}(\tau')$.

(2) If $v \in \mathcal{G}_{sa}^N$ and $\tau \wedge \tau' = 0$ then $v^{ml}(\tau \vee \tau') \geq v^{ml}(\tau) + v^{ml}(\tau')$.

(3) If $v \in \mathcal{G}_c^N$ then $v^{ml}(\tau \vee \tau') + v^{ml}(\tau \wedge \tau') \geq v^{ml}(\tau) + v^{ml}(\tau')$.

Proof (1) Let v be a monotone game. Suppose $\tau \in [0, 1]^N$ and $t \in [0, 1 - \tau(i)]$ for any player i. We get

$$\begin{aligned} v^{ml}(\tau + te^{\{i\}}) = & \sum_{\{S \subseteq N : i \in S\}} (\tau(i) + t) \prod_{j \in S \setminus \{i\}} \tau(j) \prod_{j \notin S} (1 - \tau(j)) v(S) \\ & + \sum_{\{S \subseteq N : i \notin S\}} (1 - \tau(i) - t) \prod_{j \in S} \tau(j) \prod_{j \notin S \cup \{i\}} (1 - \tau(j)) v(S) \\ & = v^{ml}(\tau) + t \sum_{\{S \subseteq N : i \in S\}} \prod_{j \in S \setminus \{i\}} \tau(j) \prod_{j \notin S \cup \{i\}} (1 - \tau(j)) [v(S) - v(S \setminus \{i\})] \\ & \geq v^{ml}(\tau), \end{aligned}$$

because, as v is monotone then $v(S) \geq v(S \setminus \{i\})$ for all coalition S. Now if $\tau \leq \tau'$ then

$$\tau' = \tau + \sum_{i \in N} [\tau'(i) - \tau(i)] e^{\{i\}}.$$

Applying sequentially the above reasoning we have $v^{ml}(\tau) \leq v^{ml}(\tau')$.
(2) Let v be a superadditive game. If $\tau, \tau' \in [0, 1]^N$ with $\tau \wedge \tau' = 0$ then for each player i one of them $\tau(i)$ or $\tau'(i)$ is null. So, we have $supp(\tau) \cap supp(\tau') = \emptyset$. We denote $supp(\tau) = N_1$ and $supp(\tau') = N_2$. Moreover $(\tau \vee \tau')(i) = \tau(i)$ if $i \in N_1$, $(\tau \vee \tau')(i) = \tau'(i)$ if $i \in N_2$ and $(\tau \vee \tau')(i) = 0$ otherwise. Each coalition into $supp(\tau \vee \tau')$ can be written as $S \cup T$ with $S \subseteq N_1$ and $T \subseteq N_2$. We obtain using the superadditivity of v and Lemma 2.1,

$$v^{ml}(\tau \vee \tau') =$$

$$= \sum_{S \subseteq N_1} \sum_{T \subseteq N_2} \left[\prod_{i \in S} \tau(i) \prod_{i \in T} \tau'(i) \prod_{i \in N_1 \setminus S} (1 - \tau(i)) \prod_{i \in N_2 \setminus T} (1 - \tau'(i)) \right] v(S \cup T)$$

$$\geq \sum_{S \subseteq N_1} \sum_{T \subseteq N_2} \left[\prod_{i \in S} \tau(i) \prod_{i \in T} \tau(i) \prod_{i \in N_1 \setminus S} (1 - \tau(i)) \prod_{i \in N_2 \setminus T} (1 - \tau'(i)) \right] (v(S) + v(T))$$

$$= \sum_{S \subseteq N_1} \prod_{i \in S} \tau(i) \prod_{i \in N_1 \setminus S} (1 - \tau(i)) \left[\sum_{T \subseteq N_2} \prod_{i \in T} \tau(i) \prod_{i \in N_2 \setminus T} (1 - \tau'(i)) \right] v(S)$$

$$+ \sum_{T \subseteq N_2} \prod_{i \in T} \tau(i) \prod_{i \in N_2 \setminus T} (1 - \tau(i)) \left[\sum_{S \subseteq N_1} \prod_{i \in S} \tau(i) \prod_{i \in N_1 \setminus S} (1 - \tau'(i)) \right] v(T)$$

$$= \sum_{S \subseteq N_1} \prod_{i \in S} \tau(i) \prod_{i \in N_1 \setminus S} (1 - \tau(i)) v(S) + \sum_{T \subseteq N_2} \prod_{i \in T} \tau(i) \prod_{i \in N_2 \setminus T} (1 - \tau(i)) v(T)$$

$$= v^{ml}(\tau) + v^{ml}(\tau').$$

(3) Now suppose v a convex game. As multilinear function v^{ml} is twice continuously differentiate for each player i it holds

$$D_i(v^{ml})(\tau) = \sum_{\{S \subseteq N : i \in S\}} \prod_{j \in S \setminus \{i\}} \tau(j) \prod_{j \notin S} (1 - \tau(j)) v(S)$$

$$- \sum_{\{S \subseteq N : i \notin S\}} \prod_{j \in S} \tau(j) \prod_{j \notin S \cup \{i\}} (1 - \tau(j)) v(S \setminus \{i\})$$

$$= \sum_{\{S \subseteq N : i \in S\}} \prod_{j \in S \setminus \{i\}} \tau(j) \prod_{j \notin S \cup \{i\}} (1 - \tau(j)) \left[v(S \cup \{i\}) - v(S) \right]$$

Therefore $D_{ii}(v^{ml})(\tau) = 0$. If $k \neq i$ then using the same reasoning in the first derivate we have

$$D_{ik}(v^{ml})(\tau) = \sum_{\{S \subseteq N : i, k \in S\}} \prod_{j \in S \setminus \{i,k\}} \tau(j) \prod_{j \notin S \cup \{i,k\}} (1 - \tau(j)) \cdot$$

$$\cdot \left[v(S \cup \{i, k\}) - v(S \cup \{i\}) - v(S \cup \{k\}) + v(S) \right].$$

Since v is convex then $v(S \cup \{i, k\}) + v(S) \geq v(S \cup \{i\}) + v(S \cup \{k\})$, thus

$$D_{ik}(v^{ml})(\tau) \geq 0.$$

Hence $D_i(v^{ml})$ is an increasing function and so are the increments regard to i, if $\tau \leq \tau'$ and $t \in [0, 1 - \tau'(i)]$ then

$$v^{ml}(\tau + te^{\{i\}}) - v(\tau) \leq v^{ml}(\tau' + te^{\{i\}}) - v(\tau').$$

Now we take any $\tau, \tau' \in [0, 1]^N$ and $R = \{i \in N : \tau'(i) < \tau(i)\} = \{i_1, \ldots, i_p\}$. We set $h_p = \tau(i_p) - \tau'(i_p)$ and then

$$\tau = \tau \wedge \tau' + \sum_{q=1}^{p} h_p e^{\{i_p\}}, \quad \tau \vee \tau' = \tau' + \sum_{q=1}^{p} h_p e^{\{i_p\}}.$$

Applying sequentially the above condition to τ', $\tau \wedge \tau'$ we get

$$v^{ml}(\tau) - v^{ml}(\tau \wedge \tau') \leq v^{ml}(\tau \vee \tau') - v^{ml}(\tau').$$

□

Remark 2.2 (1) Conditions (2) and (3) in the above proposition are true also with product and coproduct, namely a probabilistic convexity.

$$v^{ml}(\tau \otimes \tau') + v^{ml}(\tau \times \tau') \geq v^{ml}(\tau) + v^{ml}(\tau').$$

(2) A fuzzy game satisfying condition (2) in the proposition is known as a fuzzy superadditive game.
(3) A fuzzy game satisfying condition (3) is named supermodular fuzzy game. Branzei et al. [4] introduced the concept of fuzzy convex game as a fuzzy game which is supermodular and also a convex function for each component.

Obviously the multilinear extension of a game is a continuously differentiate fuzzy game as we said before. Owen [15] had showed that the Shapley value is the diagonal value of the multilinear extension before the diagonal value was defined.

Theorem 2.4 *Let $v \in \mathcal{G}^N$ be a game. It holds:*

$$\phi^d(v^{ml}) = \phi(v).$$

Proof We calculate the diagonal value of v^{ml}. If $i \in N$ then using Proposition 2.2,

$$D_i(v^{ml})(\tau) = \sum_{\{S \subseteq N : i \in S\}} \Delta_S^v \prod_{j \in S \setminus \{i\}} \tau(j). \tag{2.9}$$

Hence for each $t \in [0, 1]$ we have

$$D_i(v^{ml})(te^N) = \sum_{\{S \subseteq N : i \in S\}} \Delta_S^v t^{|S|-1}.$$

Finally we do the integral and from Theorem 1.2,

$$\int_0^1 D_i(v^{ml})(te^N)\, dt = \sum_{\{S \subseteq N : i \in S\}} \Delta_S^v \int_0^1 t^{|S|-1}\, dt$$
$$= \sum_{\{S \subseteq N : i \in S\}} \frac{\Delta_S^v}{|S|} = \phi_i(v).$$

□

Straffin [17] used the multilinear extension of a simple game (Definition 1.3) to explain the Shapley-Shubik index since a probabilistic point of view. If $v \in \mathcal{G}_s^N$ then

$$v^{ml}(\tau) = \sum_{S \in W(v)} \prod_{j \in S} \tau(j) \prod_{j \notin S} \tau(j),$$

where $W(v)$ is the set of winning coalitions. The fuzzy set τ is interpreted as a set of probability distributions, one for each player, determining the possibility to support a motion. So, $v^{ml}(\tau)$ is the probability to join a winning coalition. The Shapley-Shubik index, following the above theorem, corresponds to the diagonal value of the multilinear extension of our simple game. For each player i number

$$D_i(v^{ml})(\tau) = \sum_{S \in SW_i(v)} \prod_{j \in S \setminus \{i\}} \tau(j) \prod_{j \notin S \cup \{i\}} (1 - \tau(j))$$

(obtained in the proof of Proposition 2.3) represents the probability to get a swing (Definition 1.11) for this player. The equality

$$\phi_i(v) = \int_0^1 D_i(v^{ml})(te^N)\, dt$$

means to calculate the expectation to be the critical person in the voting under the homogeneity assumption (all the probability distributions are the same).

The Proportional Extension

We focus now on the fuzziness proposed by Butnariu [6]. Players consider the maximal level of cooperation, and the biggest coalition with this level. Then the second level and go on. This condition guarantees that each player only play once as in the classical model. If $\tau \in [0, 1]^N$ is a fuzzy coalition over N we set for each $t \in [0, 1]$

$$S_t^\tau = \{i \in N : \tau(i) = t\}. \tag{2.10}$$

Definition 2.11 Let $v \in \mathcal{G}^N$ be a game. The proportional extension of v is a fuzziness of v defined for each $\tau \in [0, 1]^N$ as

$$v^{pr}(\tau) = \sum_{t \in im(\tau)} t v\left(S_t^\tau\right).$$

Obviously family

$$pr(\tau) = (S_t^\tau, t)_{t \in im(\tau)} \tag{2.11}$$

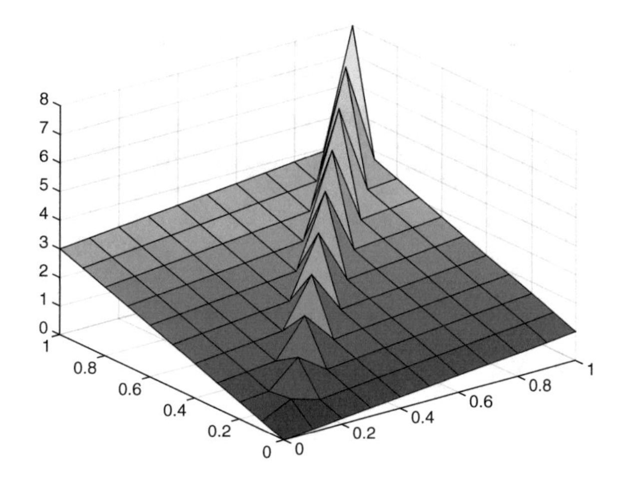

Fig. 2.2 The proportional extension v^{pr}

is a partition by levels of τ. If $\tau = e^S$ then $im(\tau) = \{1\}$ and $S_1^\tau = S$, therefore we can enunciate the following result.

Proposition 2.4 *Mapping pr is an extension.*

Example 2.10 Consider Example 2.7. The meaning now of the levels in the fuzzy coalition $\tau = (0.2, 0, 0.7, 1, 0.2)$ is complete different. In this case, for player 3, number 0.7 is the real membership and not a probability. Players look first for the maximal level of cooperation in τ, thus player 1 is interested to cooperate only with other players who are able to play at level 0.2. The same with player 3 and hence player 3 is not interested in cooperating with player 1. So, in this case,

$$v^{pr}(\tau) = v(\{4\}) + 0.7v(\{3\}) + 0.2v(\{1, 5\}).$$

Example 2.11 Suppose again the game v in Example 2.8 over $N = \{1, 2\}$ defined as $v(\{1\}) = 1$, $v(\{2\}) = 3$ and $v(\{1, 2\}) = 8$. Let $\tau = (x, y) \in [0, 1] \times [0, 1]$ be any fuzzy coalition. The proportional extension is a piecewise linear function (see Fig. 2.2) which is discontinuous on the diagonal of the square,

$$v^{pr}(x, y) = \begin{cases} 8x, & \text{if } x = y \\ x + 3y, & \text{if } x \neq y. \end{cases}$$

Example 2.12 Let $T \subseteq N$ be a non-empty set. The proportional extension of the unanimity game u_T is given for each $\tau \in [0, 1]^N$ as

$$(u_T)^{pr}(\tau) = \begin{cases} t, & \text{if } T \subseteq S_t^\tau \\ 0, & \text{otherwise.} \end{cases}$$

Using the worths of the fuzzy coalitions for unanimity games obtained in the above example we give another formula for the proportional extension.

Proposition 2.5 *For all game $v \in \mathcal{G}^N$ it holds*

$$v^{pr}(\tau) = \sum_{\{S \subseteq N: S \neq \emptyset, \exists t_S \text{ with } \tau(i)=t_S \, \forall i \in S\}} t_S \Delta_S^v.$$

Proof We get for two games v_1, v_2,

$$(av_1 + bv_2)^{pr} = av_1^{pr} + bv_2^{pr}.$$

Proposition 1.1 and Example 2.12 imply the result. □

Not always the properties of a game are transmitted to the fuzziness. Next example shows that monotonicity and convexity are not transmitted from the game to its fuzziness.

Example 2.13 Consider the game $v \in \mathcal{G}_m^N$ in Example 2.11. We have $v^{pr}(\tau) = 2.4$ with $\tau = (0.3, 0.3)$ and $v^{pr}(\tau') = 1.8$ with $\tau' = (0.3, 0.5)$, but $\tau \leq \tau'$.
Our game v is also convex (in this case, being convex coincides with being superadditive). Now take $\tau = (0.5, 0.5)$ and $\tau' = (0.4, 0.6)$. We have $\tau \vee \tau' = (0.5, 0.6)$ and $\tau \wedge \tau' = (0.4, 0.5)$. But

$$v(\tau \vee \tau') + v(\tau \wedge \tau') = 4.2 \leq 6.2 = v(\tau) + v(\tau').$$

At least superadditivity is transmitted by the proportional extension.

Proposition 2.6 *Let $v \in \mathcal{G}_{sa}^N$ be a superadditive game. For each two fuzzy coalitions τ, τ' with $\tau \wedge \tau' = 0$ it holds*

$$v^{pr}(\tau \vee \tau') \geq v^{pr}(\tau) + v^{pr}(\tau')$$

Proof Let $\tau, \tau' \in [0, 1]^N$ with $\tau \wedge \tau' = 0$. In that case $supp(\tau) \cap supp(\tau') = \emptyset$, moreover $im(\tau \vee \tau') = im(\tau) \cup im(\tau')$. This fact implies that for each $t \in im(\tau \vee \tau')$ we have

$$S_t^{\tau \vee \tau'} = S_t^{\tau} \cup S_t^{\tau'}, \quad S_t^{\tau} \cap S_t^{\tau'} = \emptyset.$$

So, using that v is superadditive

$$v^{pr}(\tau \vee \tau') = \sum_{t \in im(\tau \vee \tau')} t\, v(S_t^{\tau \vee \tau'}) = \sum_{t \in im(\tau \vee \tau')} t\, v(S_t^{\tau} \cup S_t^{\tau'})$$
$$\geq \sum_{t \in im(\tau) \cup im(\tau')} t\, v(S_t^{\tau}) + \sum_{t \in im(\tau) \cup im(\tau')} t\, v(S_t^{\tau'}) = v^{pr}(\tau) + v^{pr}(\tau').$$

Observe that if $t \in im(\tau') \setminus im(\tau)$ (or $t \in im(\tau) \setminus im(\tau')$) then $S_t^{\tau} = \emptyset$ ($S_t^{\tau'} = \emptyset$). □

The Choquet Extension

The third option uses the Choquet integral. Tsurumi et al. [18] proposed the following fuzziness. Players look for setting the largest coalition and for that they play with the smallest level.

Definition 2.12 Let $v \in \mathcal{G}^N$ be a game. The Choquet extension of v is a fuzziness of v defined for each $\tau \in [0, 1]^N$ as

$$v^{ch}(\tau) = \int_c \tau \, dv.$$

The definition of the Choquet integral in a finite set shows that the above definition is indeed a fuzziness of v,

$$v^{ch}(\tau) = \sum_{k=1}^{m} (\lambda_k - \lambda_{k-1}) v([\tau]_k),$$

if $im_0(\tau) = \{\lambda_0 < \lambda_1 < \cdots < \lambda_m\}$. So, the partition by levels is

$$ch(\tau) = (\lambda_k - \lambda_{k-1}, [\tau]_k)_{k=1}^{m}. \tag{2.12}$$

If $i \in N$ and $\tau(i) = \lambda_{k_0}$ then

$$\sum_{\{k : i \in [\tau]_k\}} (\lambda_k - \lambda_{k-1}) = \sum_{k=1}^{k_0} (\lambda_k - \lambda_{k-1}) = \tau(i).$$

Example 2.14 Following to Examples 2.7 and 2.10 we determine now the Choquet extension of v in τ. In this case, players look for the biggest coalition, so they have to use the smallest level in the image of τ, namely 0.2. Coalition $\{1, 3, 4, 5\}$ is formed. Now players still can cooperate choosing level $0.7 - 0.2 = 0.5$ and coalition $\{3, 4\}$ (using actually $(0, 0, 0.5, 0.8, 0)$) and go on.

$$v^{ch}(\tau) = (0.2 - 0)v(\{1, 3, 4, 5\}) + (0.7 - 0.2)v(\{3, 4\}) + (1 - 0.7)v(\{4\})$$
$$= 0.2v(\{1, 3, 4, 5\}) + 0.5v(\{3, 4\}) + 0.3v(\{4\}).$$

Next example shows graphically the proposed fuzziness with two players.

Example 2.15 Suppose again the game v in Examples 2.8 and 2.11 over $N = \{1, 2\}$ defined as $v(\{1\}) = 1$, $v(\{2\}) = 3$ and $v(\{1, 2\}) = 8$. Let $\tau = (x, y) \in [0, 1] \times [0, 1]$ be any fuzzy coalition. The Choquet extension is also a piece linear function (see Fig. 2.3) but it is continuously although it is not differentiate,

$$v^{ch}(x, y) = \begin{cases} x + 7y, & \text{if } x \geq y \\ 5x + 3y, & \text{if } x \leq y. \end{cases}$$

Example 2.16 Let $T \subseteq N$ be a non-empty set and $\lambda_{k_0} = \bigwedge_{i \in T} \tau(i)$. The Choquet extension of the unanimity game u_T is given for each $\tau \in [0, 1]^N$ with $im_0 = \{\lambda_0 < \lambda_1 < \cdots \lambda_m\}$ as

$$(u_T)^{ch}(\tau) = \sum_{\{k: T \subseteq [\tau]_k\}} (\lambda_k - \lambda_{k-1}) = \sum_{k=1}^{k_0} (\lambda_k - \lambda_{k-1}) = \bigwedge_{i \in T} \tau(i).$$

The above example allows again to describe the extension by the dividends of the game.

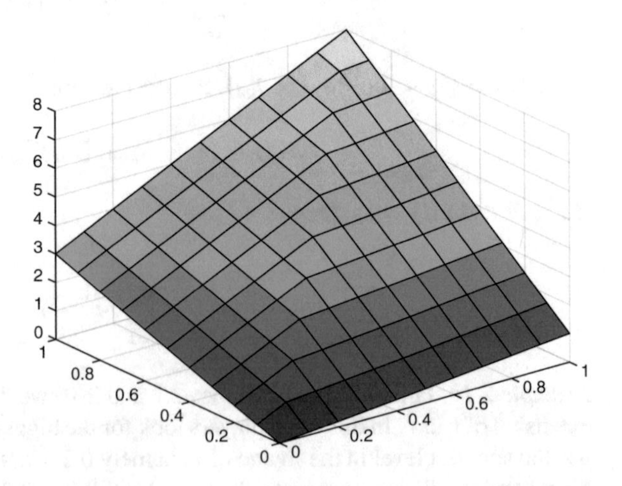

Fig. 2.3 The Choquet extension v^{ch}

Proposition 2.7 *For all game $v \in \mathcal{G}^N$ it holds*

$$v^{ch}(\tau) = \sum_{S \subseteq supp(\tau)} \Delta_S^v \bigwedge_{i \in S} \tau(i) = \sum_{k=1}^{m} \lambda_k \sum_{\{S \subseteq [\tau]_k : S \cap S_{\lambda_k}^{\tau} \neq \emptyset\}} \Delta_S^v,$$

if $im(\tau) = \{\lambda_1 < \cdots < \lambda_m\}$.

Proof We get for two games v_1, v_2,

$$(av_1 + bv_2)^{ch} = av_1^{ch} + bv_2^{ch}.$$

Proposition 1.1, Example 2.16 imply both formulas. Observe that $\bigwedge_{i \in S} \tau(i) = t$ if there exists $i \in S$ with $\tau(i) = t$. □

The Choquet extension works well with the properties of the original crisp game as the multilinear extension.

Proposition 2.8 *Let $\tau, \tau' \in [0, 1]^N$ be two fuzzy coalitions.*

(1) If $v \in \mathcal{G}_m^N$ and $\tau \leq \tau'$ then $v^{ch}(\tau) \leq v^{ch}(\tau')$.
(2) If $v \in \mathcal{G}_{sa}^N$ and $\tau \wedge \tau' = 0$ then $v^{ch}(\tau \vee \tau') \geq v^{ch}(\tau) + v^{ch}(\tau')$.
(3) If $v \in \mathcal{G}_c^N$ then $v^{ch}(\tau \vee \tau') + v^{ch}(\tau \wedge \tau') \geq v^{ch}(\tau) + v^{ch}(\tau')$.

Proof (1) The result follows since property (C8) of the Choquet integral (Sect. 2.2). (2) Let $v \in \mathcal{G}_{sa}^N$ If $\tau \wedge \tau' = 0$ then $im(\tau \vee \tau') = im(\tau) \cup im(\tau')$. Moreover $[\tau \vee \tau']_t = [\tau]_t \cup [\tau']_t$ and $[\tau]_t \cap [\tau']_t = \emptyset$ for all $t \in (0, 1]$. So, if $im_0(\tau \vee \tau') = \{0 = \lambda_0 < \lambda_1 < \cdots < \lambda_m\}$

$$\begin{aligned}
v^{ch}(\tau \vee \tau') &= \int_c \tau \vee \tau' \, dv = \sum_{k=1}^{m} (\lambda_k - \lambda_{k-1}) v([\tau]_k \cup [\tau']_k) \\
&\geq \sum_{k=1}^{m} (\lambda_k - \lambda_{k-1}) [v([\tau]_k) + v([\tau']_k)] \\
&= \sum_{k=1}^{m} (\lambda_k - \lambda_{k-1}) v([\tau]_k) + \sum_{k=1}^{m} (\lambda_k - \lambda_{k-1}) v([\tau']_k) \\
&= \int_c \tau \, dv + \int_c \tau' \, dv,
\end{aligned}$$

using (C10) in the last equality. Observe that $im(\tau), im(\tau')$ and both them contains

(3) Suppose $v \in \mathcal{G}_c^N$. Obviously $im(\tau \vee \tau') \cup im(\tau \wedge \tau') = im(\tau) \cup im(\tau')$. We denote

$$im_0(\tau) \cup im_0(\tau') = \{0 = \lambda_0 < \lambda_1 < \cdots < \lambda_m\}.$$

For each $t \in (0, 1]$ we obtain $[\tau \vee \tau']_t = [\tau]_t \cup [\tau']_t$ and $[\tau \wedge \tau']_t = [\tau]_t \cap [\tau']_t$. Now, by (C10) and convexity,

$$\begin{aligned} v^{ch}(\tau \vee \tau') + v^{ch}(\tau \wedge \tau') &= \int_c \tau \vee \tau' \, dv + \int_c \tau \wedge \tau' \, dv \\ &= \sum_{k=1}^{m} (\lambda_k - \lambda_{k-1})[v([\tau \vee \tau']_k) + v([\tau \wedge \tau']_k)] \\ &\geq \sum_{k=1}^{m} (\lambda_k - \lambda_{k-1})[v([\tau]_k) + v([\tau']_k)] \\ &= \int_c \tau \, dv + \int_c \tau\tau' \, dv \\ &= v^{ch}(\tau) + v^{ch}(\tau'). \end{aligned}$$

□

There are several ways to apply the diagonal formula to the Choquet extension by smoothing processes in Weiss [19].

References

1. Aubin, J.P.: Cooperative fuzzy games. Math. Oper. Res. **6**(1), 1–13 (1981)
2. Aumann, R.J., Shapley, L.S.: Values for Non-Atomic Games. Princeton University Press, Princeton (1974)
3. Borkotokey, S., Mesiar, R.: The Shapley value of cooperative games under fuzzy settings: a survey. Int. J. Gen. Syst. **43**(1), 75–95 (2014)
4. Branzei, R., Dimitrov, D., Tijs, S.: Convex fuzzy games and participation monotonic allocation schemes. Fuzzy Sets Syst. **139**, 267–281 (2003)
5. Branzei, R., Dimitrov, D., Tijs, S.: Models in Cooperative Game Theory. Springer, Berlin (2008)
6. Butnariu, D.: Stability and Shapley value for a n-persons fuzzy games. Fuzzy Sets Syst. **4**(1), 63–72 (1980)
7. Butnariu, D., Klement, E.P.: Triangular Norm-Based Measures and Games with Fuzzy Coalitions. Kluwer Academic Publishers, Dordrecht (1993)
8. Chalkiadakis, G., Elkind, E., Markakis, E., Polukarov, M., Jennings, N.R.: Cooperative games with overlapping coalitions. J. Artif. Intell. Res. **39**, 179–216 (2010)
9. Choquet, G.: Theory of capacities. Annales de L'Institut Fourier, vol. 5, pp. 131–295 (1953)
10. de Waegenaere, A., Wakker, P.P.: Nonmonotonic Choquet integrals. J. Math. Econ. **36**, 45–60 (2001)
11. Li, S.J., Zhang, Q.: A simplified expression of the Shapley function for fuzzy games. Eur. J. Oper. Res. **196**(1), 234–245 (2009)

12. Mares, M.: Fuzzy Cooperative Games. Studies in Fuzziness and Soft Computing vol. 72. Physica-Verlag, Heildelberg; Springer, Heildelberg (2001)
13. Meng, F.Y., Zhang, Q.: The Shapley function for fuzzy cooperative games with multilinear extension form. Appl. Math. Lett. **23**(5), 644–650 (2010)
14. Mertens, J.F.: The Shapley value in the non-differentiable case. Int. J. Game Theory **17**, 1–65 (1988)
15. Owen, G.: Multilinear extensions of games. Manag. Sci. **18**(5), 64–79 (1972)
16. Schmeidler, D.: Integral representation without additivity. Proc. Am. Math. Soc. **97**(2), 255–261 (1986)
17. Straffin, P.D.: Game Theory and Strategy. The Mathematical Association of America, vol. 36. New Mathematical Library (1993)
18. Tsurumi, M., Tanino, T., Inuiguchi, M.: A Shapley function on a class of cooperative fuzzy games. Eur. J. Oper. Res. **129**(3), 596–618 (2001)
19. Weiss, C.: Games with Fuzzy Coalitions, Concepts Based on the Choquet Extension. Bielefeld University, Germany (2003)
20. Zadeh, L.A.: Fuzzy sets. Inf. Control **8**, 338–353 (1965)

Chapter 3
Games with a Fuzzy Bilateral Relation Among the Players

3.1 Introduction

Values solve games for a particular set of players considering the cooperation among all of them. So, the solution of the cooperative game depends on the both of the elements defining it, the set of players cooperating and the mapping of the game. The classical solutions suppose as peers all the players. In real life, political, social or economic circumstances may impose certain restraints on coalition formation from the relations among the players. This idea has led several authors to develop models of cooperative games with partial cooperation. So, the classical model is modified including the information among the players. Shapley functions are values for games with this information which coincide with the Shapley value if the circumstances are harmless. In this book we analyze situations where there exists certain kind of information about the players and their relations. Now the payoff vector should be changed considering the information. There are two mean ways in the literature to do that. The first one consists in restricting the feasible coalitions by the information and then defining new values using only these coalitions. In Aumann and Dreze [3], Faigle and Kern [10], Bilbao [5], Jiménez [15] or Jiménez-Losada [16] the authors used this first way. The second one modifies the characteristic function of the game by the information defining a new classical game and later solving this new game. In Myerson [22], Derks and Peters [9], Gilles et al. [14], Algaba et al. [2], Gallardo et al. [12] the second way is followed. Fuzzy versions of some of these models were studied in Jiménez-Losada et al. [17], Meng et al. [19], Gallardo et al. [11, 12] or Gallego et al. [13].

We focus our analysis on the second way. So, we try to modify the game using some information about the players. This way means certain flexibility in the modification but, as we said in the before chapters, we work with non flexibility situations and then should respect the search of the greatest cooperations. Given a game and certain information we have to answer two different mean questions. How to assimilate the additional information in the game? depending on the situation and the interpretation

A. Jiménez-Losada, *Models for Cooperative Games with Fuzzy Relations among the Agents*, Studies in Fuzziness and Soft Computing 355,
DOI 10.1007/978-3-319-56472-2_3

of the information we can have several ways to modify the game. Once we have defined the new game we can apply the usual solutions, but formulas and axioms should be written from the original game and the information. Therefore the modified game will be only a tool and we should be able to decide which solution to apply.

In this chapter we present the general model focused on information from bilateral relations among the players. We will study several of these situations in the next chapters but now, in this one, we describe an easy particular case to illustrate how the model works.

3.2 Games with Information About the Players

Let N be our finite set of players and $v \in \mathcal{G}^N$. Suppose $I(N)$ a family of some kind of mathematical object representing different relations among the players such that there exists $K_{I(N)} \in I(N)$ harmless for them, and then $K_{I(N)}$ is identified to the classical case. For each element in $I(N)$ we have a different solution of the game. The inclusion of the new information into the game forces certain flexibility, but we try to follow the usual behavior of maximal cooperation restricted by the information. So fixed N, we have now two data points: a game $v \in \mathcal{G}^N$ and some information $K \in I(N)$. We name *game over N with information* $I(N)$ to the pair (v, K).

Definition 3.1 A value for games over N with information $I(N)$ is a mapping $f : \mathcal{G}^N \times I(N) \to \mathbb{R}^N$ such that $f(v, K)$ is interpreted as the payoff vector obtained if the relations among the players are regulated by K.

This kind of values can be also seen as uncertain solutions. So, a value f for games over N introduces for each game not just a payoff vector but a solution depending on the known information, namely for every $v \in \mathcal{G}^N$ we have a function f^v over $I(N)$ obtaining a vector $f^v(K)$ for all information set $K \in I(N)$. This other vision is named as *value function*, see Tsurumi et al. [26] for instance. We will use the first option, as in Definition 3.1, but any of both names.

The focus of our analysis is the Shapley value, therefore we look for extensions of this value to different kinds of information over the players.

Definition 3.2 A value f for games over N with information $I(N)$ is named a Shapley value if $f(v, K_{I(N)}) = \phi(v)$, where $K_{I(N)}$ represents the usual situation into $I(N)$.

During several years a lot of Shapley values have been studied using different mathematical structures explaining certain situations among the players: coalition structures [3], communication structures [22], a priori unions [24], permission structures [14], incompatibility structures [4], convex geometries [5], matroids [6], antimatroids [2], coalition configurations [1], cooperation structures [8], authorization structures [12], etc.

Following the same way it is possible to define also games with fuzzy information about the players. Now the information is certain kind of fuzzy mathematical structure $Y(N)$, containing $K_{Y(N)}$ as the classical situation.

Definition 3.3 A value for games over N with fuzzy information $Y(N)$ is a mapping $f : \mathcal{G}^N \times Y(N) \to \mathbb{R}^N$.

Observe that we always consider an usual game but we need to incorporate fuzzy information. So really we do not need to work with fuzzy games. The fuzzy information will be included into the characteristic function using fuzziness tools but always using classical games. In the literature another different approach has been also analyzed, fuzzy games with crisp information. The reader can see the studied of this other model in Tsurumi et al. [26], Meng and Zhang [19] or Meng and Zhang [20]. Usually the chosen fuzzy information structure $Y(N)$ is a generalization of a known crisp structure studied before. So we denote as $Y^{cr}(N)$ the crisp situations contained in $Y(N)$ and obviously $K_{Y(N)} = K_{Y^{cr}(N)}$. To determine a value f for games with fuzzy information $Y(N)$ we will take into account the known crisp version of the value, we denote it as f^{cr}, for games with information $Y^{cr}(N)$.

Definition 3.4 A value f for games over N with fuzzy information $Y(N)$ is named a Shapley value if f^{cr} is a Shapley value and $f(v, K) = f^{cr}(v, K)$ for all $K \in Y^{cr}(N)$.

3.3 Crisp and Fuzzy Bilateral Relationships Among Players

One kind of information system about the players is a bilateral relation among them. Equivalence relations or order relations are examples. We will treat with some of the most known of these situations in the next chapters. Let N be our finite set of players. A *binary relationship*[1] on N is a mapping $r : N \times N \to \{0, 1\}$ where $r(i, j) = 1$ if and only if player i is related with player j and $r(i, j) = 0$ otherwise. The family of bilateral relations is $R(N)$. We named *domain* of $r \in R(N)$ to the set

$$N^r = \{i \in N : r(i, i) = 1\}. \tag{3.1}$$

The relation $r \in R(N)$ is called:

[1]This way to introduce a binary relationship is not the usual one, but it allows to understand the extension to fuzzy relations.

- *Reflexive*, if $r(i,i) = 1$ for all $i \in N$, namely $N^r = N$. It is *quasi-reflexive* if $r(i,j) = 1$ implies $j \in N^r$.
- *Symmetric*, if $r(i,j) = r(j,i)$ for all $i, j \in N$, namely if player i is related with j then j is related with i. In that case we denote each pair (unordered) as ij.
- *Antisymmetric*, if $r(i,j) = r(j,i) = 1$ implies $i = j$, namely if i is related with a different player j then j is not related with i.
- *Transitive*, if $r(i,j) = r(j,k) = 1$ implies that $r(i,k) = 1$ for all different $i, j, k \in N$, namely if i is related with j and j with k then i is related with k.

Our relation r on N is an *equivalence relation* if r is reflexive, symmetric and transitive. Relation r is a *partial ordering* if r is reflexive, antisymmetric and transitive. Let $T \subseteq N$ be a coalition, the *restriction* of r to T is another binary relation $r_T \in R(N)$ such that $r_T(i,j) = r(i,j)$ for all $i, j \in T$ and $r(i,j) = 0$ otherwise. Observe that the domain of this new relation is $N^{r_T} = N^r \cap T$.

Each relation $r \in R(N)$ defines a mapping $r : N \to 2^N$ where $r(i) = \{j \in N : r(i,j) = 1\}$ for all $i \in N$. Furthermore this is another way to introduce the concept of binary relation, namely we can give r defining the players related with each one. Relation r is reflexive if and only if $i \in r(i)$ for every player i, r is symmetric if and if $j \in r(i)$ implies $i \in r(j)$, r is antisymmetric if and if $j \in r(i)$ implies $i \notin r(j)$ and r is transitive if and only if $i \in r(j)$ and $j \in r(k)$ imply $i \in r(k)$.

A binary relation r is numerically represented by a matrix using the same letter r with size $n \times n$ where $r_{ij} = r(i,j)$, moreover any matrix $n \times n$ with elements 0 or 1 is a binary relation. Relation r is quasi-reflexive if and only if one element 0 in the diagonal of the matrix means null corresponding column, r is symmetric if and only if the associated matrix is symmetric, r is antisymmetric if and only if so the matrix is and r is transitive if and only if matrix r^2 satisfies that $r^2_{ij} > 0$ implies $r_{ij} = 1$.

Graphically a relation $r \in R(N)$ is represented by a graph, and moreover any graph (directed or undirected) represents a binary relation. Suppose $L(N) = \{(i,j) : i, j \in N, i \neq j\}$. In this book we will use the following representation, If $r \in R(N)$ then we consider the directed graph r with vertices (points) labeled by N, the vertex i is in black[2] if $i \in N^r$ and white otherwise. The links are the bilateral relationships among them,

$$L(r) = \{(i,j) \in L(N) : r(i,j) = 1\},$$

namely we draw a directed line, arrow, from the first vertex to the second one if the first one is related with the second one. If the relation is symmetric we will use an undirected graph, that is we draw only lines and no arrows for the links. The usual language of graph theory is used to describe circumstances about the binary relations. Relation r is *complete* if for all $i, j \in N^r$ we have $r(i,j) = 1$. Given two players $i, j \in N$, a *path* from i to j is a finite sequence $\{i_0, \ldots, i_m\}$ of players with $i_0 = i$, $i_m = j$, $\{i_k\}_{k=1}^{m-1} \subseteq N \setminus \{i, j\}$ are two to two different and $r(i_{k-1}, i_k) = 1$ for all $k = 1, \ldots, m$. If $i = j$ and $m \geq 2$ then the path is named *cycle* for player i. An

[2]In graph theory this idea corresponds to a loop.

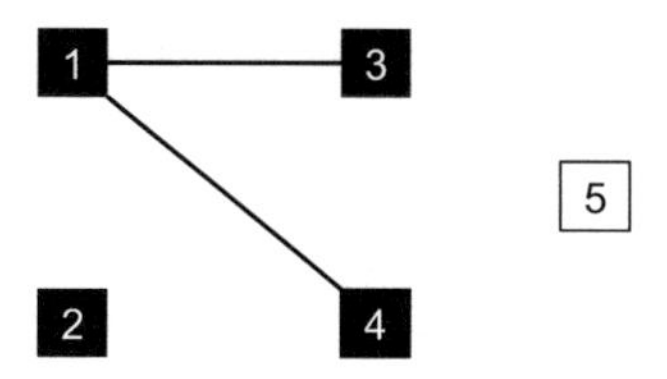

Fig. 3.1 Quasi-reflexive and symmetric relation r

acyclic relation is that without cycles. Two players $i, j \in N$ are connected if there is a path between them. Relation r is *connected*[3] if all $i, j \in N^r$ are connected. A coalition S is connected if $S \subseteq N^r$ and r_S is connected. The connected coalitions for r can be ordered by inclusion, the maximal elements are named *components*[4] of r. We denote

$$N/r = \{S \subseteq N : S \text{ is a component in } r\}. \tag{3.2}$$

Set N/r is a partition of the domain of r, namely $\bigcup_{S \in N/r} S = N^r$ and $S \cap T = \emptyset$ if $S, T \in N/r$ are different.

Example 3.1 Suppose $N = \{1, 2, 3, 4, 5\}$ and r a quasi-reflexive and symmetric relation on N given by the following matrix,

$$r = \begin{bmatrix} 1\,0\,1\,1\,0 \\ 0\,1\,0\,0\,0 \\ 1\,0\,1\,0\,0 \\ 1\,0\,0\,1\,0 \\ 0\,0\,0\,0\,0 \end{bmatrix}.$$

The domain of r is $N^r = \{1, 2, 3, 4\}$. It is not a reflexive because $r(5, 5) = 0$ and it is not transitive because $r(3, 1) = r(1, 4) = 1$ but $r(3, 4) = 0$. Relation r is not connected and its components are $N/r = \{\{1, 3, 4\}, \{2\}\}$. The undirected graph representing relation r is in Fig. 3.1.

Example 3.2 Consider again $N = \{1, 2, 3, 4, 5\}$ and r an acyclic transitive binary relation on N given by the following matrix,

$$r = \begin{bmatrix} 1\,1\,1\,1\,0 \\ 0\,1\,0\,1\,0 \\ 0\,0\,1\,0\,0 \\ 1\,1\,0\,0\,0 \\ 0\,0\,0\,0\,1 \end{bmatrix}.$$

It is not quasi-reflexive and it is not symmetric. The directed graph representing the relation is in Fig. 3.2.

[3]This concept corresponds to 1-connected for directed graphs because there are three different kinds of connection. We will only use one of them.

[4]Connected component is the usually name, but we can use component without danger of confusion.

Fig. 3.2 Transitive relation r

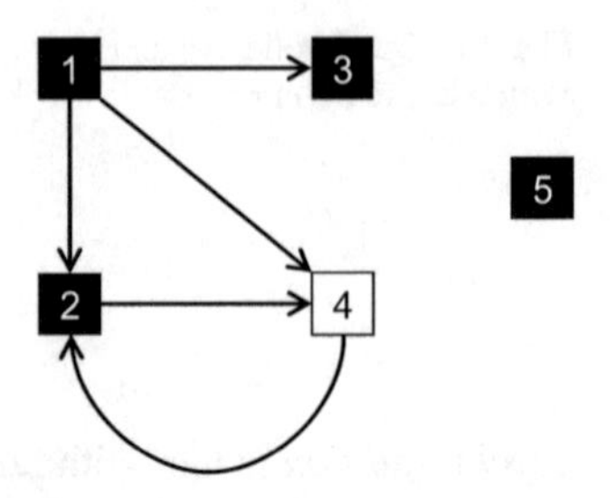

But the relations among players are not always black or white, so fuzzy bilateral relations will allow us to introduce leveled relations among them. We follow Mordeson and Nair [21] and Ovchinnikov [23] to describe the basic concepts of fuzzy binary relations and fuzzy graphs. A *fuzzy binary relation*[5] on N is a mapping $\rho : N \times N \to [0, 1]$ where $\rho(i, j)$ is the relationship level of i with j. The family of fuzzy binary relations is $[0, 1]^{N\times N}$. Any relation in $R(N)$ is particularly a fuzzy bilateral relationship. In a fuzzy context we said crisp to them and we denote $\rho \in R(N)$. Let $\rho \in [0, 1]^{N\times N}$. The *crisp version* of the fuzzy relation ρ is $r^\rho \in R(N)$ defined as $r^\rho(i, j) = 1$ if and only if $\rho(i, j) > 0$. The *domain* of ρ is the set

$$N^\rho = \{i \in N : \rho(i, i) > 0\} = N^{r^\rho}. \tag{3.3}$$

The *fuzzy domain* is a fuzzy coalition $\tau^\rho \in [0, 1]^N$ with the levels of the elements in the domain, for each $i \in N$

$$\tau^\rho(i) = \rho(i, i). \tag{3.4}$$

Hence, we have $N^\rho = supp(\tau^\rho)$. The extensions of reflexivity, symmetry, antisymmetry and transitivity have in the fuzzy case a linkage with the election of a particular T-norm (see Sect. 2.2). In this book we consider several of them. A fuzzy relation ρ is called:

- *Reflexive*, if $\rho(i, i) = 1$ for all $i \in N$. The relation is *weakly reflexive* if

$$\rho(i, j) \leq \rho(j, j)\, \forall i, j \in N,$$

  and the relation is *semi-reflexive* if

$$\rho(i, j) \leq \rho(i, i)\rho(j, j)\, \forall i, j \in N.$$

- *Symmetric*, if $\rho(i, j) = \rho(j, i)$ for all $i, j \in N$. We will use in that case $\rho(ij)$. Observe that, if ρ is symmetric then ρ is weakly reflexive if and only if $\rho(ij) \leq \rho(ii) \wedge \rho(jj)$.
- *Antisymmetric*, if $\rho(i, j) \wedge \rho(j, i) = 0$ for all $i, j \in N$, $i \neq j$. The relation ρ is *weakly antisymmetric* if $\rho(i, j) + \rho(j, i) \leq 1$.

[5]There exists a more general concept of fuzzy binary relation on a fuzzy set τ. In our case $\tau = e^N$.

- *Transitive*, if for all different $i, j, k \in N$ it holds $\rho(i, j) \wedge \rho(j, k) \leq \rho(i, k)$. We say ρ is *semi-transitive* if $\rho(i, j)\rho(j, k) \leq \rho(i, k)$.

The fuzzy relation ρ on N is a *proximity relation* if it satisfies reflexivity and symmetry. *Similarity relations* are reflexive, symmetric and transitive, they coincide in $R(N)$ with the crisp equivalence relations. As in the crisp case relation ρ is a *fuzzy partial order* if ρ is reflexive, antisymmetric and transitive, particularly it is a *fuzzy total order* if $\rho(i, j) \vee \rho(j, i) > 0$ for every $i \neq j$. Let $T \subseteq N$ be a coalition, the *restriction* of ρ to T is another fuzzy relation $\rho_T \in [0, 1]^{N\times N}$ such that $\rho_T(i, j) = \rho(i, j)$ for all $i, j \in T$ and $\rho_T(i, j) = 0$ otherwise. Also the domain of this new fuzzy relation is $N^{\rho_T} = N^{\rho} \cap T$.

A fuzzy relation ρ is a fuzzy set over $N \times N$ too. Hence we can use the usual notations and concepts from this theory in Sect. 2.2. Therefore, the support, the image, the cuts and the Choquet integral for set functions on $N \times N$ can be calculate of a fuzzy relation. Each fuzzy relation ρ defines also a mapping $\rho : N \to [0, 1]^N$ where $\rho(i)(j) = \rho(i, j)$ for all $i \in N$. So we can see the fuzzy relation as a set of n fuzzy sets.

We represent numerically a fuzzy binary relation ρ by a matrix using the same letter ρ with size $n \times n$ where $\rho_{ij} = \rho(i, j)$, moreover any matrix $n \times n$ with elements in $[0, 1]$ is a fuzzy binary relation. Relation ρ is weakly reflexive if and only if one element 0 in the diagonal of the matrix means null corresponding column, ρ is symmetric if and only if the associated matrix is symmetric, ρ is antisymmetric if and only if so the matrix is and ρ is transitive if and only if matrix ρ^2 satisfies that $\rho^2_{ij} > 0$ implies $\rho_{ij} > 0$.

Graphically a relation $\rho \in [0, 1]^{N\times N}$ can be represented by a fuzzy graph, and moreover any fuzzy graph (directed or undirected) represents a binary relation. In this book we will use the following representation, if $\rho \in [0, 1]^{N\times N}$ then we consider the directed fuzzy graph ρ with fuzzy vertices labeled by N and leveled by $\rho(i, i)$ for each vertex i, in black vertices with $i \in N^{\rho}$ and white otherwise. The fuzzy links are the links in $supp(\rho)$ between different elements where for each (i, j) the level is $\rho(i, j)$, namely we draw an arrow from the first vertex i to the second one j with level a if $\rho(i, j) = a$. If the relation is symmetric we will use an undirected fuzzy graph, that is we draw only lines and no arrows for the links. Moreover, in that case we take ij as (i, j) or (j, i). The crisp version r^{ρ} is the graph resulted to delete the numbers in the fuzzy graph ρ. The fuzzy relation ρ is *complete*, *acyclic* or *connected* if and only if the crisp version so is. The *components* of ρ are the components of r^{ρ}. We denote

$$N/\rho = N/r^{\rho}. \tag{3.5}$$

Set N/ρ is a partition of the domain of ρ. The set of fuzzy links is $L(\rho) = L(r^{\rho})$.

The *maximum level* and the *minimum level* of the fuzzy relation ρ over N is

$$\vee\rho = \bigvee_{(i,j)\in N\times N} \rho(i, j), \quad \wedge\rho = \bigwedge_{(i,j)\in supp(\rho)} \rho(i, j). \tag{3.6}$$

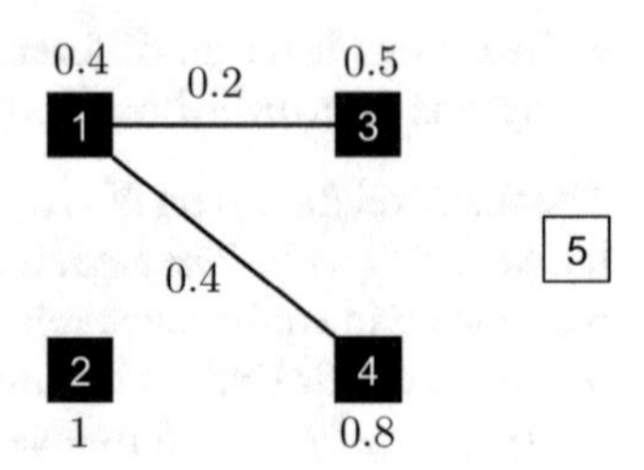

Fig. 3.3 Weakly proximity relation ρ

Example 3.3 Suppose $N = \{1, 2, 3, 4, 5\}$ and ρ given by the following matrix,

$$\rho = \begin{bmatrix} 0.4 & 0 & 0.2 & 0.4 & 0 \\ 0 & 1 & 0 & 0 & 0 \\ 0.2 & 0 & 0.5 & 0 & 0 \\ 0.4 & 0 & 0 & 0.8 & 0 \\ 0 & 0 & 0 & 0 & 0 \end{bmatrix}.$$

It is a weakly proximity relation in the sense that it is weakly reflexive and symmetric. The domain of ρ is $N^\rho = \{1, 2, 3, 4\}$. It is not a weakly similarity relation because $\rho(3, 1) \wedge \rho(1, 4) = 0.2$ but $\rho(3, 4) = 0$. Relation ρ is not connected with components $N/\rho = \{\{1, 3, 4\}, \{2\}\}$. The maximum level is $\vee\rho = 0.8$ and the minimum level is $\wedge\rho = 0.2$. The undirected fuzzy graph representing the relation is in Fig. 3.3 and the crisp version is Fig. 3.1.

Example 3.4 Consider again $N = \{1, 2, 3, 4, 5\}$ and ρ a weakly antisymmetric and transitive relation. Its matrix is

$$\rho = \begin{bmatrix} 1 & 0.3 & 0.3 & 0.6 & 0 \\ 0 & 1 & 0 & 0.7 & 0 \\ 0 & 0 & 1 & 0 & 0 \\ 1 & 0.1 & 0 & 0 & 0 \\ 0 & 0 & 0 & 0 & 1 \end{bmatrix}.$$

It is not weakly reflexive and symmetric. The directed graph representing the relation is in Fig. 3.4.

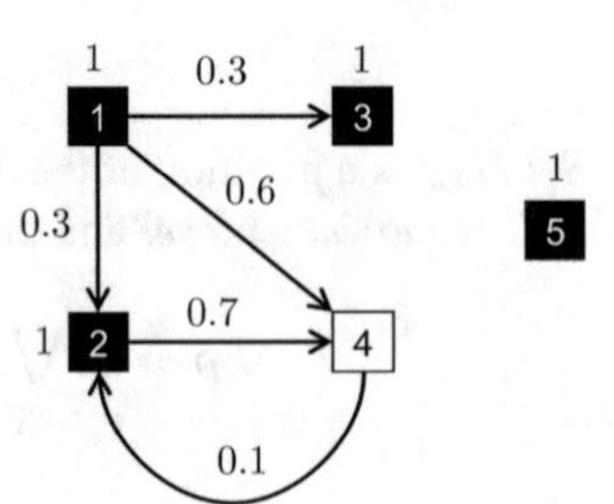

Fig. 3.4 Weakly antisymmetric and transitive relation ρ

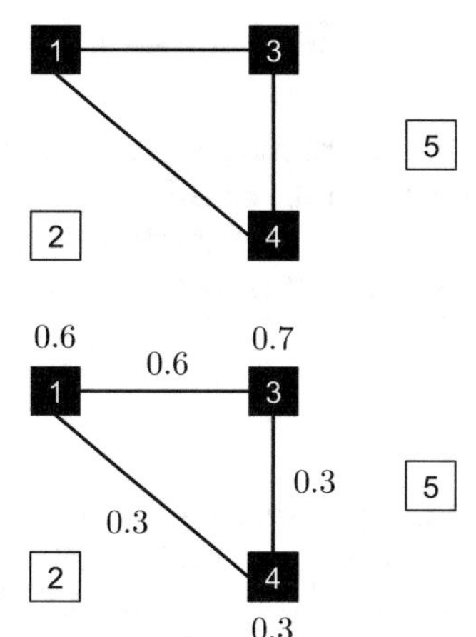

Fig. 3.5 Relation r^S

Fig. 3.6 Relation ρ^τ

In this book several Shapley values for games with information given by a particular kind of crisp and fuzzy binary relations will be studied. Namely, we will consider different models taking $I(N) \subseteq R(N)$ and $Y(N) \subseteq [0,1]^{N\times N}$ with particular interpretations. In the next two sections we explain an easy example of the model using the most simple case of crisp and fuzzy relation: the formation of an specific coalition or fuzzy coalition. Any coalition $S \subseteq N$ is identified to a quasi-equivalence relation (quasi-reflexive, symmetric and transitive) r^S where $N^{r^S} = S$. Relation r^S is defined as $r^S(i,j) = 1$ if and only if $i, j \in S$. More generally, any fuzzy coalition $\tau \in [0,1]^N$ is also identified to a weakly similarity relation (weakly reflexive, symmetric and transitive) ρ^τ as $\rho^\tau(i,j) = \tau(i) \wedge \tau(j)$ for all $i, j \in N$. Observe that $\rho^{e^S} = r^S$.

Example 3.5 Let $N = \{1,2,3,4,5\}$. In Fig. 3.5 we can see the relation r^S with coalition $S = \{1,3,4\}$ and in Fig. 3.6 the relation ρ^τ with $\tau = (0.6, 0, 0.7, 0.3, 0)$. The matrices associated to these relations are

$$r^S = \begin{bmatrix} 1\,0\,1\,1\,0 \\ 0\,0\,0\,0\,0 \\ 1\,0\,1\,1\,0 \\ 1\,0\,1\,1\,0 \\ 0\,0\,0\,0\,0 \end{bmatrix}, \quad \rho^\tau = \begin{bmatrix} 0.6 & 0 & 0.6 & 0.3 & 0 \\ 0 & 0 & 0 & 0 & 0 \\ 0.6 & 0 & 0.7 & 0.3 & 0 \\ 0.3 & 0 & 0.3 & 0.3 & 0 \\ 0 & 0 & 0 & 0 & 0 \end{bmatrix}.$$

In the next chapters several Shapley values for different situations using partitions [3], communications [22], hierarchies [14] and closeness [24] among the players. Following next sections as a sample, we will introduce an appropriate modification of games for each kind of information, we will analyze the transmission of properties for the game, we will define a Shapley value and properties for it, and finally we will provide the new value with at least an axiomatization.

3.4 The Coalitional Model

We describe in this section an easy example of games with information and a Shapley value for the situation. The classical model suppose that coalition N is formed and then $v(N)$ is allocated among all the players. We can study the situation of a game where another coalition is formed. So, $I(N) = 2^N$ and $K_{I(N)} = N$. A value function for games with this kind of information is

$$f : \mathscr{G}^N \times 2^N \to \mathbb{R}^N.$$

Each pair $(v, T) \in \mathscr{G}^N \times 2^N$ is called a *game over N with coalition*. We see this model as games with a bilateral relation among the players identifying each coalition T with r^T (see Example 3.5). At the end of Sect. 1.2 we told a way to reduce a game using subgames. But our model, as we explained in the introduction, is based on modifying the original game with the information and so obtaining a new game over N. The subgame corresponds rather to the first way.

Next definition introduces another known way to reduce a game to a coalition in our sense.

Definition 3.5 Let $v \in \mathscr{G}^N$ be a game and $T \subset N$ a coalition. The restricted game to T is a new game $v_T \in \mathscr{G}^N$ satisfying for each coalition $S \subseteq N$,

$$v_T(S) = v(S \cap T).$$

Let $(v, T) \in \mathscr{G}^N \times 2^N$. Both of the options, subgame and restricted game, get that players out of coalition T are not active in the game but with finer shades. In the first one (subgame) players out of T are not in the game, they are not really players. In the second one (restricted game) they are players but their integration into a coalition has never influence. The restricted game also allow us to extend a game defined in a set T incorporating non-active players as null players, i.e. each $i \in N \setminus T$ is a null player of v_T. In fact, if $S \subseteq N \setminus \{i\}$ then

$$v_T(S \cup \{i\}) = v((S \cup \{i\}) \cap T) = v(S \cap T) = v_T(S).$$

Hence given (v, T) we consider the restricted game $v_T \in \mathscr{G}^N$. An interesting question about the chosen modification is: does the new game preserve the properties of the original one? Next two propositions summary several of them. But not all property is preserved. For instance, the anonymous condition is not transferable by the restriction of games. Suppose $N = \{1, 2, 3\}$, $v(S) = |S|$ for all coalition S and $T = \{1, 2\}$. Game $v \in \mathscr{G}_a^N$ but $v_T(\{1, 2\}) = 2 \neq 1 = v_T(\{1, 3\})$.

Proposition 3.1 *Let $v \in \mathcal{G}^N$ and $T \subseteq N$.*

(1) If v is additive then v_T is additive.
(2) If v is superadditive (subadditive) then so is v_T.
(3) If v is convex (concave) then so is v_T.
(4) If v is monotone then v_T is monotone
(5) If v is a $\{0, 1\}$-game then v_T is also a $\{0, 1\}$-game.
(6) If v is simple then so is v_T.

Proof If $v \in \mathbb{R}^N$ then $v_T(S) = \sum_{i \in S \cap T} v_i$. Therefore v_T is the additive game associated to the vector $v^T \in \mathbb{R}^N$ with $v_i^T = v_i$ if $i \in T$ and $v_i^T = 0$ otherwise. Suppose $v \in \mathcal{G}_c^N$ and $S, R \subseteq N$. As $(S \cup R) \cap T = (S \cap T) \cup (R \cap T)$ and $(S \cap R) \cap T = (S \cap T) \cap (R \cap T)$ we get

$$v_T(S \cup R) + v_T(S \cap R) \geq v(S \cap T) + v(R \cap T) = v_T(S) + v_T(R).$$

The superadditive condition follows in the same way. If $v \in \mathcal{G}_m^N$ then $v_T(S) \leq v_T(R)$ when $S \subseteq R$ because $S \cap T \subseteq R \cap T$. The two final statements are trivial. □

More properties in the next proposition.

Proposition 3.2 *Let $v, w \in \mathcal{G}^N$ and $T \subseteq N$.*

(1) $(av + bw)_T = av_T + bw_T$ for all $a, b \in \mathbb{R}$.
(2) If v, w are strategically equivalent then so are v_T, w_T.
(3) $(-v)_T = -(v_T)$.
(4) $(v^{svg})_T = (v_T)^{svg}$.
(5) If $v, w \in \mathcal{G}_s^N$ then $(v \wedge w)_T = v_T \wedge w_T$ and $(v \vee w)_T = v_T \vee w_T$.

Proof Suppose $a, b \in \mathbb{R}$. We get

$$(av + bw)_T(S) = (av + bw)(S \cap T) = av(S \cap T) + bw(S \cap T) = av_T(S) + bw_T(S).$$

So, if v, w are strategically equivalent, namely there exist $a > 0$ and $b \in \mathbb{R}^N$ with $v = aw + b$ then

$$v_T = aw_T + b^T,$$

where $b^T \in \mathbb{R}^N$ (see the proof of the above proposition). Thus v_T, w_T are strategically equivalent. Obviously $(-v)_T = -(v_T)$. Now we determine the saving game,

$$(v^{svg})_T(S) = v^{svg}(S \cap T) = \sum_{i \in S \cap T} v(\{i\}) - v(S \cap T)$$
$$= \sum_{i \in S} v(\{i\} \cap T) - v(S \cap T) = \sum_{i \in S} v_T(\{i\}) - v_T(S) = (v_T)^{svg}(S).$$

Finally

$$(v \wedge w)_T(S) = (v \wedge w)(S \cap T) = v(S \cap T) \wedge w(S \cap T) = (v_T \wedge w_T)(S).$$

The same with operator $\vee$. □

The dual operator (1.5) has problems with the restricted game in the sense that it is not always true the equality $(v^{dual})_T = (v_T)^{dual}$, as it is showed in the next example.

Example 3.6 Consider the bankruptcy game in Examples 1.15 and 1.18. Let $T = \{2, 3\}$ and $S = \{1, 3\}$. Using Table 1.5 we obtain

$$(v^{dual})_T(S) = v^{dual}(\{3\}) = 40000.$$

On the other hand taking into account (1.5) and Table 1.3,

$$(v_T)^{dual}(S) = v_T(N) - v_T(\{2\}) = v(T) - v(\{2\}) = 25000.$$

Next we define a Shapley value for games with a coalition using the Shapley value of the modification of the game with the coalition.

Definition 3.6 The extended Shapley value is a value for games over N with coalition defined for each game $v \in \mathcal{G}^N$ and coalition $T \subseteq N$ as

$$\phi(v, T) = \phi(v_T).$$

Obviously the above mapping coincides with the Shapley value in the classical situation because $v_N = v$. We can use very similar axioms to the Shapley value for the extended on. Let $f : \mathcal{G}^N \times 2^N \to \mathbb{R}^N$ a value function for games over N with coalition.

Restricted efficiency. For all $v \in \mathcal{G}^N$ and $T \subseteq N$, $f(v, T)(N) = v(T)$.

Restricted null player. Let $i \in N$ be a null player in the game v, it holds $f_i(v, T) = 0$ for all coalition T with $i \in T$

Restricted symmetry. If $i, j \in N$ are symmetric for a game v then $f_i(v, T) = f_j(v, T)$ for all T with $i, j \in T$.

Linearity. For all $v, w \in \mathcal{G}^N$, $a, b \in \mathbb{R}$ and coalition T it holds $f(av + bw, T) = af(v, T) + bf(w, T)$.

Also we add the following axiom introduced by Shapley [25] jointly with efficiency. Players out the formed coalition have neither harm nor good in the payoff vector.

Carrier. Let $v \in \mathcal{G}^N$ and $T \subseteq N$. For all player $i \notin T$ it holds $f_i(v, T) = 0$.

Remark 3.1 Shapley [25] defined a carrier in a game $v \in \mathcal{G}^N$ as a coalition T satisfying $v(S) = v(S \cap T)$ for all coalition S. He explained the carrier axiom for a value f as

$$\sum_{i \in T} f_i(v) = v(T)$$

if T is a carrier in v. Given (v, T), coalition T is carrier in v_T. Moreover, coalition $T \cup \{i\}$ is also a carrier in v_T for each $i \in N \setminus T$. Thus, following the carrier axiom of Shapley we have

$$\sum_{j \in T \cup \{i\}} f_j(v_T) = v_T(T \cup \{i\}) = v(T) = v_T(T) = \sum_{j \in T} f_j(v_T)$$

and then $f_i(v_T) = 0$.

Theorem 3.1 *The extended Shapley value is the only value for games over N with coalition satisfying restricted efficiency, restricted null player, restricted symmetry, linearity and carrier.*

Proof Obviously the extended Shapley value satisfies the first four axioms from definition because the Shapley value verifies them over T (see Sect. 1.4). We test carrier. Let T be a non-empty coalition (if $T = \emptyset$ the result is trivial) and $i \notin T$. Observe that i is also a null player in v_T for every game v. For all $S \subset N \setminus \{i\}$

$$v_T(S \cup \{i\}) = v(S \cap T) = v_T(S).$$

Since Shapley value satisfies null player axiom we get $\phi_i(v, T) = 0$.

Let f be a value. Following the proof of Theorem 1.3, linearity implies that we only need to find out the uniqueness for unanimity games. So, let u_R with R a non-empty coalition. We consider now T another coalition. Using restricted null player and carrier we have that $f_i(v, T) = 0$ for any player i outside $T \cap R$. All the players in $T \cap R$ are symmetric and then from restricted symmetry all of them have the same payoff. Finally, using restricted efficiency as in the Theorem 1.3 again we get the uniqueness. □

Carrier is also a necessary axiom as we show in the following remark.

Remark 3.2 Suppose $n > 2$. Given a coalition $T' \subset N$ with $N \setminus T' = \{i, j\}$ we can define $f(v, T) = \phi(v, T)$ for all $T \neq T'$ and

$$f_k(v, T') = \begin{cases} \phi_k(v, T'), & \text{if } k \neq i, j \\ v(\{i\}) - v(\{j\}), & \text{if } k = i \\ v(\{j\}) - v(\{i\}), & \text{if } k = j. \end{cases}$$

Our value $f \neq \phi$ when $v(\{i\}) \neq v(\{j\})$ but f satisfies all the axioms in the above theorem except carrier.

Restricted symmetry tells about players in a determined coalition. Next proposition analyzes anonymity and symmetry in more general cases.

Proposition 3.3 *Let $T \subsetneq N$ be a non-empty coalition and $i \in T$. If $\theta \in \Theta^N$ is a permutation then the extended Shapley value satisfies*

$$\phi_{\theta(i)}(\theta v, \theta T) = \phi_i(v, T).$$

Moreover:

(1) If $\theta T = T$ then $\phi_{\theta(i)}(\theta v, T) = \phi_i(v, T)$,
(2) If i, j are symmetric players with $j \notin T$ for $v \in \mathcal{G}^N$ then

$$\phi_i(v, T) = \phi_j(v, T \setminus \{i\} \cup \{j\}).$$

Proof If $\theta \in \Theta^N$ then

$$(\theta v)_{\theta T}(\theta S) = (\theta v)(\theta(S \cap T)) = v(S \cap T) = v_T(S).$$

Hence $(\theta v)_{\theta T} = \theta(v_T)$. As Shapley value verifies anonymity (Proposition 1.10) then

$$\phi_{\theta(i)}(\theta v, \theta T) = \phi_{\theta(i)}(\theta(v_T)) = \phi_i(v_T).$$

Obviously if $\theta T = T$ the equality in (1) is true. Suppose now i, j symmetric players. For the last equality we take θ^{ij} defined as $\theta^{ij}(i) = j$, $\theta^{ij}(j) = i$ and $\theta^{ij}(k) = k$ for another $k \neq i, j$. Hence $\theta^{ij} T = T \setminus \{i\} \cup \{j\}$. As i, j are symmetric for v then we can test that $\theta^{ij} v = v$. If $i, j \notin S$ or $i, j \in S$ then $\theta^{ij} S = S$ and $(\theta^{ij} v)(S) = (\theta^{ij} v)(\theta^{ij} S) = v(S)$. If $i \in S$ but $j \notin S$ then $\theta^{ij} S = S \setminus \{i\} \cup \{j\}$ and applying the symmetric condition to $S \setminus \{i\}$ we get

$$(\theta^{ij} v)(S) = (\theta^{ij} v)(\theta^{ij}\theta^{ij} S) = v(\theta^{ij} S) = v(S \setminus \{i\} \cup \{j\}) = v(S).$$

□

Suppose now the other model described in the introduction. In that case given (v, T) we use directly the function v restricted to the chosen coalition T, i.e. the subgame (Sect. 1.2) over the coalition T. Then we apply the specific Shapley value for the coalition, namely the payoff vector would be $\phi^T(v)$. Both ways are very closed in this case. If $x \in \mathbb{R}^T$ with $T \subset N$ then $x^0 \in \mathbb{R}^N$ denotes the extension of x by zeros, namely $x_i^0 = x_i$ if $i \in T$ and $x_i^0 = 0$ otherwise.

Proposition 3.4 *Let $v \in \mathcal{G}^N$ be a game and $T \subseteq N$ a non-empty coalition. It holds*

$$\phi(v, T) = [\phi^T(v)]^0.$$

Proof Since Theorem 3.1 if we test that the value function $f(v, T) = [\phi^T(v)]^0$ ($f(v, \emptyset) = 0$) verifies all the five axioms then the equality requested is true. Obviously f satisfies carrier. As the Shapley value is efficient for games over T then $\phi^T(v)(T) = v(T)$, jointly with carrier we get efficiency. If i is a null player for v then i is a null player for each game $v|_T$ with $i \in T$ (the other side is not true). Also if $i, j \in T$ are symmetric for v then i, j are symmetric for the subgame. So, as the Shapley value satisfies null player, symmetry and linearity over $\mathcal{G}^T$ then f too. □

The extended Shapley value $\phi(v, T)$ and the original Shapley value $\phi^T(v)$ are slightly different as the above proposition suggests. Players out of T do not have any payoff for the Shapley value but these payoffs are zero for the extended one. Next we show other properties of the Shapley value for this model.

Proposition 3.5 *The extended Shapley value satisfies the following properties.*

(1) For all $v \in \mathcal{G}^N$ and $i, j \in N$ different players it holds for any T with $i, j \in T$

$$\phi_i(v, T) - \phi_i(v, T \setminus \{j\}) = \phi_j(v, T) - \phi_j(v, T \setminus \{i\}).$$

(2) If $v, w \in \mathcal{G}^N$ with $v <_T w$ for any $T \subseteq N$ non empty then $\phi(v, T) \leq \phi(w, T)$.
(3) If $v, w \in \mathcal{G}^N$ with $v <_i w$ for any $i \in N$ then $\phi_i(v, T) \leq \phi_i(w, T)$ for all $T \subseteq N$.
(4) If $i \in T$ is a necessary player for $v \in \mathcal{G}_m^N$ then $\phi_j(v, T) \leq \phi_i(v, T)$.
(5) If $i \in T$ and $v \in \mathcal{G}_{sa}^N$ then $\phi_i(v, T) \geq v(\{i\})$.
(6) If $S \subseteq T$ and $v \in \mathcal{G}_c^N$ then $\phi(v, T)(S) \geq v(S)$.
(7) $\phi(-v, T) = -\phi(v, T)$.
(8) For all $i \in T$ and $v \in \mathcal{G}^N$ it holds $\phi_i(v, T) = v(\{i\}) - \phi_i(v^{svg}, T)$.
(9) If $v, w \in \mathcal{G}_s^N$ then for all coalition T it holds

$$\phi_i(v \wedge w, T) + \phi(v \vee w, T) = \phi_i(v, T) + \phi_i(w, T).$$

Proof (1) The Shapley value in general satisfies balanced contributions (see Proposition 1.16). Let $i, j \in N$ and $T \subseteq N$ with $i, j \in T$. If $v \in \mathcal{G}^N$ then ϕ^T verifies balanced contributions for the subgame $v \in \mathcal{G}^T$, thus by Proposition 3.4

$$\begin{aligned}\phi_i(v, T) - \phi_j(v, T) &= \phi_i^T(v) - \phi_j^T(v) = \phi_i^{T\setminus\{j\}}(v) - \phi_j^{T\setminus\{i\}}(v) \\ &= \phi_i(v, T \setminus \{j\}) - \phi_j(v, T \setminus \{i\}).\end{aligned}$$

(2) If $i \notin T$ then $\phi_i(v, T) = \phi_i(w, T) = 0$. Also the subgames over T verifies $v <_T w$ and hence Proposition 1.12 implies $\phi_i^T(v) \leq \phi_i^T(w)$ for all $i \in T$. Now Proposition 3.4 says if $i \in T$

$$\phi_i(v, T) = \phi^T(v) \leq \phi_i^T(w) = \phi_i(w, T).$$

(3) If $v <_i w$ then $v_T <_i w_T$. In fact, for each $S \subseteq N \setminus \{i\}$ we have two cases: if $i \in T$

$$\begin{aligned}v_T(S \cup \{i\}) - v_T(S) &= v((S \cup \{i\}) \cap T) - v(S \cap T) = v((S \cap T) \cup \{i\}) - v(S \cap T) \\ &\leq w((S \cap T) \cup \{i\}) - w(S \cap T) = w_T(S \cup \{i\}) - w_T(S),\end{aligned}$$

and if $i \notin T$ then $v_T(S \cup \{i\}) - v_T(S) = 0 = w_T(S \cup \{i\}) - w_T(S)$. Therefore

$$\phi_i(v, T) = \phi_i(v_T) \leq \phi_i(w_T) = \phi_i(w, T).$$

(4) Game v_T is monotone if $v \in \mathcal{G}_m^N$ since Proposition 3.1. If $i \in T$ is a necessary player for v then $v_T(S) = v(T \cap S) = 0$ for all $S \subseteq N \setminus \{i\}$, i.e. i is a necessary player also for v_T. As Shapley value satisfies the necessary player axiom (Proposition 1.11) we have for any $j \in N \setminus \{i\}$

$$\phi_j(v, T) = \phi_j(v_T) \leq \phi_i(v_T) = \phi_i(v, T).$$

(5) Proposition 3.1 implies that if $v \in \mathcal{G}_{sa}^N$ then also $v_T \in \mathcal{G}_{sa}^N$ for each non-empty coalition T. Let $i \in T$. Since the Shapley value is individually stable (Proposition 1.13) we get $\phi_i(v, T) = \phi_i(v_T) \geq v_T(\{i\}) = v(\{i\})$.
(6) Proposition 3.1 again says that $v_T \in \mathcal{G}_c^N$ if $v \in \mathcal{G}_c^N$. So, if $S \subseteq T$ then by Proposition 1.14

$$\phi(v, T)(S) = \phi_i(v_T)(S) \geq v_T(S) = v(S).$$

(7) Linearity of the Shapley value and Proposition 3.2 imply the condition.
(8) If $i \in T$ then $v_T(\{i\}) = v(\{i\})$. From Propositions 1.15 and 3.2 it holds

$$\begin{aligned}\phi_i(v, T) &= \phi_i(v_T) = v_T(\{i\}) - \phi_i((v_T)^{svg}) = v(\{i\}) - \phi_i((v^{svg})_T) \\ &= v(\{i\}) - \phi_i(v^{svg}, T).\end{aligned}$$

(9) The result is trivial from Proposition 3.2 (5). □

In this context balanced contributions property (condition (1) in the above proposition) does not modify the set of players. The modification of the game can also modifies the stability properties of the solution. In this case the extended value keeps the conditions internally into the chosen coalition (see steps (5), (6) in the above proposition). Out of T actually payoffs are zero from Proposition 3.4. Payoffs in T of a subadditive game can be calculated from the saving game in the same way of the classical value. But, following Example 3.6 we will see that it is not true that the dual game has the same payoff vector of the original game.

Example 3.7 The Shapley value of the bankruptcy game in Table 1.3 was calculated in Example 1.15 and we test that using the dual game in Table 1.5 the payoff vector was the same in Example 1.18. Following Example 3.6 we calculate for that game

$$\phi_3(v, T) = \phi_3^T(v) = \frac{1}{2}[v(\{2, 3\}) - v(\{2\})] + \frac{1}{2}v(\{3\}) = 15000,$$

with $T = \{2, 3\}$. Now the payoff of player 3 in the dual game is

$$\phi_3(v^{dual}, T) = \phi_3^T(v^{dual}) = \frac{1}{2}[v^{dual}(\{2, 3\}) - v^{dual}(\{2\})] + \frac{1}{2}v^{dual}(\{3\}) = 35000.$$

It is possible to solve this downside by thinking of ad hoc concept of dual. Given a coalition T and a game v, the T-*dual* game is $v^{Tdual} \in \mathcal{G}^N$ where

$$v^{Tdual}(S) = v(T) - v(T \setminus S). \tag{3.7}$$

This new definition is more consistent with the information and it gets the same result for the Shapley value. But you need to analyze each coalition one by one.

Proposition 3.6 *Let $T \subseteq N$ be a coalition. For all game v it holds*

$$\phi(v^{Tdual}, T) = \phi(v, T).$$

Proof Formula (3.7) is given looking for the equality which failed in Example 3.6, $(v^{Tdual})_T = (v_T)^{dual}$. In fact, for all coalition S we have

$$\begin{aligned}(v^{Tdual})_T(S) &= v^{Tdual}(S \cap T) = v(T) - v(T \setminus (S \cap T)) \\ &= v(T) - v((N \setminus S) \cap T) = v_T(N) - v_T(N \setminus S) = (v_T)^{dual}(S).\end{aligned}$$

Now, Proposition 1.15 (3) implies

$$\phi(v^{Tdual}, T) = \phi((v^{Tdual})_T) = \phi((v_T)^{dual}) = \phi(v_T) = \phi(v, T).$$

□

Propositions 1.1 and 1.6 showed the interest of the unanimity games to analyze games and also particularly simple games. The next proposition evaluates the restricted game of a unanimity game and its extended Shapley value.

Proposition 3.7 *Let $R \subseteq N$ be a non empty coalition. For another non empty coalition $T \subseteq N$ it holds $(u_R)_T = u_R$ if $R \subseteq T$ and $(u_R)_T = 0$ otherwise. Furthermore*

$$\phi(u_R, T) = \begin{cases} \phi(u_R), & \text{if } R \subseteq T \\ 0, & \text{otherwise.} \end{cases}$$

Proof Suppose $S \subseteq N$, we have $(u_R)_T(S) = u_R(S \cap T)$. So, when $R \subseteq T$ then $R \subseteq S \cap T$ if and only if $R \subseteq S$, and $u_R(S \cap T) = u_R(S)$. But otherwise $S \cap T$ cannot contain R and $u_R(S \cap T) = 0$. Definition 3.6 implies the result about the index. □

Next we see that it is possible to get the extended Shapley value by the dividends of the game without determining the restricted game.

Theorem 3.2 *For each $v \in \mathcal{G}^N$ and non-empty coalition $T \subseteq N$ the extended Shapley value is*

$$\phi_i(v, T) = \sum_{i \in S \subseteq T} \frac{\Delta_S^v}{|S|}.$$

Proof Propositions 1.1 and 3.2 obtain

$$\sum_{\{S \subseteq N : S \neq \emptyset\}} \Delta_S^{v_T} u_S = v_T = \sum_{\{S \subseteq N : S \neq \emptyset\}} \Delta_S^v (u_S)_T.$$

The above proposition gets

$$\sum_{\{S \subseteq N : S \neq \emptyset\}} \Delta_S^{v_T} u_S = \sum_{\{S \subseteq T : S \neq \emptyset\}} \Delta_S^v u_S.$$

But Proposition 1.1 assures the uniqueness of the coefficients reward to the unanimity games thus

$$\Delta_S^{v_T} = \begin{cases} \Delta_S^v, & \text{if } S \subseteq T \\ 0, & \text{otherwise.} \end{cases} \tag{3.8}$$

So, Theorem 1.2 implies the expected formula. □

Example 3.8 Consider the bankruptcy game in Example 1.15. We got the dividends of this game in Example 1.19, see Table 1.7. For instance, if $T = \{2, 3\}$ and $i = 3$ the above theorem calculates the extended Shapley value as

Table 3.1 Extended Shapley values of game v

T	$\{1\}$	$\{2\}$	$\{3\}$	$\{1, 2\}$	$\{1, 3\}$	$\{2, 3\}$
$\phi(v, T)$	$(0, 0, 0)$	$(0, 0, 0)$	$(0, 0, 5000)$	$(5000, 5000, 0)$	$(12500, 0, 17500)$	$(0, 10000, 15000)$

$$\phi_3(v, T) = \Delta^v_{\{3\}} + \frac{1}{2}\Delta^v_{\{2,3\}} = 5000 + \frac{1}{2}20000 = 15000.$$

If $T = N$ then $\phi(v, N) = \phi(v) = (14166.6, 13333.3, 22500)$. Table 3.1 shows the payoff vectors for all the rest of coalitions.

The extended Shapley value as index measures the power of each agent when a determined coalition is formed. Restricted simple games can be described by unanimity games into the lattice.

Proposition 3.8 *Let $v \in \mathcal{G}_s^N$. For each non empty coalition T it holds*

$$v_T = \bigvee_{\{S \in W_m(v): S \subseteq T\}} u_S.$$

Particularly, if $T \in W_m(v)$ then $v_T = u_T$ and if $T \notin W(v)$ then $v_T = 0$. Moreover,

$$W_m(v_T) = \{S \in W_m(v) : S \subseteq T\}.$$

Proof We saw in Proposition 3.1 that if $v \in \mathcal{G}_s^N$ then $v_T \in \mathcal{G}_s^N$ too. We know since Propositions 3.2 and 3.7 that

$$v_T = \bigvee_{S \in W_m(v)} (u_S)_T = \bigvee_{\{S \in W_m(v): S \subseteq T\}} u_S.$$

Obviously if $T \notin W(v)$, namely it is a losing coalition, then $\{S \in W_m(v) : S \subseteq T\} = \emptyset$, and if $T \in W_m(v)$ then $\{S \in W_m(v) : S \subseteq T\} = \{T\}$. Now we will prove the claim about the minimal winning coalitions. Suppose $S \in W_m(v)$ with $S \subseteq T$. We have $v_T(S) = v(S) = 1$, i.e. $S \in W(v_T)$. Also, if $R \subsetneq S$ then $v_T(R) = v(R) = 0$. Hence $S \in W_m(v_T)$. On the other hand, let $S \in W_m(v_T)$. The above formula implies that there exists $S' \subseteq T$ with $S' \in W_m(v)$ and $u_{S'}(S) = 1$. As we have proved that $S' \in W_m(v_T)$ and $S' \subseteq S$ then $S' = S$. □

Example 3.9 We take the voting situation in Example 1.20 given by $[26; 20, 15, 6, 5]$. We had calculated the swings of the players for v in Table 1.8. But when we take a winning coalition (if it is a losing coalition the game is zero as we saw before) the swings can increase or decrease. Suppose $i = 2$. For instance, given $T = \{1, 2, 3\} \in W(v)$, $\{3, 4\} \in SW_2(v)$ but $\{3, 4\} \notin SW_2(v_T)$ because $v_T(\{2, 3, 4\}) = v(\{2, 3\}) = 0$. Now if we take $T = \{2, 3, 4\}$ we have that $\{1, 3, 4\} \notin SW_2(v)$ but $\{1, 3, 4\} \in SW_2(v_T)$ because $v_T(N) = v(T) = 1$ and $v_T(\{1, 3, 4\}) = v(\{3, 4\}) = 0$. For the last T, as it is a minimal winning coalition, then $v_T = u_T$ by Proposition 3.8. So, if coalition $\{2, 3, 4\}$ is formed, for instance in a government convention, the power index is

$$\phi(v, T) = (0, 1/3, 1/3, 1/3).$$

3.5 The Fuzzy Coalitional Model

The example developed in the above section is extended in a fuzzy way. So, in this case the objects of information are the different fuzzy coalitions, namely $Y(N) = [0, 1]^N$ and $K_{Y(N)} = e^N$. Each fuzzy coalition τ is identified to a fuzzy relation ρ^τ (see Example 3.5). A value for games over N with a fuzzy coalition is

$$f : \mathcal{G}^N \times [0, 1]^N \to \mathbb{R}^N.$$

In order to incorporate the fuzzy information in the crisp cooperative game we will use the fuzziness methods studied in the above chapter. In general, for each extension independent of the game (Definition 2.9) it is possible to define a different Shapley value by different restricted games. Now we specify more the kind of partition function that we need.

Definition 3.7 A partition function pl is inherited if for all $\tau \in [0, 1]^N$ and $S \subseteq N$ with $supp(\tau) \cap S \neq \emptyset$ it holds

$$pl(\tau \times e^S) = \left\{ \left(R, \sum_{\{(T,t) \in pl(\tau) : T \cap S = R\}} t \right) : R \neq \emptyset, \exists (T, t) \in pl(\tau) \text{ with } T \cap S = R \right\}.$$

In an inherited partition function the partition by levels of the crisp restriction of a given fuzzy coalition, i.e. reducing the support but keeping the levels into the new support, is calculated restricting the coalitions in the partition and adding all the levels that obtains the same restriction. Observe that an inherited partition function pl is an extension if and only if $pl(e^N) = \{(N, 1)\}$. We will consider from now on inherited extensions. Furthermore the three extensions studied in Sect. 2.4 are inherited.

Proposition 3.9 *The multilinear, proportional and Choquet extensions are inherited.*

Proof Let $\tau \in [0, 1]^N$, $\tau \neq 0$, and $S \subseteq N$ with $S \cap supp(\tau) \neq \emptyset$.

MULTILINEAR. The partition of τ is taken following (2.8),

$$ml(\tau) = \left\{ \left(T, \prod_{i \in T} \tau(i) \prod_{i \notin T} (1 - \tau(i)) \right) \right\}_{[\tau]_1 \subseteq T \subseteq supp(\tau)}.$$

Since (2.8) again we have $(R, r) \in ml(\tau \times e^S)$ if and only if,

- $[\tau \times e^S]_1 \subseteq R \subseteq supp(\tau \times e^S)$, in other words $[\tau]_1 \cap S \subseteq R \subseteq supp(\tau) \cap S$,
- and

$$r = \prod_{i \in R} (\tau \times e^S)(i) \prod_{i \notin R} (1 - (\tau \times e^S)(i)) = \prod_{i \in R} \tau(i) \prod_{i \in S \setminus R} (1 - \tau(i))$$

  because $R \subseteq S$ and if $i \notin S$ then $(\tau \times e^S)(i) = 0$.

We set

$$\{T : [\tau]_1 \subseteq T \subseteq supp(\tau), T \cap S = R\} = \{R \cup T : [\tau]_1 \setminus S \subseteq T \subseteq supp(\tau) \setminus S\} \neq \emptyset.$$

Using Lemma 2.1 with coalition $N \setminus S$ we obtain

$$\begin{aligned}
&\sum_{\{R \cup T : [\tau]_1 \setminus S \subseteq T \subseteq supp(\tau) \setminus S\}} \prod_{i \in R \cup T} \tau(i) \prod_{i \notin R \cup T} (1 - \tau(i)) \\
&= \prod_{i \in R} \tau(i) \prod_{i \in S \setminus R} (1 - \tau(i)) \sum_{[\tau]_1 \setminus S \subseteq T \subseteq supp(\tau) \setminus S} \prod_{i \in T} \tau(i) \prod_{i \in (N \setminus S) \setminus T} (1 - \tau(i)) \\
&= \prod_{i \in R} \tau(i) \prod_{i \in S \setminus R} (1 - \tau(i)) \sum_{T \subseteq N \setminus S} \prod_{i \in T} \tau(i) \prod_{i \in (N \setminus S) \setminus T} (1 - \tau(i)) = r.
\end{aligned}$$

PROPORTIONAL Suppose for this case the partition by levels following (2.11) of τ,

$$pr(\tau) = \{(S_t^\tau, t)\}_{t \in im(\tau)}.$$

For our coalition S we have $pr(\tau \times e^S) = \{(S_r^{\tau \times e^S}, r)\}_{r \in im(\tau \times e^S)}$. For each number $r \in im(\tau \times e^S) \subset im(\tau)$,

$$S_r^{\tau \times e^S} = \{i \in N : (\tau \times e^S)(i) = r\} = S_r^\tau \cap S.$$

As $S_t^\tau \cap S_{t'}^\tau = \emptyset$ then S_r^τ is the only coalition in the partition of τ with $S_t^\tau \cap S = S_r^{\tau \times e^S}$.
CHOQUET. Following (2.12) we have

$$ch(\tau) = \{(\lambda_k - \lambda_{k-1}, [\tau]_k)\}_{k=1}^m$$

with $im_0(\tau) = \{\lambda_0 < \lambda_1 < \cdots < \lambda_m\}$. For coalition S we denote as $im_0(\tau \times e^S) = \{\lambda'_0 < \lambda'_1 < \cdots < \lambda'_{m'}\} \subseteq im_0(\tau)$. Let $p \in \{1, \ldots, m'\}$ and $(\lambda'_p - \lambda'_{p-1}, [\tau \times e^S]_p)$

$\in ch(\tau \times e^S)$. There exist $q, q' \in \{1, \ldots, m\}$ with $q' < q, \lambda'_p = \lambda_q$ and $\lambda'_{p-1} = \lambda_{q'}$. Furthermore,

$$[\tau]_{q''} \cap S = [\tau \times e^S]_p.$$

for all $q' < q'' \leq q$. Consider $\{q' + 1 = q_1 < \cdots < q_r = q\}$, we obtain

$$\sum_{l=1}^{r} \lambda_{q_l} - \lambda_{q_l - 1} = \lambda_{q_r} - \lambda_{q_1 - 1} = \lambda'_p - \lambda'_{p-1}.$$

□

We define a restricted game for each inherited extension with good properties as we will see later.

Definition 3.8 Let $v \in \mathcal{G}^N$, $\tau \in [0,1]^N$ and pl an inherited extension. The pl-restricted game is defined for all $S \subseteq N$ as

$$v_\tau^{pl}(S) = v^{pl}(\tau \times e^S).$$

The inherited condition permits to describe these games from the crisp restricted game as says the following lemma.

Lemma 3.1 *Let pl be an inherited extension. If $\tau \in [0,1]^N$ with $pl(\tau) = \{(R_k, r_k)\}_{k=1}$ then*

$$v_\tau^{pl} = \sum_{k=1}^{m} r_k v_{R_k}.$$

Proof Let pl be an inherited extension. Suppose $\tau \in [0,1]^N$ with $pl(\tau) = \{(R_k, r_k)\}_{k=1}^m$. If $S \cap supp(\tau) = \emptyset$ then $v_\tau^{pl}(S) = 0 = v_{R_k}(S)$ for all k, because always $R_k \subseteq supp(\tau)$. Let $S \cap supp(\tau) \neq \emptyset$. We get that if $(R, r) \in pl(\tau \times e^S)$ then

$$r = \sum_{\{k \in \{1,\ldots,m\} : R_k \cap S = R\}} r_k,$$

and we have

$$rv(R) = \sum_{\{k \in \{1,\ldots,m\} : R_k \cap S = R\}} r_k v(R_k \cap S).$$

If $R_k \cap S = \emptyset$ then $v(R_k \cap S) = 0$ and we obtain

$$v_\tau^{pl}(S) = v^{pl}(\tau \times e^S) = \sum_{(R,r)\in pl(\tau\times e^S)} rv(R) = \sum_{(R,r)\in pl(\tau\times e^S)} \sum_{\{k\in\{1,\ldots,m\}:R_k\cap S=R\}} r_k v(R_k \cap S)$$

$$= \sum_{k=1}^{m} r_k v(R_k \cap S) = \sum_{k=1}^{m} r_k v_{R_k}(S).$$

□

Example 3.10 Consider $N = \{1, 2, 3\}$ and $v(S) = |S|^2$ for all S. Let $\tau = (0.4, 0.7, 0.4)$. Our game is anonymous but τ introduced asymmetry among the players involved into the support. We determine in this example the pl-restricted games for $pl = ml, pr, ch$. Take for instance $S = \{2, 3\}$. We have $[\tau \times e^S]_1 = \emptyset$ and $supp\,(\tau \times e^S) = S$, so

$$v_\tau^{ml}(\{2,3\}) = \tau(2)(1-\tau(3))v(\{2\}) + \tau(3)(1-\tau(2))v(\{3\}) + \tau(2)\tau(3)v(\{2,3\}) = 1.54.$$

As $im\,(\tau \times e^S) = \{0.4, 0.7\}$ and $S^{0.4}_{\tau\times e^S} = \{3\}$, $S^{0.7}_{\tau\times e^S} = \{2\}$,

$$v_\tau^{pr}(\{2,3\}) = 0.4v(\{3\}) + 0.7v(\{2\}) = 1.1.$$

Finally,

$$v_\tau^{ch}(\{2,3\}) = (0.4-0)v(\{2,3\}) + (0.7-0.4)v(\{2\}) = 1.9.$$

Table 3.2 represents the worths of the three games.

The inherited condition permits to obtain the same transference of properties of the game than in the crisp case (see Proposition 3.1).

Proposition 3.10 *Let $v \in \mathcal{G}^N$ and $\tau \in [0,1]^N$. Let pl be an inherited extension.*

(1) If $\tau = e^T$ then $v_\tau^{pl} = v_T$.
(2) If v is additive then v_τ^{pl} is additive.
(3) If v is superadditive (subadditive) then so is v_τ^{pl}.
(4) If v is convex (concave) then so is v_τ^{pl}.
(5) If v is monotone then v_τ^{pl} is monotone

Table 3.2 pl-restricted games

S	{1}	{2}	{3}	{1, 2}	{1, 3}	{2, 3}	N
$v_\tau^{ml}(S)$	0.4	0.7	0.4	1.54	1.12	1.54	3.698
$v_\tau^{pr}(S)$	0.4	0.7	0.4	1.1	1.6	1.1	2.3
$v_\tau^{ch}(S)$	0.4	0.7	0.4	1.9	1.6	1.9	3.9

Proof (1) Suppose $\tau = e^T$. For each coalition S, as pl is an extension

$$v_\tau^{pl}(S) = v^{pl}(e^T \times e^S) = v^{pl}(e^{S\cap T}) = v(S \cap T) = v_T(S).$$

(2) Suppose $v \in \mathbb{R}^N$. We will prove that $v_\tau^{pl} = v \times \tau = (v_i\tau(i))_{i\in N} \in \mathbb{R}^N$. If $S \subseteq N$ then from Example 2.6

$$v_\tau^{pl}(S) = v^{pl}(\tau \times e^S) = v \cdot (\tau \times e^S) = \sum_{i\in S} v_i\tau(i).$$

(3) Suppose $v \in \mathcal{G}_{sa}^N$ and two coalitions $S, T \subseteq N$ with $S \cap T = \emptyset$. Consider that $pl(\tau) = \{(R_k, r_k)\}_{k=1}^m$. As pl is inherited we get since Lemma 3.1

$$\begin{aligned} v_\tau^{pl}(S) + v_\tau^{pl}(T) &= \sum_{k=1}^m r_k v_{R_k}(S) + \sum_{k=1}^m r_k v_{R_k}(T) \\ &= \sum_{k=1}^m r_k[v_{R_k}(S) + v_{R_k}(T)] \leq \sum_{k=1}^m r_k v_{R_k}(S \cup T) = v_\tau^{pl}(S \cup T), \end{aligned}$$

using also that v_{R_k} is superadditive for any k from Proposition 3.1.
(4) Suppose $v \in \mathcal{G}_c^N$ and two coalitions $S, T \subseteq N$. Consider that $pl(\tau) = \{(R_k, r_k)\}_{k=1}^m$. Using again Lemma 3.1 in a similar way of the above point

$$\begin{aligned} v_\tau^{pl}(S) + v_\tau^{pl}(T) &= \sum_{k=1}^m r_k[v_{R_k}(S) + v_{R_k}(T)] \\ &\leq \sum_{k=1}^m r_k[v(R_k)(S \cup T) + v_{R_k}(S \cap T) = v_\tau^{pl}(S \cup T) + v_\tau^{pl}(S \cap T), \end{aligned}$$

using the convexity of v_{R_k} from Proposition 3.1.
(5) Suppose $v \in \mathcal{G}_m^N$. Let $S \subseteq T$ and $pl(\tau) = \{(R_k, r_k)\}_{k=1}^m$. Following again Lemma 3.1 and Proposition 3.1 we have

$$v_\tau^{pl}(S) = \sum_{k=1}^m r_k v_{R_k}(S) \leq \sum_{k=1}^m r_k v_{R_k}(T) = v_\tau^{pl}(T).$$

□

Obviously if v is a $\{0, 1\}$-game then v_τ^{pl} is not in general a $\{0, 1\}$-game. Moreover it is only true if $\tau = e^S$ for any coalition S. We have the same problem with simple games.

Remark 3.3 Observe that the inherited condition was used in the proof of all the points in the above proposition except for the first two ones. We obtain an additive

Table 3.3 pl-restricted game in Example 3.11

R	$\{1\}$	$\{2\}$	$\{3\}$	$\{1,2\}$	$\{1,3\}$	$\{2,3\}$	N
$v_\tau^{pl}(R)$	0	0	0	5	0	0	0

game actually for any partition function, even if it is not an extension. For the second one it is only necessary to be an extension.

Example 3.11 We use the game v over $N = \{1, 2, 3\}$ in Example 1.17 which is superadditive and monotone. Consider the fuzzy coalition $\tau_0 = (0.2, 0.5, 0.4)$. We take $S = \{1, 2\}$. Now we define a mixed extension between the Choquet and the proportional extensions,

$$pl(\tau) = \begin{cases} pr(\tau), & \text{if } 3 \in supp(\tau) \\ ch(\tau), & \text{otherwise.} \end{cases}$$

The partition function pl is an extension but it is not inherited, for instance $pl(\tau_0) = \{(\{1\}, 0.2), (\{2\}, 0.5), (\{3\}, 0.4)\}$ however $pl(\tau_0 \times e^S) = \{(\{1, 2\}, 0.2), (\{2\}, 0.3)\}$. Table 3.3 determines game v_τ^{pl} which is not supeadditive and it is not monotone.

Proposition 3.11 *Let $v, w \in \mathscr{G}^N$ and $T \subseteq N$. For any inherited extension pl it holds*

(1) $(av + bw)_\tau^{pl} = av_\tau^{pl} + bw_\tau^{pl}$ for all $a, b \in \mathbb{R}$.
(2) If v, w are strategically equivalent then so are v_τ^{pl}, w_τ^{pl}.
(3) $(-v)_\tau^{pl} = -(v_\tau^{pl})$.
(4) $(v^{svg})_\tau^{pl} = (v_\tau^{pl})^{svg}$.

Proof Suppose $a, b \in \mathbb{R}$. We get for each coalition S if $pl(\tau \times e^S) = \{(S_k, s_k)\}_{k=1}^m$,

$$(av + bw)_\tau^{pl}(S) = \sum_{k=1}^{m} s_k(av + bw)(S_k) = av_\tau^{pl}(S) + bw_\tau^{pl}(S).$$

So, if v, w are strategically equivalent, namely there exist $a > 0$ and $b \in \mathbb{R}^N$ with $v = aw + b$ then

$$v_\tau^{pl} = aw_\tau^{pl} + b_\tau^{pl},$$

where $b_\tau^{pl} \in \mathbb{R}^N$ (see the above proposition). Thus v_τ^{pl}, w_τ^{pl} are strategically equivalent. Obviously $(-v)_\tau^{pl} = -(v_\tau^{pl})$. Now we determine the saving game, let $pl(\tau) = \{(T_k, t_k)\}_{k=1}^m$. Hence, by Lemma 3.1 and Proposition 3.2

$$(v^{svg})^{pl}_{\tau}(S) = \sum_{k=1}^{m} t_k (v^{svg})_{T_k}(S) = \sum_{k=1}^{m} t_k (v_{T_k})^{svg}(S)$$
$$= \sum_{k=1}^{m} t_k \sum_{i \in S} v_{T_k}(\{i\}) - \sum_{k=1}^{m} t_k v_{T_k}(S) = (v^{pl}_{\tau})^{svg}(S).$$

□

Since Proposition 3.10 (1) any property does not satisfies by the crisp restricted game is not verified by the fuzzy versions either. So, generally $(v^{dual})^{pl}_{\tau} \neq (v^{pl}_{\tau})^{dual}$ from Example 3.6.

Now, we introduce a game identified with a fuzzy coalition and each particular extension which allows us to describe formulas. Let $\tau \in [0, 1]^N$ and pl an inherited extension. The *pl-game* is defined for every coalition $S \subseteq N$ as

$$\tau^{pl}(S) = \sum_{\{(R,r) \in pl(\tau): S \subseteq R\}} r. \tag{3.9}$$

This game explains the membership of the coalitions in the fuzzy coalition depending on the chosen extension. So, we get from Lemma 2.1

$$\tau^{ml}(S) = \sum_{\{R \subseteq N: S \subseteq R\}} \prod_{i \in R} \tau(i) \prod_{i \in N \setminus R} (1 - \tau(i))$$
$$= \prod_{i \in S} \tau(i) \sum_{R \subseteq N \setminus S} \prod_{i \in R} \tau(i) \prod_{i \in (N \setminus S) \setminus R} (1 - \tau(i)) = \prod_{i \in S} \tau(i).$$

Thus, we have that $\tau^{ml}(S)$ represents the probability to obtain a coalition containing S according to τ,

$$\tau^{ml}(S) = \prod_{i \in S} \tau(i). \tag{3.10}$$

In the proportional case, coalition S is formed when all the players has the same level, and we get

$$\tau^{pr}(S) = \begin{cases} t, & \text{if } \tau(i) = t \ \forall i \in S \\ 0, & \text{otherwise.} \end{cases} \tag{3.11}$$

Finally, in the Choquet case the worth of S in the ch-game is the measure of the interval where coalition is ensured,

$$\tau^{ch}(S) = \vee(\tau \times e^S). \tag{3.12}$$

In fact, if $im_0(\tau) = \{0 = \lambda_0 < \lambda_1 < \cdots < \lambda_m\}$ with $\vee(\tau \times e^S) = \lambda_p$ we know that $S \subseteq [\tau]_k$ when $1 \leq k \leq p$. So,

$$\tau^{ch}(S) = \sum_{k=1}^{p}(\lambda_k - \lambda_{k-1}) = \lambda_p - \lambda_0 = \vee(\tau \times e^S).$$

For instance, in the next proposition we give a formula using the pl-game for the pl-restricted unanimity games. The pl-game determines the constant of proportionality between an unanimity game and its pl-restriction.

Proposition 3.12 *Let $\tau \in [0, 1]^N$ be a fuzzy coalition and pl be an inherited extension. For all $T \subseteq N$ be a non-empty coalition the pl-restricted game of the unanimity game u_T is*

$$(u_T)^{pl}_\tau = \tau^{pl}(T)u_T.$$

Proof Let $T \subseteq N$ a non-empty coalition. Since Lemma 3.1 and Proposition 3.7 we have

$$(u_T)^{pl}_\tau = \sum_{(R,r)\in pl(\tau)} r(u_T)_R = \left[\sum_{\{(R,r)\in pl(\tau):T\subseteq R\}} r\right] u_T.$$

So, from (3.9) we get the desired result. □

Following the crisp version (Sect. 3.4) we can define an extended value for each inherited extension.

Definition 3.9 Let pl be an inherited extension. The pl-extended Shapley value is a value for games over N with fuzzy coalition defined for each game $v \in \mathcal{G}^N$ and fuzzy coalition $\tau \in [0, 1]^N$ as

$$\phi^{pl}(v, \tau) = \phi(v^{pl}_\tau).$$

The above value coincides with the extended Shapley value if we take $\tau = e^T$. If $\tau = e^N$ then $\phi^{pl}(v, e^N) = \phi(v) = \phi^{cr}(v^{pl})$ from Theorem 2.3. Using Lemma 3.1 we write the pl-extended values according to the extended Shapley value.

Lemma 3.2 *Let pl be an inherited extension. It holds for all game $v \in \mathcal{G}^N$ and $\tau \in [0, 1]^N$ with $pl(\tau) = \{(R_k, r_k)\}$ that*

$$\phi^{pl}(v, \tau) = \sum_{k=1}^{m} r_k\phi(v, R_k).$$

Proof We use Lemma 3.1 and the linearity of the Shapley value (Proposition 1.8),

$$\phi^{pl}(v,\tau) = \phi(v_\tau^{pl}) = \phi\left(\sum_{k=1}^{m} r_k v_{R_k}\right) = \sum_{k=1}^{m} r_k \phi(v_{R_k}).$$

□

Next theorem gives a formula to determine the *pl*-extended values by dividends using also the *pl*-game (3.8).

Theorem 3.3 *Let pl be an inherited extension. For each* $(v,\tau) \in \mathcal{G}^N \times [0,1]^N$ *it holds for all player* $i \in N$ *that*

$$\phi_i^{pl}(v,\tau) = \sum_{\{S \subseteq N : i \in S\}} \tau^{pl}(S) \frac{\Delta_S^v}{|S|}.$$

Proof Following (3.8), we obtain

$$\begin{aligned} v_\tau^{pl} &= \sum_{(R,r)\in pl(\tau)} r v_R = \sum_{(R,r)\in pl(\tau)} r \sum_{\{S\subseteq R : S\neq\emptyset\}} \Delta_S^v u_S \\ &= \sum_{\{S\subseteq N : S\neq\emptyset\}} \left[\sum_{\{(R,r)\in pl(\tau) : S\subseteq R\}} r\right] \Delta_S^v u_S. \end{aligned}$$

So, by the uniqueness of the coefficients, the dividend for a non-empty coalition S of the *pl*-restricted game is

$$\Delta_S^{v_\tau^{pl}} = \tau^{pl}(S)\Delta_S^v. \tag{3.13}$$

Now, Theorem 1.2 implies

$$\phi_i^{pl}(v,\tau) = \phi_i(v_\tau^{pl}) = \sum_{\{S\subseteq N : i\in S\}} \frac{\Delta_S^{v_\tau^{pl}}}{|S|} = \sum_{\{S\subseteq N : i\in S\}} \tau^{pl}(S)\frac{\Delta_S^v}{|S|}.$$

□

Taking into account the above lemma and expressions (2.8), (2.11) and (2.12), we describe three interesting fuzzy extended Shapley values.

Definition 3.10 Let $v \in \mathcal{G}^N$ be a game and $\tau \in [0,1]^N$ be a fuzzy coalition.

- The multilinear Shapley value is defined as

$$\phi^{ml}(v,\tau) = \sum_{S \subseteq N} \left[\prod_{i \in S} \tau(i) \prod_{i \notin S} (1-\tau(i)) \right] \phi(v,S).$$

- The proportional Shapley value is defined as

$$\phi^{pr}(v,\tau) = \sum_{t \in im(\tau)} t\phi(v, S_t^\tau).$$

- The Choquet-Shapley value is defined as

$$\phi^{ch}(v,\tau) = \sum_{k=1}^{m} (\lambda_k - \lambda_{k-1})\phi(v,[\tau]_k),$$

with $im_0(\tau) = \{\lambda_0 < \lambda_1 < \cdots < \lambda_m\}$.

The multilinear and Choquet values can be expressed as integrals in a similar way of the values of these extensions in the above chapter.

Theorem 3.4 *Let $(v,\tau) \in \mathcal{G}^N \times [0,1]^N$. For each $i \in N$ it holds*

(1) $\phi_i^{ml}(v,\tau) = \tau(i)\int_0^1 D_i v^{ml}(t\tau)\,dt$.
(2) $\phi_i^{ch}(v,\tau) = \int_c \tau\, d\phi_i^v$ where $\phi_i^v(T) = \phi_i(v,T)$ for all $T \subseteq N$.

Proof (1) Suppose $i \in supp(\tau)$, otherwise the result follows because from Lemma 3.1 if $i \notin supp(\tau)$ then $\phi_i^{ml}(v,\tau) = 0$. Expression (2.9) describes the partial derivate of the multilinear extension and then

$$D_i v^{ml}(t\tau) = \sum_{\{S \subseteq supp(\iota): i \in S\}} t^{|S|-1} \Delta_S^v \prod_{j \in S \setminus \{i\}} \tau(j).$$

Integrating,

$$\tau(i)\int_0^1 D_i v^{ml}(t\tau)\,dt = \tau(i) \sum_{\{S \subseteq supp(\tau): i \in S\}} \Delta_S^v \prod_{j \in S \setminus \{i\}} \tau(j) \int_0^1 t^{|S|-1}\,dt$$

$$= \tau(i) \sum_{\{S \subseteq supp(\tau): i \in S\}} \prod_{j \in S \setminus \{i\}} \tau(j) \frac{\Delta_S^v}{|S|} = \sum_{\{S \subseteq supp(\tau): i \in S\}} \prod_{j \in S} \tau(j) \frac{\Delta_S^v}{|S|}$$

$$= \sum_{\{S \subseteq N: i \in S\}} \tau^{ml}(S) \frac{\Delta_S^v}{|S|} = \phi_i^{ml}(v, \tau),$$

from the concept of ml-game (3.10) and Theorem 3.3.
(2) For each player i the mapping $\phi_i^v : 2^N \to \mathbb{R}$ with $\phi_i^v(T) = \phi_i(v, T)$ if $T \subseteq N$ is a signed capacity because $\phi_i^v(\emptyset) = 0$. Definition 3.10 (3) and the definition of Choquet integral imply

$$\phi_i^{ch}(v, \tau) = \sum_{k=1}^{m} (\lambda_k - \lambda_{k-1}) \phi_i^v([\tau]_k) = \int_c \tau \, d\phi_i^v,$$

with $im_0(\tau) = \{\lambda_0 = 0 < \lambda_1 < \cdots < \lambda_m\}$. □

Butnariu [7] studied the proportional Shapley value in the context of fuzzy games, namely as a value for a particular class of fuzzy games, those proportional extensions of crisp games. In a similar way, Tsurumi et al. [26] analyzed the Choquet-Shapley value, namely for the fuzzy games which are Choquet extensions of crisp games. In both of these papers the authors obtained particular axiomatizations in the fuzzy context. Later Li and Zhang [18] got a common axiomatization for both of the families of fuzzy games and useful for all the fuzzy games. Now we find out an axiomatization in our context for all the inherited extensions. We can use very similar axioms to the Shapley value for the pl-extended one. But we substitute symmetry by necessary player axiom, the motive will be explained later. Let $f : \mathcal{G}^N \times [0, 1]^N \to \mathbb{R}^N$ a value for games over N with fuzzy coalition.

***pl*-Efficiency**. For all $v \in \mathcal{G}^N$ and $\tau \in [0, 1]^N$, $f(v, \tau)(N) = v^{pl}(\tau)$.

Support null player. Let $i \in N$ be a null player in the game v, it holds $f_i(v, \tau) = 0$ for all fuzzy coalition τ with $i \in supp(\tau)$

Support necessary player. Let $\tau \in [0, 1]^N$. If $i \in supp(\tau)$ is a necessary player for a monotone game v then $f_i(v, \tau) \geq f_j(v, \tau)$ for all $j \in N \setminus \{i\}$.

Linearity. For all $v, w \in \mathcal{G}^N$, $a, b \in \mathbb{R}$ and fuzzy coalition τ it holds that $f(av + bw, \tau) = af(v, \tau) + bf(w, \tau)$.

Support carrier. Let $v \in \mathcal{G}^N$ and $\tau \in [0, 1]^N$. For all player $i \notin supp(\tau)$ it holds $f_i(v, \tau) = 0$.

Theorem 3.5 *The pl-extended Shapley value is the only value for games over N with fuzzy coalition satisfying pl-efficiency, support null player, support necessary player, linearity and support carrier.*

Proof Since Theorem 3.1 the extended Shapley value verifies restricted efficiency, restricted null player, linearity and carrier. Proposition 3.5 (4) implies that this value also satisfies restricted necessary player. Using Lemma 3.2 we test that the pl-extended Shapley value satisfies the five axioms of the statement. Let $\tau \in [0, 1]^N$ with $pl(\tau) = \{(R_k, r_k)\}_{k=1}^m$. We have

$$\sum_{i \in N} \phi_i^{pl}(v, \tau) = \sum_{i \in N} \sum_{k=1}^m r_k \phi_i(v, R_k) = \sum_{k=1}^m r_k \sum_{i \in N} \phi_i(v, R_k) = \sum_{k=1}^m r_k v(R_k) = v^{pl}(\tau).$$

If $i \in N$ is a null player for v then $\phi_i^{pl}(v, \tau) = \sum_{k=1}^m r_k \phi_i(v, R_k) = 0$ because if $i \in R_k$ we use restricted null player for the crisp version and if $i \notin R_k$ we use carrier. The carrier axiom of the crisp version implies also the support carrier axiom for the pl-version because if $i \notin supp(\tau)$ then $i \notin R_k$ for all k. Linealidad follows since

$$\phi^{pl}(av + bw, \tau) = \sum_{k=1}^m r_k \phi(av + bw, R_k) = a\phi^{pl}(v, \tau) + b\phi^{pl}(w, \tau).$$

Finally we take i a necessary player for a monotone game v. If for any k we get $i \notin R_k$ then $v_{R_k} = 0$, thus $\phi(v, R_k) = \phi(v_{R_k}) = 0$. Otherwise, if $i \in R_k$ then i is a necessary player for game v_{R_k}. Therefore

$$\phi_i^{pl}(v, \tau) = \sum_{k=1}^m r_k \phi_i(v, R_k) \geq \sum_{k=1}^m r_k \phi_j(v, R_k) = \phi_j^{pl}(v, \tau).$$

Now let f be a value for games with fuzzy coalition. Linearity implies that we only need to test the uniqueness for unanimity games. So, let u_R with R a non-empty coalition and $\tau \in [0, 1]^N$. If $i \notin supp(\tau)$ then support carrier axiom implies $f_i(u_R, \tau) = 0$. Players out $R \cup supp(\tau)$ are null players and $f_i(u_R, \tau) = 0$. If $R \cap supp(\tau) = \emptyset$ we have finished the proof. Suppose $R \cap supp(\tau) \neq \emptyset$. All the players in $R \cap supp(\tau)$ are necessary players and hence $f_i(u_R, \tau) = f_j(u_R, \tau)$ if $i, j \in R \cap supp(\tau)$. We obtain

$$\sum_{i \in N} f_i(u_R, \tau) = |R \cap supp(\tau)| f_{i_0}(u_R, \tau) = u_R^{pl}(\tau),$$

for certain $i_0 \in R \cap supp(\tau)$. Therefore

$$f_{i_0}(u_R, \tau) = \frac{u_R^{pl}(\tau)}{|R \cap supp(\tau)|}.$$

□

Several of the properties analyzed in the crisp version, those related more directly to only the game, are extended to the pl-extensions. So, as in Proposition 3.5 we get the following properties.

Proposition 3.13 *Let pl be an inherited extension. The pl-extended Shapley value satisfies the following properties.*

(1) If $v, w \in \mathcal{G}^N$ with $v <_{supp(\tau)} w$ for $\tau \in [0, 1]^N$ then $\phi^{pl}(v, \tau) \leq \phi^{pl}(w, \tau)$.
(2) If $v, w \in \mathcal{G}^N$ with $v <_i w$ for $i \in N$ then $\phi_i^{pl}(v, \tau) \leq \phi_i^{pl}(w, \tau)$ for all $\tau \in [0, 1]^N$.
(3) If $i \in supp(\tau)$ and $v \in \mathcal{G}_{sa}^N$ then $\phi_i(v, \tau) \geq \tau(i)v(\{i\})$.
(4) If $S \subseteq supp(\tau)$ and $v \in \mathcal{G}_c^N$ then $\phi(v, \tau)(S) \geq v_\tau^{pl}(S)$.
(5) $\phi(-v, \tau) = -\phi(v, \tau)$.
(6) If $i \in supp(\tau)$ and $v \in \mathcal{G}^N$ then $\phi_i(v, \tau) = \tau(i)v(\{i\}) - \phi_i(v^{svg}, \tau)$.

Proof All these properties follow from Proposition 3.5 and Lemma 3.2. Given $\tau \in [0, 1]^N$, $\tau \neq 0$, we denote as $pl(\tau) = \{(R_k, r_k)\}_{k=1}^m$ its partition by levels by pl.
(1) If $i \notin supp(\tau)$ then $\phi_i(v, \tau) = \phi_i(w, \tau) = 0$. For all k we have $R_k \subseteq supp(\tau)$. If $R_k \subsetneq supp(\tau)$ then $v_{R_k} = w_{R_k}$, thus $\phi(v, R_k) = \phi(w, R_k)$. If $R_k = supp(\tau)$ then Proposition 3.5 (2) implies $\phi_i(v, R_k) \leq \phi_i(w, R_k)$ for all $i \in supp(\tau)$. Since Lemma 3.2 we get $\phi^{pl}(v, \tau) \leq \phi^{pl}(w, \tau)$.
(2) If $v <_i w$ then $\phi_i(v, R_k) \leq \phi_i(w, R_k)$ for all k from Proposition 3.5 (3). The result is obtained using Lemma 3.2.
(3) Let $i \in supp(\tau)$. Proposition 3.5 (5) implies that if $v \in \mathcal{G}_{sa}^N$ then $\phi_i(v, R_k) \geq v(\{i\})$ when $i \in R_k$. Hence (2.6) says

$$\phi_i(v, \tau) \geq v(\{i\}) \sum_{\{k: i \in R_k\}} r_k = \tau(i)v(\{i\}).$$

(4) Proposition 3.10 (4) again says that $v_\tau^{pl} \in \mathcal{G}_c^N$ if $v \in \mathcal{G}_c^N$. So, if $S \subseteq T$ then by Proposition 1.14

$$\phi^{pl}(v, \tau)(S) = \phi_i(v_\tau^{pl})(S) \geq v_\tau^{pl}(S).$$

(5) It is trivial.
(6) If $i \in R_k$ then $v_{R_k}(\{i\}) = v(\{i\})$. From Proposition 3.5 (8), (2.6) and Lemma 3.2 it holds

$$\begin{aligned}\phi_i(v, \tau) &= \sum_{\{k: i \in R_k\}} r_k \phi_i(v, R_k) = \left[\sum_{\{k: i \in R_k\}} r_k\right] v(\{i\}) - \sum_{\{k: i \in R_k\}} r_k \phi_i(v^{svg}, R_k) \\ &= \tau(i)v(\{i\}) - \phi_i(v^{svg}, \tau).\end{aligned}$$

□

Those properties related with the structure depend on the extension the most. The determination of a fuzzy coalition implies an asymmetry among the players, therefore the symmetry for the game is not enough to get two players with the same payoff. It is possible to find general conditions including certain symmetry in the extension.

Definition 3.11 An extension pl is named anonymous if for all $\tau \in [0,1]^N$ and all permutation $\theta \in \Theta^N$ it happens

$$pl(\theta\tau) = \{(\theta R, r) : (R, r) \in pl(\tau)\},$$

where $\theta\tau(i) = \tau(\theta^{-1}(i))$.

All the interesting proposed extensions are anonymous.

Proposition 3.14 *The multilinear, proportional and Choquet extensions are anonymous.*

Proof Let $\tau \in [0,1]^N$ be a non-zero fuzzy coalition, and $\theta \in \Theta^N$ a permutation over N.
MULTILINEAR. The partition of τ is taken following (2.8),

$$ml(\tau) = \left\{\left(T, \prod_{i \in T} \tau(i) \prod_{i \notin T} (1 - \tau(i))\right)\right\}_{[\tau]_1 \subseteq T \subseteq supp(\tau)}.$$

Since (2.8) again we have $(R, r) \in ml(\theta\tau)$ if and only if,

- $[\theta\tau]_1 \subseteq R \subseteq supp(\theta\tau)$. But $[\theta\tau]_1 = \theta[\tau]_1$ and $supp(\theta\tau) = \theta\, supp(\tau)$,
- and

$$r = \prod_{i \in R} (\theta\tau)(i) \prod_{i \notin R} (1 - (\theta\tau)(i)) = \prod_{i \in R} \tau(\theta^{-1}(i)) \prod_{i \notin R'} (1 - \tau(\theta^{-1}(i))).$$

So, we take $[\tau]_1 \subseteq R' = \theta^{-1} R \subseteq supp(\tau)$ with

$$r = \prod_{i \in R'} \tau(i) \prod_{i \notin R'} (1 - \tau(i)).$$

PROPORTIONAL Suppose for this case the partition by levels following (2.11) of τ,

$$pr(\tau) = \{(S_t^\tau, t)\}_{t \in im(\tau)}.$$

For our permutation θ we have $pr(\theta\tau) = \{(S_r^{\theta\tau}, r)\}_{r \in im(\theta\tau)}$. But $im(\theta\tau) = im(\tau)$ and

$$S_r^{\theta\tau} = \{i \in N : (\theta\tau)(i) = r\} = \{i \in N : \tau(\theta^{-1}(i)) = r\} = \theta S_r^{\tau}.$$

CHOQUET. Following (2.12) we have

$$ch(\tau) = \{(\lambda_k - \lambda_{k-1}, [\tau]_k)\}_{k=1}^{m}$$

with $im_0(\tau)=\{\lambda_0 < \lambda_1 < \cdots < \lambda_m\}$. For permutation θ we have $im_0(\theta\tau) = im_0(\tau)$. Let $p \in \{1, \ldots, m'\}$ and $(\lambda'_p - \lambda'_{p-1}, [\tau \times e^S]_p) \in ch(\tau \times e^S)$. We also get

$$[\theta\tau]_k = \{i \in N : \tau(\theta^{-1}(i)) = \lambda_k\} = \theta[\tau]_k.$$

□

Inherited property and anonymity are not related for extensions. Next examples show this fact.

Example 3.12 (1) Let $N = \{1, 2\}$. Consider the extension

$$pl(\tau) = \begin{cases} pr(\tau), & \text{if } \tau(1) > \tau(2) \\ ch(\tau), & \text{otherwise.} \end{cases}$$

Extension pl is inherited because for each player $i = 1, 2$ we have $pr(\tau \times e^{\{i\}}) = ch(\tau \times e^{\{i\}})$. But pl is not anonymous. If we take for instance $\tau = (0.5, 0.3)$ and $\theta(1) = 2, \theta(2) = 1$ then $pl(\tau) = \{(\{1\}, 0.5), (\{2\}, 0.3)\}$ and

$$pl(\theta\tau) = \{(\{1, 2\}, 0.3), (\{2\}, 0.2)\}.$$

(2) Now take $N = \{1, 2, 3\}$. Suppose the extension

$$pl(\tau) = \begin{cases} pr(\tau), & \text{if } supp(\tau) = N \\ ch(\tau), & \text{otherwise.} \end{cases}$$

Condition $supp(\theta\tau) = N$ is equivalent to $supp(\tau) = N$. As pr and ch are anonymous (Proposition 3.14) we deduce that pl too. But pl is not inherited. Consider $\tau = (0.1, 0.2, 0.3)$, we have $pl(\tau) = \{(\{1\}, 0.1), (\{2\}, 0.2), (\{3\}, 0.3)\}$. If we take $S = \{2, 3\}$ then

$$pl(\tau \times e^S) = \{(\{2, 3\}, 0.2), (\{3\}, 0.1)\}.$$

Adding the anonymity for the extension we get some properties about symmetry for our values.

Proposition 3.15 *Let pl be an anonymous and inherited extension. The pl-extended Shapley value satisfies the following properties.*

(1) For all permutation $\theta \in \Theta^N$, *player* $i \in N$ *and* $(v, \tau) \in \mathcal{G}^N \times [0, 1]^N$,

$$\phi^{pl}_{\theta(i)}(\theta v, \theta\tau) = \phi^{pl}_i(v, \tau).$$

(2) If i, j *are symmetric players for a game* v *and* $\tau \in [0, 1]^N$ *then*

$$\phi^{pl}_i(v, \tau) = \phi^{pl}_j(v, \tau^{ij}),$$

where $\tau^{ij} = \tau + (\tau(i) - \tau(j))e^j + (\tau(j) - \tau(i))e^i$.

Proof (1) As pl is anonymous, if $pl(\tau)=\{(R_k, r_k)\}_{k=1}^m$ then $pl(\theta\tau) = \{(\theta R_k, r_k)\}_{k=1}^m$ for all permutation θ. Lemma 3.2 and Proposition 3.5 (9) implies that

$$\phi^{pl}_{\theta(i)}(\theta v, \theta\tau) = \sum_{k=1}^{m} r_k \phi_{\theta(i)}(\theta v, \theta R_k) = \sum_{k=1}^{m} r_k \phi_i(v, R_k) = \phi^{pl}_i(v, \tau).$$

(2) Let $\theta^{ij} \in \Theta^N$ with $\theta^{ij}(i) = j, \theta^{ij}(j) = i$ and $\theta^{ij}(i') = i'$ for all $i' \in N \setminus \{i, j\}$. As i, j are symmetric players then $\theta^{ij} v = v$ because $\theta^{ij} S = S$ if $S \subseteq N \setminus \{i, j\}$ or $i, j \in S$, and if $i, j \notin S$ then

$$\theta^{ij} v(S \cup \{i\}) = v(S \cup \{j\}) = v(S \cup \{i\}).$$

Moreover $\theta^{ij}\tau = \tau^{ij}$. Following the above step we obtain

$$\phi^{pl}_j(v, \tau^{ij}) = \phi^{pl}_{\theta^{ij}(i)}(\theta^{ij} v, \theta^{ij}\tau) = \phi_i(v, \tau).$$

□

As consequence of the second step in the above proposition we get that symmetric players with the same level must have the same payoff. If $\tau(i) = \tau(j)$ for two players i, j symmetric for a game v then $\tau^{ij} = \tau$ and

$$\phi^{pl}_i(v, \tau) = \phi^{pl}_j(v, \tau).$$

As we said before the pl-restricted game of a simple game is not another simple in general. This question introduces a difference with regard to the crisp version. But the pl-extended values applied to simple games can be used as power indices. The ml-extended index $\phi^{ml}_i(v, \tau)$ determines the expected power of the players following the probability distribution τ. Extended indices with pr, ch are interesting to determine the power of groups with different feels into them.

References

1. Albizuri, M.J., Aurrekoetxea, J., Zarzuelo, J.M.: Configuration values: extensions of the coalitional Owen value. Games Econ. Beh. **57**, 1–17 (2006)
2. Algaba, E., Bilbao, J.M., van den Brink, R., Jiménez-Losada, A.: Cooperative games on antimatroids. Discret. Math. **282**, 1–15 (2004)
3. Aumann, R.J., Dreze, J.H.: Cooperative games with coalition structures. Int. J. Game Theory **3**(4), 217–237 (1974)
4. Bergantiños, G., Carreras, F., García-Jurado, I.: Cooperation when some players are incompatible. Math. Methods Oper. Res. **38**(2), 187–201 (1993)
5. Bilbao, J.M.: Axioms for the Shapley value on convex geometries. Eur. J. Oper. Res. **110**(2), 368–376 (1998)
6. Bilbao, J.M., Driessen, T.S.H., Jiménez-Losada, A., Lebrón, E.: The Shapley value for games on matroids: the static model. Math. Methods Oper. Res. **53**(2), 333–348 (2001)
7. Butnariu, D.: Stability and Shapley value for a n-persons fuzzy games. Fuzzy Sets Syst. **4**(1), 63–72 (1980)
8. Casajus, A.: Beyond basic structures in game theory. Ph.D. thesis. University of Leipzig. Germany (2007)
9. Derks, J., Peters, H.: A Shapley value for games with restricted coalitions. Int. J. Game Theory **21**(4), 351–360 (1993)
10. Faigle, U., Kern, W.: The Shapley value for cooperative games under precedence constraints. Int. J. Game Theory **21**(3), 249–266 (1992)
11. Gallardo, J.M., Jiménez, N., Jiménez-Losada, A., Lebrón, E.: Games with fuzzy permission structure: a conjunctive approach. Inf. Sci. **278**, 510–519 (2014)
12. Gallardo, J.M., Jiménez, N., Jiménez-Losada, A., Lebrón, E.: Games with fuzzy authorization structure: a Shapley value. Fuzzy Sets Syst. **272**(1), 115–125 (2015)
13. Gallego, I., Fernández, J.R., Jiménez-Losada, A., Ordóñez, M.: Cooperation among agents with a proximity relation. Eur. J. Oper. Res. **250**(2), 555–565 (2016)
14. Gilles, R.P., Owen, G., van den Brink, R.: Games with permission structures: the conjunctive approach. Int. J. Game Theory **20**(3), 277–293 (1992)
15. Jiménez, N.: Solution concepts for games on closure spaces. Ph.D. thesis. University of Seville. Spain (1998)
16. Jiménez-Losada, A.: Values for games on combinatorial structures. Ph.D. thesis. University of Seville (in spanish). Spain (1998)
17. Jimínez-losada, A., Fernández, J.R., Ordóñez, M.: Myerson values for games with fuzzy communication structure. Fuzzy Sets Syst. **213**, 74–90 (2013)
18. Li, S.J., Zhang, Q.: A simplified expression of the Shapley function for fuzzy games. Eur. J. Oper. Res. **196**(1), 234–245 (2009)
19. Meng, F., Zhang, Q.: The Shapley function for cooperative fuzzy games on convex geometries. J. Syst. Eng. Electron. **33**, 1305–1309 (2011)
20. Meng, F., Zhang, Q.: The symmetric Banzhaf value for fuzzy games with a coalition structure. Int. J. Autom. Comput. **9**(6), 600–608 (2012)
21. Mordeson, J.N., Nair, P.S.: Fuzzy Graphs and Fuzzy Hypergraphs. Studies in Fuzzines and Soft Computing. Springer, Berlin (2000)
22. Myerson, R.B.: Graphs and cooperation in games. Math. Oper. Res. **2**(3), 225–229 (1977)
23. Ovchinnikov, S.: On fuzzy preference relations. Int. J. Intell. Syst. **6**(2), 225–234 (1991)
24. Owen, G.: Values of games with a priori unions. In: Henn, R., Moeschlin, O. (eds.) Mathematical Economics and Game Theory. Lecture Notes in Economics and Mathematical, vol. 141, pp. 76–88. Springer, Berlin (1977)
25. Shapley, L.S.: A value for n-person games. In: Kuhn, H.W., Tucker, A.W. (eds.) Contributions to the Theory of Games, Volume II. Annals of Mathematical Studies, vol. 28, pp. 307–317. Princeton University Press, Princeton (1953)
26. Tsurumi, M., Tanino, T., Inuiguchi, M.: A Shapley function on a class of cooperative fuzzy games. Eur. J. Oper. Res. **129**(3), 596–618 (2001)

Chapter 4
Fuzzy Communication

4.1 Introduction

In the first chapter we saw as the classical cooperative game theory and particularly the construction of the Shapley value is thinking of the great coalition. The model analyzed in the last chapter considered that a particular coalition, non necessary the great coalition is formed. We took the decided coalition as an additional information over the classical model and supposed players out the coalition as players out of game (carrier axiom implies payoff zero for this kind of players).

But the first model in this sense was given by Aumann and Dreze [1] in 1974. In this case they considered that a partition of the set of players is formed. Hence all the players play but they are organized in several different final groups. Obviously the classical model is a particular case of their model. This option can be seen as a model in both of the lines explained in the introduction of the above chapter. Although there exist different extensions of the Shapley value to the model, we are interesting in describing as games with a bilateral relation among the agents: each player is related with those in the same set of the partition.

Three years later, Myerson [9] introduced communication structures among the agents as an additional information about the players. A communication structure consideres a graph to represent the communication options of the players. Vertices represent the agents and each link represents a feasible bilateral communication. The final formed coalitions are the connected components of the graph. Hence the Myerson model supposes also the construction of a partition of the set of players, but in this case not any pair of agents into one of the elements in the partition are related between them. This model establishes a clear bilateral relation among the players in the sense of our general model (see Sects. 3.2 and 3.3). The Aumann and Dreze [1] model is seen then as a particular case where the subgraph defined into each element of the partition is a complete graph (all the bilateral relations are

A. Jiménez-Losada, *Models for Cooperative Games with Fuzzy Relations among the Agents*, Studies in Fuzziness and Soft Computing 355,
DOI 10.1007/978-3-319-56472-2_4

feasible). Moreover the classical model is represented as a communication structure by the trivial partition and therefore by the complete graph over all the players. The reader can find a very extensive study of the Myerson model in Slikker and van den Nouweland [13]. The own Myerson extended his model in [10] by conference structures. In that case communication relations are not necessarily bilateral and then this kind of information goes out of our general model.

Communication in the real life is clearly an information for leveling. Hence Jiménez-Losada et al. [6] introduced fuzzy communication structures using fuzzy graphs. Fuzzy graphs allow to describe leveled membership and leveled communication. In Jiménez-Losada et al. [7] several different models to extend the Myerson value are shown based on the proportional and the Choquet fuzziness. Calvo et al. [4] analyzed games with probabilistic graphs using the multilinear extension.

In this chapter we study all these extensions. First the coalition structures and communication structures are explained. Later several fuzzy models are introduced and studied.

4.2 Coalition Structures. The Aumann-Dreze Model

Aumann and Dreze [1] proposed a partition of the set of players as the final distribution in coalitions. We explain their model in a more general way, in order to involve this idea in the fuzzy communication situations that we will study later. The coalitional model in Sect. 3.3 took any coalition as the final argument of cooperation. Now players can form several coalitions. Let N be a finite set.

Definition 4.1 A coalition structure for N is a non-empty family of coalitions $\mathscr{B} \subset 2^N$ such that $B \cap B' = \emptyset$ for all $B, B' \in \mathscr{B}$. The set of coalition structures over N is denoted as P_0^N.

Each coalition structure[1] $\mathscr{B}$ is associated to a quasi-equivalence relation (quasi-reflexive, symmetric and transitive) $r_{\mathscr{B}}$ where the domain is the active set of players,

$$N^{r_{\mathscr{B}}} = \bigcup_{B \in \mathscr{B}} B = N^{\mathscr{B}},$$

and the equivalence classes or components are the coalitions in $\mathscr{B}$, i.e. $\mathscr{B} = N/r_{\mathscr{B}}$. Coalition structures have been a crucial tool to analyze coalition formation (see for instance [3]). Obviously if $|\mathscr{B}| = 1$ then we are in the situation of Sect. 3.4.

Example 4.1 Suppose $N = \{1, 2, 3, 4, 5, 6, 7, 8\}$ a finite set of eight sellers of determined product. They decide cooperate but only among those working in the same

[1] A coalition structure $\mathscr{B}$ for Aumann y Dreze [1] satisfies $\bigcup_{B \in \mathscr{B}} B = N$ in addition.

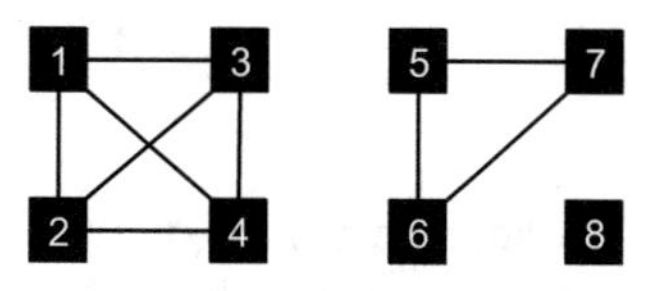

Fig. 4.1 Coalition structure

city. They form the coalitional structure $\mathscr{B} = \{\{1, 2, 3, 4\}, \{5, 6, 7\}, \{8\}\}$. Each coalition B in the structure is associated to the relation r^B as we say in Example 3.5. The aggregation of these relations establishes a new relation $r_{\mathscr{B}}$ in Fig. 4.1 given by

$$r_{\mathscr{B}} = \begin{bmatrix} 1\,1\,1\,1\,0\,0\,0\,0 \\ 1\,1\,1\,1\,0\,0\,0\,0 \\ 1\,1\,1\,1\,0\,0\,0\,0 \\ 1\,1\,1\,1\,0\,0\,0\,0 \\ 0\,0\,0\,0\,1\,1\,1\,0 \\ 0\,0\,0\,0\,1\,1\,1\,0 \\ 0\,0\,0\,0\,1\,1\,1\,0 \\ 0\,0\,0\,0\,0\,0\,0\,1 \end{bmatrix}$$

Taking into account the concept of restricted game (Definition 3.5) we introduce the instrumental game for this model.

Definition 4.2 Let $v \in \mathscr{G}^N$ be a game and $\mathscr{B} \in P_0^N$ a coalition structure. The coalitional game associated to $\mathscr{B}$ is the new game $v_{\mathscr{B}} \in \mathscr{G}^N$ with

$$v_{\mathscr{B}} = \sum_{B \in \mathscr{B}} v_B.$$

Almost all the properties verified by the restricted game in Sect. 3.4 are also satisfied by the coalitional game.

Proposition 4.1 *Let $v \in \mathscr{G}^N$ and $\mathscr{B} \in P_0^N$.*

(1) If v is additive then $v_{\mathscr{B}}$ is additive, moreover $v_{\mathscr{B}} = v_{N^{\mathscr{B}}}$.
(2) If v is superadditive (subadditive) or convex (concave) or monotone then so is $v_{\mathscr{B}}$.
(3) $(av + bw)_{\mathscr{B}} = av_{\mathscr{B}} + bw_{\mathscr{B}}$ for all $a, b \in \mathbb{R}$. Hence $(-v)_{\mathscr{B}} = -(v_{\mathscr{B}})$.
(4) If v, w are strategically equivalent then so are $v_{\mathscr{B}}, w_{\mathscr{B}}$.
(5) $(v^{svg})_{\mathscr{B}} = (v_{\mathscr{B}})^{svg}$.

Proof It is very easy to prove all these properties using Propositions 3.1 and 3.2. We only show the first and the last one. If $v \in \mathbb{R}^N$ then

$$v_{\mathscr{B}}(S) = \sum_{B\in\mathscr{B}} \sum_{i\in S\cap B} v_i = \sum_{i\in S\cap N^{\mathscr{B}}} v_i = v_{N^{\mathscr{B}}}(S).$$

The additive game $v_{N^{\mathscr{B}}}$ was defined on Proposition 3.1. Now we determine the saving game, for each coalition S

$$(v^{svg})_{\mathscr{B}}(S) = \sum_{B\in\mathscr{B}} (v^{svg})_B(S) = \sum_{B\in\mathscr{B}} (v_B)^{svg}(S) = \sum_{B\in\mathscr{B}} \sum_{i\in S} v_B(\{i\}) - v_B(S)$$
$$= \sum_{i\in S} \sum_{B\in\mathscr{B}} v_B(\{i\}) - v_B(S) = \sum_{i\in S} v_{\mathscr{B}}(\{i\}) - v_{\mathscr{B}}(S).$$

□

A value for games with coalition structure is

$$f : \mathscr{G}^N \times P_0^N \to \mathbb{R}^N.$$

Each pair $(v, \mathscr{B}) \in \mathscr{G}^N \times P_0^N$ is called a *game over N with coalition structure*. We define a Shapley value for games with a coalition structure using the extended Shapley value (Definition 3.6) for each coalition in the structure.

Definition 4.3 The coalitional value is a value for games over N with coalition structure defined for each game $v \in \mathscr{G}^N$ and each coalition structure $\mathscr{B} \in P_0^N$ as

$$\mu(v, \mathscr{B}) = \phi(v_{\mathscr{B}}).$$

The above mapping coincides with the Shapley value in the classical situation, namely $\mathscr{B} = \{N\}$. Additivity of the Shapley value implies the equality

$$\mu(v, \mathscr{B}) = \sum_{B\in\mathscr{B}} \phi(v, B)$$

Since the extended Shapley values verifies carrier and the elements in the coalition structure are disjoint we have that for each $B \in \mathscr{B}$ and $i \in B$

$$\mu_i(v, \mathscr{B}) = \phi_i(v, B). \tag{4.1}$$

Aumann and Dreze [1] defined the coalitional value using the subgame over each coalition in the structure. Proposition 3.4 says that both are the same,

$$\mu_i(v, \mathscr{B}) = \phi_i^B(v)$$

for all $i \in B \in \mathscr{B}$. Hence $\mu_i(v, \mathscr{B}) = 0$ if $i \notin N^{\mathscr{B}}$.

Similar axioms to the extended Shapley value can be used for the coalitional value. Let $f : \mathscr{G}^N \times P_0^N \to \mathbb{R}^N$ a value for games over N with coalition structure.

Efficiency by components. For all $v \in \mathscr{G}^N$ and $B \in \mathscr{B}$, $f(v, \mathscr{B})(B) = v(B)$.

Restricted null player. Let $i \in N$ be a null player in the game v, it holds $f_i(v, \mathscr{B}) = 0$ if $i \in N^{\mathscr{B}}$.

Symmetry by components. If $i, j \in N$ are symmetric for a game v and $\mathscr{B} \in P_0^N$ with $i, j \in B \in \mathscr{B}$ then $f_i(v, \mathscr{B}) = f_j(v, \mathscr{B})$.

Linearity. For all $v, w \in \mathscr{G}^N$, $a, b \in \mathbb{R}$ and $\mathscr{B} \in P_0^N$ it holds $f(av + bw, \mathscr{B}) = af(v, \mathscr{B}) + bf(w, \mathscr{B})$.

Carrier. Let $v \in \mathscr{G}^N$ and $\mathscr{B} \in P_0^N$. For any non-active player i, namely $i \notin N^{\mathscr{B}}$, it holds $f_i(v, \mathscr{B}) = 0$.

Restricted null player is the same axiom given in Sect. 3.4 for coalition $N^{\mathscr{B}}$. Efficiency and symmetry by components are the usual axioms relative to each component in the binary relation.[2] If $|\mathscr{B}| = 1$ then the axiomatization coincides with the axiomatization of the extended Shapley value.

Theorem 4.1 *The coalitional value is the only value for games over N with coalition structure satisfying efficiency by components, restricted null player, symmetry by components, linearity and carrier.*

Proof Obviously, from (4.1) the coalitional value satisfies the first four axioms. As the extended Shapley value verifies carrier, if $i \notin N^{\mathscr{B}}$ then $\phi_i(v, B) = 0$ for every $B \in \mathscr{B}$.

Let f be a value. Following the proof of Theorem 1.3, linearity implies that we only need to find out the uniqueness for unanimity games. So, let u_R with R a non-empty coalition. We consider now $\mathscr{B}$ a coalition structure. Using carrier we have $f_i(u_R, \mathscr{B}) = 0$ if $i \notin N^{\mathscr{B}}$. Suppose $B \in \mathscr{B}$, restricted null player says that $f_i(u_R, \mathscr{B}) = 0$ for all $i \in B \setminus R$. If $B \cap R \neq \emptyset$ then all the players in this intersection are symmetric, thus there exists $i_0 \in B \cap R$ with $f_i(u_R, \mathscr{B}) = f_{i_0}(u_R, \mathscr{B})$ for all $i \in B \cap R$. Efficiency by components implies

$$\sum_{i \in B} f_i(u_R, \mathscr{B}) = |B \cap R| f_{i_0}(u_R, \mathscr{B}) = u_R(B).$$

So we get the uniqueness for player i_0. □

[2] Aumann and Dreze [1] named them relative efficiency and relative symmetry.

We show more results about anonymity and symmetry in this context in the next proposition. Given $\mathscr{B} \in P_0^N$ a coalition structure and $\theta \in \Theta^N$ a permutation over N, we introduce the new coalition structure

$$\theta\mathscr{B} = \{\theta B : B \in \mathscr{B}\}.$$

Proposition 4.2 *Let $\mathscr{B} \in P_0^N$ be a coalition structure and $\theta \in \Theta^N$. It holds for any game v and player i,*

$$\mu_{\theta(i)}(\theta v, \theta\mathscr{B}) = \mu_i(v, \mathscr{B}).$$

Moreover,

(1) If $\theta\mathscr{B} = \mathscr{B}$ then $\mu_{\theta i}(\theta v, \mathscr{B}) = \mu_i(v, \mathscr{B})$,
(2) If $B, B' \in \mathscr{B}$ with $B \neq B'$, and $i \in B$, $j \in B'$ are two symmetric players for v then the coalitional value satisfies $\mu_i(v, \mathscr{B}) = \mu_j(v, \mathscr{B}')$, where

$$\mathscr{B}' = \mathscr{B} \setminus \{B, B'\} \cup \{B \setminus \{i\} \cup \{j\}, B' \setminus \{j\} \cup \{i\}\}.$$

(3) If $B \in \mathscr{B}$ and $i \in B$, $j \notin N^{\mathscr{B}}$ are symmetric for v then $\mu_i(v, \mathscr{B}) = \mu_j(v, \mathscr{B}')$, where

$$\mathscr{B}' = \mathscr{B} \setminus \{B\} \cup \{B \setminus \{i\} \cup \{j\}\}.$$

Proof We use Theorem 3.1. Suppose $i \in N^{\mathscr{B}}$, there exists only one $B \in \mathscr{B}$ with $i \in B$ and

$$\mu_{\theta(i)}(\theta v, \theta\mathscr{B}) = \phi_{\theta(i)}(v, \theta B) = \phi_i(v, B) = \mu_i(v, \mathscr{B}).$$

Otherwise, namely $i \notin N^{\mathscr{B}}$ we get $\theta(i) \notin N^{\theta\mathscr{B}}$ and by the carrier axiom of the coalitional value both payoffs are zero.

Property (1) is a trivial consequence from the above equality. For the others equalities we follow the proof in Theorem 3.1 taking permutation θ^{ij}. As i, j are symmetric then $\theta v = v$ and $\theta\mathscr{B} = \mathscr{B}'$ in both cases. □

Property (1) in the above proposition is taken as an axiom by Aumann and Dreze [1]. It is easy to prove that this property implies symmetry by components.

Next we summarize other properties of the coalitional value. The proofs are simples using (4.1) and following the proofs in Proposition 3.5.

Proposition 4.3 *The coalitional value satisfies the following properties for a coalition structure $\mathscr{B}$.*

(1) Let $i, j \in B \in \mathscr{B}$ be two different players. If $\mathscr{B}_i, \mathscr{B}_j \in P_0^N$ with $B \setminus \{i\} \in \mathscr{B}_i$ and $B \setminus \{j\} \in \mathscr{B}_j$ then for all $v \in \mathscr{G}^N$ it holds

$$\mu_i(v, \mathscr{B}) - \mu_i(v, \mathscr{B}_j) = \mu_j(v, \mathscr{B}) - \mu_j(v, \mathscr{B}_i).$$

(2) If $v, w \in \mathscr{G}^N$ with $v <_B w$ for all $B \in \mathscr{B}$ then $\mu(v, \mathscr{B}) \leq \mu(w, \mathscr{B})$.
(3) If $v, w \in \mathscr{G}^N$ with $v <_i w$ for any $i \in N$ then $\mu_i(v, \mathscr{B}) \leq \mu_i(w, \mathscr{B})$.
(4) If $i \in N^{\mathscr{B}}$ is a necessary player for $v \in \mathscr{G}_m^N$ then $\mu_j(v, \mathscr{B}) \leq \mu_i(v, \mathscr{B})$.
(5) If $i \in N^{\mathscr{B}}$ and $v \in \mathscr{G}_{sa}^N$ then $\mu_i(v, \mathscr{B}) \geq v(\{i\})$.
(6) Let $v \in \mathscr{G}_c^N$. If $S \subseteq N$ then $\mu(v, \mathscr{B})(S) \geq v_{\mathscr{B}}(S)$. Particularly if $S \subseteq B \in \mathscr{B}$ then $\mu(v, \mathscr{B})(S) \geq v(S)$.
(7) $\mu(-v, \mathscr{B}) = -\mu(v, \mathscr{B})$.
(8) For all $i \in N^{\mathscr{B}}$ and $v \in \mathscr{G}^N$ it holds $\mu_i(v, \mathscr{B}) = v(\{i\}) - \mu_i(v^{svg}, \mathscr{B})$.

Example 3.7 showed that it is not true that the dual game has the same payoff vector of the original game for the extended Shapley value. Another concept of dual was proposed in (3.7) and now we extend the idea to coalition structures. Given a coalition structure $\mathscr{B}$ and a game v, the *$\mathscr{B}$-dual* game is $v^{\mathscr{B}dual} \in \mathscr{G}^N$ where

$$v^{\mathscr{B}dual}(S) = \sum_{B \in \mathscr{B}} v(B) - v(B \setminus S). \tag{4.2}$$

This new definition is also consistent with the information of the coalition structure as the next proposition says.

Proposition 4.4 *Let $\mathscr{B} \in P_0^N$ be a coalition structure. For all game v it holds*

$$\mu(v^{\mathscr{B}dual}, \mathscr{B}) = \mu(v, \mathscr{B}).$$

Proof Suppose $i \in N^{\mathscr{B}}$, otherwise the equality is true from the carrier axiom. Let $B \in \mathscr{B}$ the only coalition with $i \in B$ in the structure. For any coalition S if $B' \in \mathscr{B} \setminus \{B\}$ then $B' \setminus (B \cap S) = B'$ because they are disjoint. We use expressions (4.2) and the definition of B-dual game (3.7) to get

$$(v^{\mathscr{B}dual})_B(S) = v^{\mathscr{B}dual}(S \cap B) = v(B) - v(B \setminus (S \cap B)) = (v^{Bdual})_B(S).$$

It implies

$$\phi_i(v^{\mathscr{B}dual}, B) = \phi_i((v^{\mathscr{B}dual})_B) = \phi_i((v^{Bdual})_B) = \phi_i(v^{Bdual}, B).$$

Now, Proposition 3.6 and expression (4.1) obtain

$$\mu_i(v^{\mathscr{B}dual}, \mathscr{B}) = \phi_i(v^{\mathscr{B}dual}, B) = \phi_i(v^{Bdual}, B) = \phi_i(v, B) = \mu_i(v, \mathscr{B}).$$

□

Now we explain the coalitional value as an index. But, given a simple game, the coalitional game is not always simple.

Proposition 4.5 *Let $v \in \mathscr{G}_s^N$ be a simple game. Game $v_{\mathscr{B}}$ is simple for all $\mathscr{B} \in P_0^N$ if and only if v is also superadditive.*

Proof Let $v \in \mathscr{G}_s^N$ be a simple game. As v is monotone then Proposition 4.1 guarantees that $v_{\mathscr{B}}$ is also monotone. The problem is to be $\{0, 1\}$-game. Observe that a simple game is superadditive if and only if there are not two disjoint winning coalitions. If we have two disjoint winning coalitions S, T then we can define $\mathscr{B}$ such that $S, T \in \mathscr{B}$ and then

$$v_{\mathscr{B}}(S \cup T) = v(S) + v(T) = 2.$$

If there are not two disjoint winning coalitions then for any coalition structure $\mathscr{B}$ and any coalition S we obtain at most one $B \in \mathscr{B}$ such that $S \cap B$ is a winning coalition for v, therefore $S \in W(v_B)$. Suppose there is one B with this condition, then

$$v_{\mathscr{B}}(S) = \sum_{B \in \mathscr{B}} v_B(S) = 1.$$

Otherwise $v_{\mathscr{B}}(S) = 0$. Thus $v_{\mathscr{B}}$ is a $\{0, 1\}$-game. □

The next proposition evaluates the coalitional game of a unanimity game which is a superadditive simple game, and its coalitional index.

Proposition 4.6 *Let $R \subseteq N$ be a non empty coalition. For a coalition structure $\mathscr{B}$ it holds $(u_R)_{\mathscr{B}} = u_R$ if there exists $B \in \mathscr{B}$ with $R \subseteq B$ and $(u_R)_{\mathscr{B}} = 0$ otherwise. Furthermore*

$$\mu(u_R, \mathscr{B}) = \begin{cases} \phi(u_R), & \text{if there is } B \in \mathscr{B} \text{ with } R \subseteq B \\ 0, & \text{otherwise.} \end{cases}$$

Proof We have that there is at most one $B \in \mathscr{B}$ with $R \subseteq B$. We get $(u_R)_{\mathscr{B}} = (u_R)_B$. Proposition 3.7 implies the result. □

We can describe the coalitional value from the dividends using the above proposition.

Theorem 4.2 *For each $v \in \mathcal{G}^N$ and a coalition structure $\mathcal{B}$ the coalitional value of a player $i \in N_{\mathcal{B}}$ is*

$$\mu_i(v, \mathcal{B}) = \sum_{i \in S \subseteq B} \frac{\Delta_S^v}{|S|},$$

where $B \in \mathcal{B}$ with $i \in B$.

Proof We get the goal using that $\mu_i(v, \mathcal{B}) = \phi_i(v, B)$ and by (3.8). □

Remark 4.1 It is possible to modify the concept of coalitional game for simple games in order to get always a new simple game. Let $\mathcal{B} \in P_0^N$. If $v \in \mathcal{G}_s^N$ then the *simple coalitional game* is defined for each coalition S as

$$v_s^{\mathcal{B}}(S) = \bigvee_{B \in \mathcal{B}} v(S \cap B). \tag{4.3}$$

Obviously $v_s^{\mathcal{B}} \in \mathcal{G}_s^N$ and if $v \in \mathcal{G}_s^N \cap \mathcal{G}_{sa}^N$ then $v_s^{\mathcal{B}} = v_{\mathcal{B}}$. But this concept of simple coalitional game is not perfect. The definition does not work correctly with the lattice structure of simple games. It is true that $(v \vee w)_s^{\mathcal{B}} = v_s^{\mathcal{B}} \vee w_s^{\mathcal{B}}$ for all $\mathcal{B} \in P_0^N$ but next example shows what happens with the minimum. Let $N = \{1, 2, 3\}$ and the coalition structure $\mathcal{B} = \{\{1, 2\}, \{3\}\}$. Suppose the unanimity games $v = u_{\{1,2\}}$ and $w = u_{\{3\}}$. The minimum of these games is $v \wedge w = u_N$ and then $(v \wedge w)_s^{\mathcal{B}} = 0$. On the other hand $v_s^{\mathcal{B}} = u_{\{1,2\}}$ and $w_s^{\mathcal{B}} = u_{\{3\}}$. Thus $v_s^{\mathcal{B}} \wedge w_s^{\mathcal{B}} = u_N \neq 0$.

4.3 Communication Structures. The Myerson Model

Coalition structures over N can be seen as a partition of the set of players in several final coalitions through the bilateral communication between players but without communication among them. If $\mathcal{B} \in P_0^N$ then each $B \in \mathcal{B}$ represents a family of players without communication with the players out of B and where all the communications between players in B are feasible. Myerson [9] introduced a new step in the analysis of the partial cooperation in this sense. Now players are organized in a coalition structure but into each coalition of the structure not all the bilateral communications are feasible. In a coalition of the structure enough communication is supposed in order to connect the set. Myerson used a undirected graph to represent this situation. In this graph the vertices are the players and the links are the feasible bilateral coalitions. A undirected graph is really an special case of binary relation

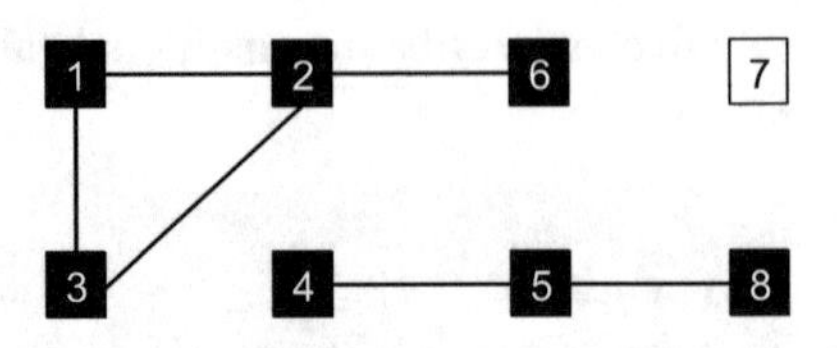

Fig. 4.2 Communication structure r

over N. The sense of the relation of a player with herself is being active, a black vertex. As we are thinking of bilateral communication we consider that every player with a link to another one is active in the game, namely is quasi-reflexive. Following with the usual reasoning in the classical cooperative games, players form the maximal coalitions that they get connecting. So, the final coalition structure is the set of components of the relation.

Definition 4.4 A communication structure over N is a bilateral relation r satisfying two conditions:

(1) r is quasi-reflexive, $r(ij) = 1$ implies $i \in N^r$,
(2) r is symmetric, $r(i, j) = r(j, i) = r(ij)$ if $i, j \in N$ with $i \neq j$.

The family of communication structures is denoted as G^N

Example 4.2 Suppose $N = \{1, 2, 3, 4, 5, 6, 7, 8\}$. We consider the next communication situation. Player 7 is not active. Player 1 is able to communicate with players 2, 3, player 2 with 3 and player 2 with 6. There exists also communication of 5 with 4, 8 (see Fig. 4.2). The relation is represented by the matrix

$$r = \begin{bmatrix} 1\,1\,1\,0\,0\,0\,0\,0 \\ 1\,1\,1\,0\,0\,1\,0\,0 \\ 1\,1\,1\,0\,0\,0\,0\,0 \\ 0\,0\,0\,1\,1\,0\,0\,0 \\ 0\,0\,0\,1\,1\,0\,0\,1 \\ 0\,1\,0\,0\,0\,1\,0\,0 \\ 0\,0\,0\,0\,0\,0\,0\,0 \\ 0\,0\,0\,0\,1\,0\,0\,1 \end{bmatrix}.$$

Following Sect. 3.3 we have

$$L(r) = \{12, 13, 23, 26, 58, 45\}.$$

Connecting among them, players form the coalition structure of the components in the relation,

$$N/r = \{\{1, 2, 3, 6\}, \{4, 5, 8\}\}.$$

But although $\{4, 5, 8\}$ is a final coalition, communication between 4 and 8 is not feasible.

Coalition structures as in the Aumann-Dreze model are transitive communication structures, i.e. into each final coalition every pair of players are related. Each pair $(v, r) \in \mathcal{G}^N \times G^N$ is called a *game over N with communication structure*.

A "measure" of the benefit obtained in a each game with communication structure is defined by Myerson. He supposed that this benefit is the sum of the worth of the components in the relation.

Definition 4.5 Let $v \in \mathcal{G}^N$. The graph-worth over N is the function $g : \mathcal{G}^N \times G^N \to \mathbb{R}$ given by

$$g(v, r) = \sum_{S \in N/r} v(S)$$

for all $v \in \mathcal{G}^N$ and $r \in G^N$.

Next we introduce an operation of graphs interesting in several moments of the book.

Definition 4.6 If r is a communication structure and $ij \in L(r)$ then $r_{-ij} \in G^N$ with $r_{-ij}(i'j') = r(i'j')$ for all $i'j' \neq ij$ and $r_{-ij}(ij) = 0$.

Example 4.3 Consider r in Example 4.2. Suppose $ij = 45 \in L(r)$. Relation r_{-45} is represented by the matrix

$$r_{-45} = \begin{bmatrix} 1\,1\,1\,0\,0\,0\,0\,0 \\ 1\,1\,1\,0\,0\,1\,0\,0 \\ 1\,1\,1\,0\,0\,0\,0\,0 \\ 0\,0\,0\,1\,0\,0\,0\,0 \\ 0\,0\,0\,0\,1\,0\,0\,1 \\ 0\,1\,0\,0\,0\,1\,0\,0 \\ 0\,0\,0\,0\,0\,0\,0\,0 \\ 0\,0\,0\,0\,1\,0\,0\,1 \end{bmatrix}.$$

Hence now $L(r_{-45}) = \{12, 13, 23, 26, 58\}$. If we delete a link in a graph the set of component can change, now $N/r_{-45} = \{\{1, 2, 3, 6\}, \{4\}, \{5, 8\}\}$ (Fig. 4.3).

The graph-worth g satisfies the following conditions,

(1) Connection, if r is connected then $g(v, r) = v(N^r)$,
(2) Component additivity, $g(v, r) = \sum_{S \in N/r} g(v, r_S)$,
(3) Link monotonicity, if $v \in \mathcal{G}_{sa}^N$ then $g(v, r) \geq g(v, r_{-ij})$ for all $ij \in L(r)$.

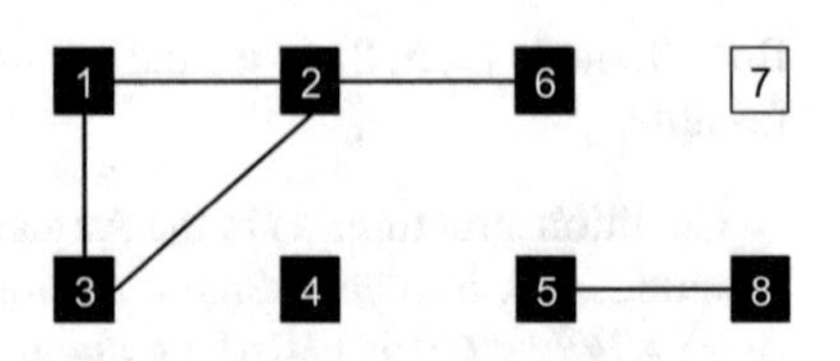

Fig. 4.3 Communication structure r_{-45}

The first property, following the Myerson idea says that in a communication structure the most important thing is connecting people. The second one implies that if two groups are not connected then they cannot improve their profits for cooperating and the worth is the sum of the separated benefits. Remember that in a superadditive game cooperation is a good chance for players, therefore the graph-worth must respect this idea rewarding the communication. Last condition expresses this fact. Observe that the first two properties implies the existence of only one function, the graph-worth.

Myerson [9] used the graph-worth to describe a model to study communication structures. Following our general model a new game is defined introducing the information about the communication.

Definition 4.7 Let $v \in \mathcal{G}^N$ and r a communication structure. The vertex game is $v/r \in \mathcal{G}^N$ with

$$v/r(S) = g(v, r_S)$$

for any coalition S.

Obviously, if the relation r is also transitive, $r = r_{\mathcal{B}}$ with a coalition structure $\mathcal{B}$, then $v/r = v_{\mathcal{B}}$.

Example 4.4 Consider again the communication situation r in Fig. 4.2. Now let $v \in \mathcal{G}_a^N$ with $v(S) = |S| - 1$ for all non-empty coalition. We calculate the worth of coalition $S = \{1, 3, 6, 5, 8\}$ in the vertex game. We have $v(S) = 4$ but

$$v/r(S) = g(v, r_S) = \sum_{T \in N/r_S} v(T) = v(\{1, 3\}) + v(\{6\}) + v(\{5, 8\}) = 2,$$

because $N/r_S = \{\{1, 3\}, \{6\}, \{5, 8\}\}$. The graph representing r_S is in Fig. 4.4.

Almost all the properties of those verified by the restricted game are also satisfied by the coalitional game.

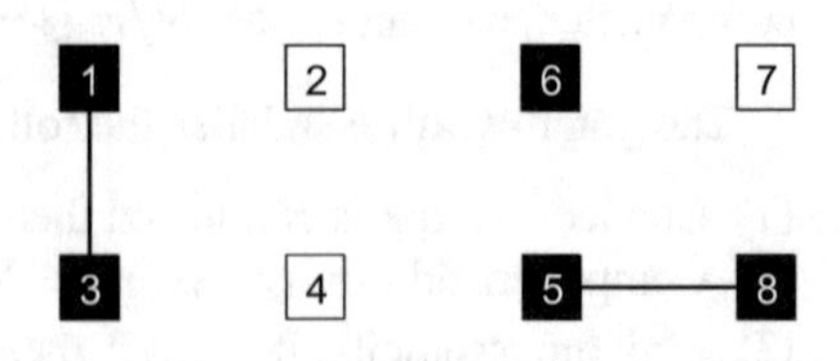

Fig. 4.4 Communication structure r_S

Proposition 4.7 *Let $v \in \mathcal{G}^N$ and $r \in G^N$.*

(1) If v is additive then v/r is additive, moreover $v/r = v_{N^r}$.
(2) If v is superadditive (subadditive) then so is v/r.
(3) $(av + bw)/r = av/r + bw/r$ for all $a, b \in \mathbb{R}$. Hence $(-v)/r = -(v/r)$.
(4) If v, w are strategically equivalent then so are $v/r, w/r$.
(5) $(v^{svg})/r = (v/r)^{svg}$.

Proof (1) Remember that the components form a partition of the domain of a relation. If $v \in \mathbb{R}^N$ then following Definition 4.7,

$$v/r(S) = \sum_{T \in N/r_S} v(T) = \sum_{T \in N/r_S} \sum_{i \in T} v_i = \sum_{i \in N^r \cap S} v_i.$$

(2) Suppose $v \in \mathcal{G}^N_{sa}$. Let S, T be coalitions with $S \cap T = \emptyset$. For each $R \in N/r_{S \cup T}$ there exists a partition of it using components in N/r_S and N/r_T. Thus, superadditivity of v implies

$$v/r(S \cup T) = \sum_{R \in N/r_{S \cup T}} v(R) \geq \sum_{R \in N/r_S} v(R) + \sum_{R \in N/r_T} v(R) = v/r(S) + v/r(T).$$

(3) A simple exercise for the reader.
(4) If v, w are strategically equivalent then there are $a \in \mathbb{R}$ and $b \in \mathbb{R}^N$ such that $v = aw + b$.The first and the third steps imply

$$v/r(v) = aw/r(v) + b_{N^r}.$$

(5) We determine the saving game, for each coalition S. If $i \in N$ then $v/r(\{i\}) = v(\{i\})$ when $i \in N^r$ and $v/r(\{i\}) = 0$ otherwise. Let $S \subseteq N$,

$$\begin{aligned}(v^{svg})/r(S) &= \sum_{T \in N/r_S} v^{svg}(T) = \sum_{T \in N/r_S} \left[\sum_{i \in T} v(\{i\}) - v(T) \right] \\ &= \sum_{i \in N^{r_S}} v(\{i\}) - \sum_{T \in N/r_S} v(T) = \sum_{i \in S} v/r(\{i\}) - v/r(S) = (v/r)^{svg}(S).\end{aligned}$$

□

But in this situation monotonicity and convexity are not inheritable. Next example shows that superadditivity plays an important role to get the inherence of the monotonicity.

Example 4.5 Suppose the monotone game v over $N = \{1, 2, 3\}$ given by $v(\{i\}) = 4$, $v(\{i, j\}) = 5$ and $v(N) = 6$. This game is not superadditive because we obtain

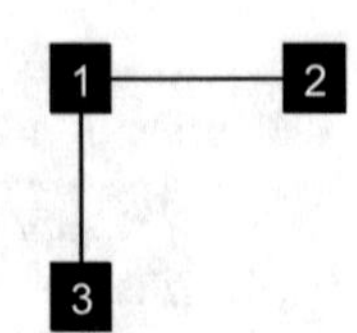

Fig. 4.5 Monotonicity is not inheritable

$v(\{1\}) + v(\{3\}) = 8 > v(\{1,3\}) = 5$. Now we consider the graph in Fig. 4.5. As the graph is connected we obtain $v/r(N) = 6$. But then, $v/r(\{1,3\}) = v(\{1\}) + v(\{3\}) = 8$. Observe that if v was superadditive then $v(\{1\}) + v(\{3\}) \leq v(\{1,3\}) \leq v(N)$.

We can remove from the above example the following fact. For any communication structure r containing the graph in Fig. 4.5[3] we can find a monotone game v such that v/r is not monotone. Supperadditivity is an usual property that it ensures the inherence of monotonicity. Remember from the first chapter that any monotone game must be non-negative, moreover if $v \in \mathcal{G}_{sa}^N$ and $v \geq 0$ then $v \in \mathcal{G}_m^N$. This kind of monotone games guarantees the monotonicity of the vertex game.

Proposition 4.8 *If $v \in \mathcal{G}_{sa}^N$ and $v \geq 0$ then v/r is monotone for all communication structure r.*

Proof Let $v \in \mathcal{G}_{sa}^N$. Proposition 4.7 (2) implies that v/r is also superadditive. Game v/r is also non-negative, for all coalition S

$$v/r(S) = g(v, r_S) = \sum_{R \in N/r_S} v(R) \geq 0.$$

Hence, v/r is superadditive and non-negative, thus it is monotone. □

Convexity is not inheritable either. Next example was given by Slikker and van den Nouweland [13] to demonstrate this fact.

Example 4.6 Let $N = \{1, 2, 3, 4\}$. Consider the reflexive communication structure r and with the set of links

$$L(r) = \{12, 24, 34, 13\},$$

the graph is in Fig. 4.6. We take again $v(S) = |S| - 1$ for any non-empty coalition. It holds $v(S) + v(T) = |S \cup T| + |S \cap T| - 2$, thus

$$v(S) + v(T) = v(S \cup T) + v(S \cap T).$$

[3]There is a coalition S with r_S drawing the same graph for the vertices in black but with different labels.

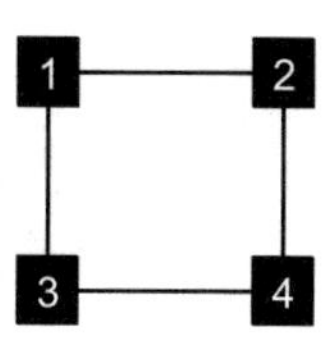

Fig. 4.6 Convexity is not inheritable

Our game is convex. Now we construct the vertex game, as the relation is reflexive

$$v/r(S) = g(v, r_S) = \sum_{T \in N/r_S} (|T| - 1) = |S| - |N/r_S|.$$

Coalitions $S = \{1, 2, 3\}$ and $T = \{2, 3, 4\}$ are connected, therefore they verifies $v/r(S) = v(S) = 2 = v(T) = v/r(T)$. As $S \cup T = N$ and the graph is connected then $v/r(N) = v(N) = 3$. But the intersection, $S \cap T = \{2, 3\}$, is not connected, and its worth is $v/r(S \cap T) = 0$. So,

$$v/r(S) + v/r(T) = 4 > 3 = v/r(S \cup T) + v/r(S \cap T).$$

Game v/r is not convex.

Convexity is a restrictive condition enough to think increase it. Perhaps non-negativity can be a good idea, but the game in the above example is non-negative. So, Slikker and van den Nouweland [13] thought of conditions for the communication structure in order to get the inheritance of convexity.

Definition 4.8 A communication structure r is cycle-complete if any cycle $\{i_0, \ldots, i_m\}$ in r satisfies $r_{\{i_0,\ldots,i_m\}}$ is complete.

In [13] the authors showed that this family is enough extensive. For instance all the acyclic communication structures are cycle-complete, but there a lot more (see Fig. 4.7).

van Nouweland and Borm [14] proved the following nice relation between this kind of graphs and the convexity of the vertex game.

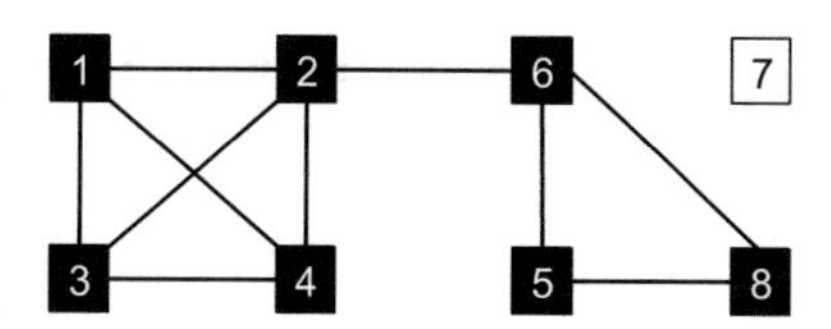

Fig. 4.7 Cycle-complete communication structure

Proposition 4.9 *A communication structure r is cycle-complete if and only if the vertex game v/r is convex for all $v \in \mathcal{G}_c^N$.*

Proof Consider r communication structure cycle-complete and $v \in \mathcal{G}_c^N$. We use Proposition 1.3 to get the convexity of the vertex game v/r. Suppose $i \in N^r$, otherwise all the marginal contributions are zero. For all coalition S with $i \notin S$, we denote the set of components in r_S containing some player with a link to i as

$$C_i^S(r) = \{C \in N/r_S : \exists j \in C,\ ij \in L(r)\}. \tag{4.4}$$

Next we prove the following claims:

- $N/r_{S\cup\{i\}} = \left[(N/r_S) \setminus C_i^S(r)\right] \cup \left\{\{i\} \cup \bigcup_{C \in C_i^S(r)} C\right\}$. In fact, any component in $r_{S\cup\{i\}}$ without player i is also a component in r_S, and, by definition, the component in $r_{S\cup\{i\}}$ containing player i connects all the components in $C_i^S(r)$.
- If $S \subseteq T$ with $i \notin T$ then for all $C \in C_i^S(r)$ there exists only one $D_C \in C_i^T(r)$ with $C \subseteq D_C$, moreover if $C \neq C'$ then $D_C \neq D'_C$. Obviously as $S \subseteq T$ for each $C \in C_i^S(r)$ must exist one $D \in C_i^T(r)$ with $C \subseteq D$, and as D is a component then D is unique. Now suppose one $D \in C_i^T(r)$ with two different $C, C' \in C_i^S(r)$ verifying $C, C' \subseteq D$. There are $j \in C$ and $j' \in C'$ such that: $\{j, i, j'\}$ is a path between j and j' (by definition of C and C'), $jj' \notin L(r)$ (they are in different components in r_S), and finally j and j' are connected in r_T (they are in the same component in r_T) by another path without using player i (she is not in T). Therefore we have a cycle in r containing j, j' but $jj' \notin L(r)$. This is not possible because r is cycle-complete.

Suppose coalitions $S \subsetneq T$ and $i \notin T$. Using the first claim,

$$v/r(T \cup \{i\}) - v/r(T) = v\left(\{i\} \cup \bigcup_{D \in C_i^T(r)} D\right) - \sum_{D \in C_i^T(r)} v(D), \tag{4.5}$$

and obviously the same with S. Now we apply the another claim and convexity for v,

$$\begin{aligned} v/r(T \cup \{i\}) - v/r(T) &\geq v\left(\{i\} \cup \bigcup_{C \in C_i^S(r)} D_C\right) - \sum_{C \in C_i^S(r)} v(D_C) \\ &\geq v\left(\{i\} \cup \bigcup_{C \in C_i^S(r)} C\right) - \sum_{C \in C_i^S(r)} v(C) \\ &= v/r(S \cup \{i\}) - v/r(S). \end{aligned}$$

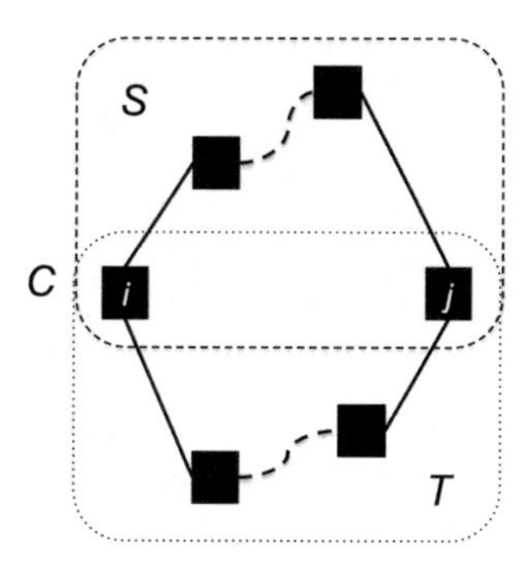

Fig. 4.8 Coalitions S, T in C

The first inequality follows from superadditivity of v because

$$\{i\} \cup \bigcup_{D \in C_i^T(r)} D = \left(\{i\} \cup \bigcup_{C \in C_i^S(r)} D_C\right) \cup \bigcup_{\{D \in C_i^T(r): D \neq D_C \forall C \in C_i^S(r)\}} D,$$

and then

$$v\left(\{i\} \cup \bigcup_{D \in C_i^T(r)} D\right) = v\left(\{i\} \cup \bigcup_{C \in C_i^S(r)} D_C\right) + \sum_{\{D \in C_i^T(r): D \neq D_C \forall C \in C_i^S(r)\}} v(D).$$

The second inequality uses convexity sequentially by Proposition 1.3, because if $C' \in C_i^S(r)$ then

$$\bigcup_{C \in C_i^S(r) \setminus C'} D_C \cup C' \subseteq \bigcup_{C \in C_i^S(r) \setminus C'} D_C \cup D_{C'}.$$

On the other hand, suppose that r is not cycle-complete. There exists a cycle in r, we denote C the set of players in this cycle, containing i, j such that $ij \notin L(r)$. Consider again the game $v(S) = |S| - 1$ for any non-empty coalition S. At least there are two players in $C \setminus \{i, j\}$ because C is a cycle and $ij \notin L(r)$. We take coalitions S, T as in Fig. 4.7 satisfying: S, T are connected, $S \cap T = \{i, j\}$ (it is not connected) and $S \cup T = C$ (connected). Thus, as $|S| + |T| = |C| + 2$ (Fig. 4.8)

$$v/r(S) + v/r(T) = v(S) + v(T) = |C| > |C| - 1 = v/r(S \cup T) + v(S \cap T).$$

□

A value for games with communication structure is

$$f : \mathcal{G}^N \times G^N \to \mathbb{R}^N.$$

Myerson [9] proposed to define a value applying the Shapley value to the vertex game.

Definition 4.9 The Myerson value is a value for games over N with communication structure defined for each game $v \in \mathscr{G}^N$ and communication structure r as

$$\mu(v, r) = \phi(v/r).$$

Obviously, the Myerson value coincides with the coalitional value if the communication structure is transitive, and then with the Shapley value in the classical situation. In the above section we saw that the coalitional value of a player only depends on the component where she is (4.1). The Myerson value also satisfies this component decomposability, but the result is not trivial. Van den Nouweland [15] proved this fact. In our context we get the next result.

Proposition 4.10 *Let $(v, r) \in \mathscr{G}^N \times G^N$ be a game with communication structure. The Myerson value of a player $i \in N^r$ satisfies*

$$\mu_i(v, r) = \mu_i(v, r_T) = \phi_i\,(v/r, T) = \phi_i\,(v/r, N/r)\,,$$

where $T \in N/r$ and $i \in T$.

Proof Given a player $i \in N^r$, we take again for each coalition S not containing the player, the set $C_i^S(r)$ (4.4) of components in r_S with a link to i. Expression (4.5) showed that the marginal contribution of player i to a coalition S only depends on $C_i^S(r)$. As there exists only one component $T \in N/r$ with $i \in T$ we have $C_i^S(r) = C_i^S(r_T)$ because all the elements in $C_i^S(r)$ are contained in T. Hence,

$$v/r(S \cup \{i\}) - v/r(S) = v/r_T(S \cup \{i\}) - v/r_T(S).$$

Now, by Theorem 1.1

$$\mu_i(v, r) = \phi_i(v/r) = \sum_{S \subseteq N \setminus \{i\}} c_S^n[v/r(S \cup \{i\}) - v/r(S)] = \phi_i(v/r_T) = \mu_i(v, r_T).$$

Finally, following Definition 3.6, we see that $v/r_T = (v/r)_T$. In fact, for any coalition S we have $N/(r_T)_S = N/r_{S \cap T}$, and then since Definition 3.5

$$v/r_T(S) = \sum_{R \in N/(r_T)_S} v(R) = \sum_{R \in N/r_{T \cap S}} v(R) = v/r(T \cap S) = (v/r)_T(S).$$

The last equality in the proposition is given by (4.1). □

In order to find an axiomatization of the Myerson value we take into account that players are asymmetric in the structure because in each component the situation of the players contained in it is different, for instance the position of player 1 in Fig. 4.5 is not the same of players 2, 3. Therefore equal treatment of players in the same component (symmetry) is not possible. Myerson [9] used then an axiomatization based in a fairness axiom, thinking about the deletion of a link by Definition 4.6. Let $f : \mathscr{G}^N \times G^N \to \mathbb{R}^N$ a value for games over N with coalition structure.

Efficiency by components. If $v \in \mathscr{G}^N$ and $r \in G^N$ then $f(v, r)(T) = v(T)$ for all $T \in N/r$.

Fairness. Let $v \in \mathscr{G}^N$ and $r \in G^N$, it holds for all $ij \in L(r)$ that

$$f_i(v, r) - f_i(v, r_{-ij}) = f_j(v, r) - f_j(v, r_{-ij}).$$

Carrier. Let $v \in \mathscr{G}^N$ and $r \in G^N$. For all player $i \notin N^r$ it holds $f_i(v, r) = 0$.

But fairness is an axiom which only is possible to use in the context of graphs, thus next axiomatization is not valid for the family of usual games. Similar to the following axiomatization was given in [9].

Theorem 4.3 *The Myerson value is the only value for games over N with communication structure satisfying efficiency by components, fairness and carrier.*

Proof First we test that the Myerson value verifies the three axioms.
CARRIER. Suppose $i \notin N^r$. For all coalition $S \subseteq N \setminus \{i\}$, player i does not change the components in the coalition, namely $N/r_{S\cup\{i\}} = N/r_S$. Thus i is a null player for v/r because $v/r(S \cup \{i\}) = v/r(S)$. As Shapley value verifies null player property (Proposition 1.11) then $\mu_i(v, r) = 0$.
EFFICIENCY BY COMPONENTS. Let $T \in N/r$ be a component in a communication structure r. The above proposition says

$$\sum_{i\in T} \mu_i(v, r) = \sum_{i\in T} \mu_i(v, r_T).$$

Carrier axiom of the Myerson value (we have just proved it) and the efficiency of the Shapley value (Proposition 1.9) imply

$$\sum_{i\in T} \mu_i(v, r) = \sum_{i\in N} \mu_i(v, r_T) = \sum_{i\in N} \phi_i(v/r_T) = v/r_T(N).$$

But $N/r_T = \{T\}$ because T is connected and then $v/r_T(N) = g(v, r_T) = v(T)$.
FAIRNESS. Let $ij \in L(r)$. We consider $w^{ij} \in \mathscr{G}^N$ defined as

$$w^{ij} = v/r - v/r_{-ij}. \tag{4.6}$$

If one of the players, i or j (or both of them) are not in a coalition S then $w^{ij}(S) = 0$ because the components in r and r_{-ij} are the same. Hence $w^{ij}(S \cup \{i\}) = w^{ij}(S \cup \{j\}) = 0$ for all $S \subseteq N \setminus \{i, j\}$. Players i, j are symmetric for w^{ij}. As the Shapley value satisfies symmetry (Proposition 1.10) then

$$\phi_i(w^{ij}) = \phi_j(w^{ij}).$$

From the linearity of the Shapley value (Proposition 1.8) we get

$$\mu_i(v, r) - \mu_i(v, r_{-ij}) = \phi_i(v/r) - \phi_i(v/r_{-ij}) = \phi_i(w^{ij}) = \mu_j(v, r) - \mu_j(v, r_{-ij}).$$

Take now two values f^1, f^2 for games with communication structure satisfying the three axioms. We will prove that $f^1 = f^2$. If $i \notin N^r$ then carrier implies $f_i^1(v, r) = f_i^2(v, r) = 0$. For the rest of players we work by induction in $|L(r)|$. If $|L(r)| = 0$ then each player $i \in N^r$ verifies $\{i\} \in N/r$ and by efficiency by components $f_i^1(v, r) = f^2(v, r) = v(\{i\})$. Suppose that the equality of both values is true if $|L(r)| = k - 1$ and consider r with $|L(r)| = k$. Let $T \in N/r$ and $i \in T$. If $j \in T$ is such $ij \in L(r)$ then fairness and induction say

$$\begin{aligned} f_i^1(v, r) - f_j^1(v, r) &= f_i^1(v, r_{-ij}) - f_j^1(v, r_{-ij}) \\ &= f_i^2(v, r_{-ij}) - f_j^2(v, r_{-ij}) = f_i^2(v, r) - f_j^2(v, r). \end{aligned}$$

Otherwise, $j \in T$ but $ij \notin L(r)$, there is a path between i and j, and applying the above reasoning in each link in the path we get also the same equality. Hence, for all $j \in T$, $f_j^1(v, r) - f^2(v, r) = A$, being $A = f_i^1(v, r) - f_i^2(v, r)$. So, using efficiency in component T,

$$0 = \sum_{j \in T} f_j^1(v, r) - f_j^2(v, r) = |T|A.$$

We get $A = 0$ □

Example 4.7 Let $N = \{1, 2, 3, 4, 5\}$. Suppose the anonymity game $v(S) = |S|^2 - 1$ and the graph r in Fig. 4.9. Using Proposition 4.10 we can obtain the Myerson value by components, $N/r = \{\{1, 2, 3\}, \{4, 5\}\}$. From Proposition 3.4 we can also use the subgame in each component,

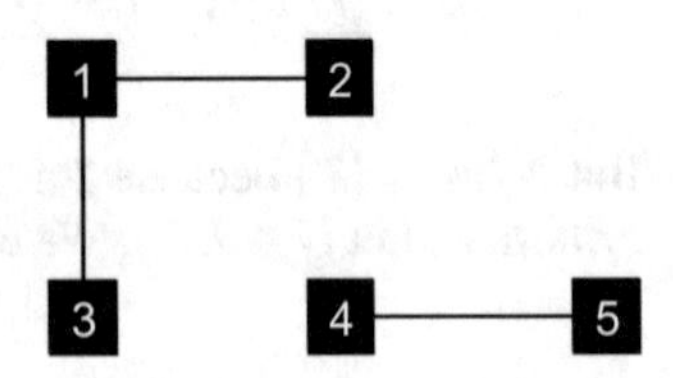

Fig. 4.9 Communication structure Example 4.7

$$\mu(v, r_T) = \phi((v/r)_T) = \left[\phi^T((v/r)|_T)\right]^0.$$

Game $(v/r)|_{\{1,2,3\}}$ is 0-normalized and $(v/r)|_{\{1,2,3\}}(\{1, 2\} = (v/r)|_{\{1,2,3\}}(\{1, 3\}) = 3$, $(v/r)|_{\{1,2,3\}}(\{2, 3\}) = 0$, $(v/r)|_{\{1,2,3\}}(\{1, 2, 3\} = 8$. So, $\mu_1(v, r) = 22/6$ and $\mu_2(v, r) = \mu_3(v, r) = 13/6$. In the another component the graph is complete and players are symmetric. As $v(\{4, 5\}) = 3$ then $\mu_4(v, r) = \mu_5(v, r) = 3/2$. We get

$$\mu(v, r) = \left(\frac{11}{3}, \frac{13}{6}, \frac{13}{6}, \frac{3}{2}, \frac{3}{2}\right).$$

Next we see some classical properties of the Shapley value extended to the Myerson value.

Proposition 4.11 *The Myerson value satisfies the following properties for a communication structure r.*

(1) If $v, w \in \mathcal{G}^N$ then $\mu(av + bw, r) = a\mu(v, r) + b\mu(w, r)$.
(2) If $i \in N^r$ is a necessary player for $v \in \mathcal{G}^N_{sa}$ and $v \geq 0$ then $\mu_j(v, r) \leq \mu_i(v, r)$ for any another player j.
(3) If $i \in N^r$ and $v \in \mathcal{G}^N_{sa}$ then $\mu_i(v, r) \geq v(\{i\})$.
(4) If $S \subseteq N$, r is cycle-complete, and $v \in \mathcal{G}^N_c$ then $\mu(v, r)(S) \geq v/r(S)$.
(5) $\mu(-v, r) = -\mu(v, r)$.
(6) For all $i \in N^r$ and $v \in \mathcal{G}^N$ it holds $\mu_i(v, r) = v(\{i\}) - \mu_i(v^{svg}, r)$.

Proof (1) The linearity is obtained because $(av + bw)/r = a(v/r) + b(w/r)$.
(2) Proposition 4.8 implies that v/r is monotone. As Proposition 1.11 says that the Shapley value satisfies necessary player then for all $j \neq i$

$$\mu_j(v, r) = \phi_j(v/r) \leq \phi_i(v/r) = \mu_i(v, r).$$

(3) From Proposition 4.7 the vertex game is superadditive and as the Shapley value satisfies individual stability (Proposition 1.13) then

$$\mu_i(v, r) = \phi_i(v/r) \geq v/r(\{i\}) = v(\{i\}).$$

Observe that if $i \in N^r$ then $N/r_{\{i\}} = \{\{i\}\}$.
(4) Proposition 4.9 obtains a convex vertex game, so

$$\mu(v, r)(S) = \phi(v/r)(S) \geq v/r(S)$$

since the Shapley value verifies coalitional stability (Proposition 1.14).
(5) Follows from (2).
(6) Let $i \in N^r$, using Propositions 1.15 and 4.7

$$\mu_i(v,r) = \phi_i(v/r) = v/r(\{i\}) - \phi_i\left((v/r)^{svg}\right) = v(\{i\}) - \phi_i\left(v^{svg}/r\right)$$
$$= v(\{i\}) - \mu_i(v^{svg}, r).$$

□

It is not possible to improve the result of the coalitional convexity as next example shows. Furthermore Slikker and van den Nouweland [13] proved that the reasoning in this example can be extended to all graph non cycle-complete.

Example 4.8 Consider the communication structure in Fig. 4.6 which is not cycle-complete. Now take the unanimity game $u_{\{1,4\}}$ which is convex. We have that the vertex game is

$$u_{\{1,4\}}/r = u_{\{1,2,4\}} \vee u_{\{1,3,4\}}.$$

which is not convex. The Myerson value is

$$\mu(u_{\{1,4\}}, r) = \left(\frac{5}{12}, \frac{1}{12}, \frac{1}{12}, \frac{5}{12}\right),$$

satisfying

$$\mu(u_{\{1,4\}}, r)(\{1,2,4\}) = \frac{11}{12} < v/r(\{1,2,4\} = 1.$$

There exist also properties of the value specific to communication situations. The following proposition summarizes several of them (the reader can find more in Slikker and van den Nouweland [13]). We need to introduce a new operation for communication structures

Definition 4.10 Let $r \in G^N$ and $k \in N^r$. The isolation of a vertex r_{-k} is defined as $r_{-k}(ij) = r(ij)$ if $i, j \neq k$, $r_{-k}(kk) = 1$ and otherwise $r_{-k}(ij) = 0$.

Proposition 4.12 *The Myerson value satisfies the following properties for a communication structure r.*

1. *(Stability by links) If $ij \in L(r)$ and $v \in \mathcal{G}^N_{sa}$ then $\mu_i(v,r) \geq \mu_i(v, r_{-ij})$.*
2. *(Balanced contributions) If $v \in \mathcal{G}^N$ and $i, j \in N^r$ then*

$$\mu_i(v,r) - \mu_i(v, r_{-j}) = \mu_j(v,r) - \mu_j(v, r_{-i}).$$

3. *(Superfluous link) If $ij \in L(r)$ and $v \in \mathcal{G}^N$ satisfies $v/r = v/r_{-ij}$ then $\mu(v,r) = \mu(v, r_{-ij})$.*

Proof (1) We use game w^{ij} in (4.6). Remember that $w^{ij}(S) = 0$ if $i \notin S$ or $j \notin S$, hence

$$\phi_i(w^{ij}) = \sum_{\{S \subseteq N \setminus \{i\}: j \in S\}} c_S^n w^{ij}(S \cup \{i\}).$$

Observe that as the graph-worth is monotone by links when the game is superadditive,

$$w^{ij}(S) = v/r(S) - v/r_{-ij}(S) = g(v,r) - g(v, r_{-ij}) \geq 0.$$

Now, from the above formula we get $\phi_i(w^{ij}) \geq 0$. As Shapley value is linear $\phi_i(v/r) \geq \phi_i(v/r_{-ij})$.
(2) Using Theorem 1.1 we get

$$\begin{aligned}\mu_i(v,r) - \mu_i(v, r_{-j}) &= \phi_i(v/r) - \phi_i(v/r_{-j}) \\ &= \sum_{S \subseteq N \setminus \{i\}} c_S^n [v/r(S \cup \{i\}) - v/r(S) - v/r_{-j}(S \cup \{i\}) + v/r_{-j}(S)]\end{aligned}$$

If $j \notin S$ then the summand is zero. If T is a coalition with $j \in T$ then $v/r_{-j}(T) = v/r(T \setminus \{j\}) + v(\{j\})$. Hence

$$\mu_i(v,r) - \mu_i(v, r_{-j}) = \sum_{S \subseteq N \setminus \{i,j\}} c_S^n [v/r(S \cup \{i,j\}) - v/r(S \cup \{j\}) - v/r(S \cup \{i\}) + v/r(S)].$$

Observe that this number is the same if we consider $\mu_j(v,r) - \mu_j(v, r_{-i})$.
(3) It is trivial from Definition 4.9. □

We need to define an appropriate dual game for communication structures. Given a graph r and a game v, the r-*dual* game is $v^{rdual} \in \mathcal{G}^N$ where

$$v^{rdual}(S) = \sum_{\{T \in N/r : T \cap S \neq \emptyset\}} \frac{1}{|N/r_{T \cap S}|} \left[v(T) - \sum_{R \in N/r_{T \setminus S}} v(R) \right]. \quad (4.7)$$

Proposition 4.13 *Let $r \in G^N$. For all $v \in \mathcal{G}^N$ it holds $(v^{rdual})_r = (v_r)^{dual}$ and*

$$\mu(v^{rdual}, r) = \mu(v, r).$$

Proof Suppose $i \in N^r$, otherwise the equality is true from the carrier axiom. We will prove that $v^{rdual}/r = v/r$. Observe that if R is connected there exists only one component $T_R \in N/r$ with $R \subseteq T_R$. For any coalition S we get

$$v^{rdual}/r(S) = \sum_{R\in N/r_S} v^{rdual}(R) = \sum_{R\in N/r_S} \frac{1}{|N/r_{T_R\cap S}|}\left[v(T_R) - \sum_{T\in N/r_{T_R\setminus S}} v(T)\right]$$

$$= \sum_{H\in N/r}\left[v(H) - \sum_{T\in N/r_{H\setminus S}} v(T)\right] = \sum_{H\in N/r} v(H) - \sum_{T\in N/r_{N\setminus S}} v(T)$$

$$= (v/r)(N) - (v/r)(N\setminus S) = (v/r)^{dual}(S),$$

because

$$N/r_{N\setminus S} = \bigcup_{H\in N/r} N/r_{H\setminus S}.$$

Thus, from Proposition 1.15

$$\mu\left(v^{rdual}, r\right) = \phi\left(v^{rdual}/r\right) = \phi\left((v/r)^{dual}\right) = \phi(v/r) = \mu(v, r).$$

□

Given a simple game, the vertex game is not always a simple game as the coalitional game. Moreover, it is also true the following relation with the same proof.

Proposition 4.14 *Let $v \in \mathcal{G}_s^N$. Game v/r is simple for all $r \in G^N$ if and only if $v \in \mathcal{G}_{sa}^N$.*

The next proposition evaluates the vertex game of a unanimity game. For each $S \subseteq N$ and r communication structure we denote as $S|r$ is the set of minimal connected coalitions containing S, in the sense that $T \in S|r$ is connected and any coalition $S \subset R \subset T$ is not connected. Observe that if S is connected then $S|r = \{S\}$ and if there is not any component containing S then $S|r = \emptyset$.

Proposition 4.15 *Let $R \subseteq N$ be a non-empty coalition. For a communication structure r it holds*

$$u_R/r = \bigvee_{T\in R|r} u_T.$$

Proof We see that $W_m(u_R/r) = R|r$ and the result follows from Proposition 1.6.
Let $T \in R|r$. If $S \subsetneq T$ then

$$u_R/r(S) = \sum_{C\in N/r_S} u_R(C).$$

But as $C \subsetneq T$ is connected then C does not contain R by definition of $R|r$, hence $u_R(C) = 0$. Furthermore, $u_R/r(T) = u_R(T) = 1$.

On the other hand, let $T \in W_m(u_R/r)$. If there is $R \subseteq S \subsetneq T$ with S connected in r then $u_R/r(S) = u_R(S) = 1$, but this fact implies that T is not a winning coalition. We prove now that T is connected. Suppose it is not true,

$$u_R/r(T) = \sum_{S \in N/r_T} v(S) = 1,$$

and then, as $|N/r| > 1$, we find S winning coalition with $S \subsetneq T$. □

We can describe the Myerson value, determining the dividends of the vertex game from the original game following Fernández [5]. Owen [12] also proposed another formula, but it is more complex.

Theorem 4.4 *For each $v \in \mathcal{G}^N$ and communication structure r the Myerson value of a player $i \in N^r$ is*

$$\mu_i(v, r) = \sum_{\{i \in S \subseteq N : S \text{ connected}\}} \frac{\Delta_S^{v/r}}{|S|},$$

where the dividends are calculated by

$$\Delta_S^{v/r} = v(S) - \sum_{\{T \subsetneq S : T \text{ connected}\}} \Delta_T^{v/r},$$

and $\Delta_{\{i\}}^{v/r} = v(\{i\})$.

Proof We only need to prove that the dividends of non-connected coalitions are zero taking into account Theorem 1.2 and (1.2). Let $S \subseteq N$ non-connected in the communication structure r. By induction in $|S|$, if $\{i\}$ is not connected then $i \notin N^r$ and $\Delta_{\{i\}}^{v/r} = v/r(\{i\}) = 0$. Suppose the claim true if $|S| < k$ and take $|S| = k$. Using (1.2) and Proposition 1.1 we obtain

$$\Delta_S^{v/r} = v/r(S) - \sum_{T \subsetneq S} \Delta_T^{v/r} = \sum_{R \in N/r_S} v(R) - \sum_{R \in N/r_S} \sum_{T \subseteq R} \Delta_T^{v/r}$$

$$= \sum_{R \in N/r_S} \left[v(R) - \sum_{T \subseteq R} \Delta_T^{v/r} \right] = \sum_{R \in N/r_S} \left[v/r(R) - \sum_{T \subseteq R} \Delta_T^{v/r} \right] = 0.$$

□

4.4 Fuzzy Communication Structures and Games

Communication is an information about the players which is susceptible to level. Furthermore, communication may be uncertain in the sense that it depends on several variables. The Myerson model allowed to get an allocation of benefits adjustable to the real communications in each moment from a priori characteristic function, but in each one of these moments the communication between two players is feasible or not. Jiménez-Losada et al. [6, 7] proposed to introduce fuzziness in communication structures following the concept non-probabilistic of fuzzy graph (see Moderson and Nair [8]).

Definition 4.11 A fuzzy communication structure over N is a (undirected) fuzzy graph ρ, namely a fuzzy bilateral relation over N which satisfies:

(1) ρ is weakly reflexive, $\rho(i, j) \leq \rho(ii) \wedge \rho(jj)$ for all players $i, j \in N$,
(2) ρ is symmetric, $\rho(i, j) = \rho(j, i) = \rho(ij)$ for all $i \neq j$.

The family of fuzzy communication structures over N is denoted as FG^N.

The interpretation of $\rho(ij)$ for each pair ij is communication level between i and j or the capacity of cooperation of these players. So, if $i = j$ we understand $\rho(ii)$ as the level of participation of i in the game.

Example 4.9 Let $N = \{1, 2, 3, 4, 5, 6, 7, 8\}$. We consider the next fuzzy communication situation. Player 7 is active at level 0.6 but she has no communication with the other agents. Player 1 is able to communicate with players 2, 3 at maximum level 1, player 2 with 3 only at level 0.1 and player 2 with 6 at 0.5, taking into account that 6 plays at level 0.8. There exists also communication of 5 with 4, 8 at level 0.2 while they play at level 0.4 (see Fig. 4.10). The relation is represented by the matrix

$$\rho = \begin{bmatrix} 1 & 1 & 1 & 0 & 0 & 0 & 0 & 0 \\ 1 & 1 & 0.1 & 0 & 0 & 0.5 & 0 & 0 \\ 1 & 0.1 & 1 & 0 & 0 & 0 & 0 & 0 \\ 0 & 0 & 0 & 0.4 & 0.2 & 0 & 0 & 0 \\ 0 & 0 & 0 & 0.2 & 0.4 & 0 & 0 & 1 \\ 0 & 0.5 & 0 & 0 & 0 & 0.8 & 0 & 0 \\ 0 & 0 & 0 & 0 & 0 & 0 & 0.6 & 0 \\ 0 & 0 & 0 & 0 & 0.2 & 0 & 0 & 0.4 \end{bmatrix}.$$

Following Sect. 3.3 we have

$$L(\rho) = \{12, 13, 23, 26, 58, 45\}, \quad im\,(\rho) = \{0.1, 0.2, 0.4, 0.5, 0.6, 0.8, 1\}.$$

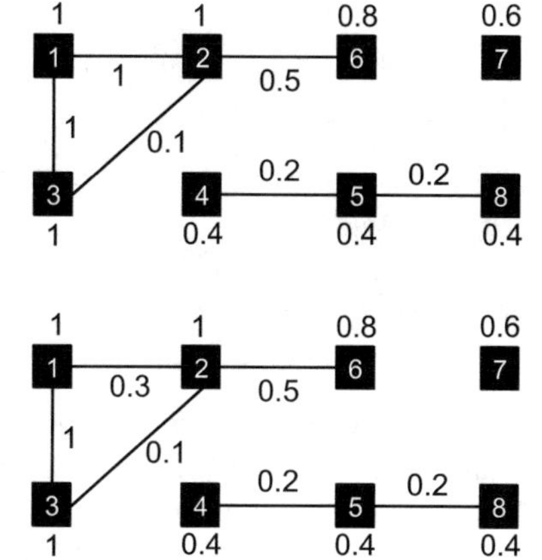

Fig. 4.10 Fuzzy communication structure ρ

Fig. 4.11 Fuzzy communication structure $\rho^{0.7}_{-12}$

Connecting among them, players form the fuzzy coalition structure of the components in the relation,

$$N/\rho = \{\{1, 2, 3, 6\}, \{4, 5, 8\}, \{7\}\}.$$

We extend the concept of graph-worth g in Definition 4.5. But in the fuzzy case we have a lot of feasible functions satisfying similar conditions. First we need to define the following operation for fuzzy graphs.

Definition 4.12 If ρ is a fuzzy communication structure, $ij \in L(\rho)$ and $t \in [0, \rho(ij)]$ then $\rho^t_{-ij} \in FG^N$ with $\rho^t_{-ij}(i'j') = \rho(i'j')$ for all $i'j' \neq ij$ and $\rho^t_{-ij}(ij) = \rho(ij) - t$.

Example 4.10 Suppose the fuzzy communication structure in Example 4.9. We reduce the level of communication between players 1 and 2 in $t = 0.7$. The new fuzzy relation is represented by the fuzzy graph in Fig. 4.11 Now we get $L(\rho^{0.7}_{-12}) = L(\rho)$ and $im\ (\rho^{0.7}_{-12}) = \{0.1, 0.2, 0.3, 0.4, 0.5, 0.6, 0.8, 1\}$. Furthermore, $N/\rho^{0.7}_{-12} = N/\rho$. The matrix represented the fuzzy graph is now

$$\rho^{0.7}_{-12} = \begin{bmatrix} 1 & 0.3 & 1 & 0 & 0 & 0 & 0 & 0 \\ 0.3 & 1 & 0.1 & 0 & 0 & 0.5 & 0 & 0 \\ 1 & 0.1 & 1 & 0 & 0 & 0 & 0 & 0 \\ 0 & 0 & 0 & 0.4 & 0.2 & 0 & 0 & 0 \\ 0 & 0 & 0 & 0.2 & 0.4 & 0 & 0 & 1 \\ 0 & 0.5 & 0 & 0 & 0 & 0.8 & 0 & 0 \\ 0 & 0 & 0 & 0 & 0 & 0 & 0.6 & 0 \\ 0 & 0 & 0 & 0 & 0.2 & 0 & 0 & 0.4 \end{bmatrix}.$$

Definition 4.13 A fuzzy graph-worth over N is a function $\gamma : \mathcal{G}^N \times FG^N \rightarrow \mathbb{R}$ satisfying

(1) Extension, if $r \in G^N$ then $\gamma(v, r) = g(v, r)$,
(2) Component additivity, $\gamma(v, \rho) = \sum_{S \in N/\rho} \gamma(v, \rho_S)$,
(3) Link monotonicity, if $v \in \mathcal{G}_{sa}^N$ then $\gamma(v, \rho) \geq \gamma(v, \rho_{-ij}^t)$ for all $ij \in L(\rho)$ and $t \in [0, \rho(ij)]$.

A trivial example of fuzzy graph-worth is $\gamma(v, \rho) = g(v, r^\rho)$ for all $\rho \in FG^N$, using the crisp version (see Sect. 3.3) of the fuzzy communication structure ρ and the Myerson graph-worth (Definition 4.5). But, obviously, this option is not reasonable. As we will see later, there are a lot of fuzzy graph-worths. Hence we think of fuzziness of the Myerson measure in the sense of Sect. 2.4. Jiménez-Losada et al. [7] gave a method to get fuzzy graph-worth as fuzziness (extensions) of the Myerson graph-worth. They needed to introduce the following operations between fuzzy graphs: sum, subtraction and product with a scalar.

Definition 4.14 Let $\rho, \rho' \in FG^N$ be two fuzzy communication structures.

(1) The sum $\rho + \rho'$ is defined as fuzzy sets over $N \times N$, namely for all $i, j \in N$
$$(\rho + \rho')(ij) = 1 \wedge [\rho(ij) + \rho'(ij)].$$
(2) If $\rho' \leq \rho$ then the subtraction for each $i, j \in N$ is
$$(\rho - \rho')(ij) = [\rho(ij) - \rho'(ij)] \wedge [\rho(ii) - \rho'(ii)] \wedge [\rho(jj) - \rho'(jj)].$$
(3) If $t \in [0, 1]$ then $(t\rho)(ij) = t\rho(ij)$ for every $i, j \in N$.

The special definition of the subtraction of fuzzy graphs is based on maintaining the weakly reflexivity, the level of a link must never be greater than the level of its vertices. The subtraction is explained in the next example.

Example 4.11 In Fig. 4.12 we see the subtraction between a fuzzy communication structure and one of its subgraphs. Observe that the usual subtraction in link 12 is $\rho(12) - \rho'(12) = 0.5$ but $\rho(22) - \rho'(22) = 0.4$. That is why we take $(\rho - \rho')(12) = 0.4$.

The reader can see that sum and subtraction of fuzzy graphs are not opposite operations, but they satisfies the next result.

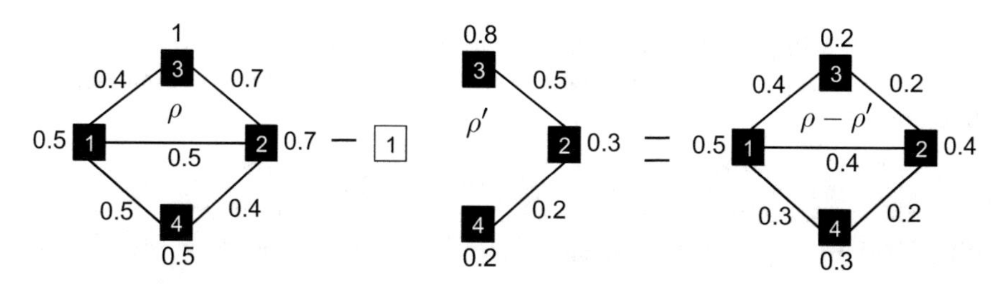

Fig. 4.12 Fuzzy graph subtraction

Proposition 4.16 *If $\rho, \rho', \rho'' \in FG^N$ with $\rho' \leq \rho$ and $\rho'' \leq \rho - \rho'$ then:*

(1) $(\rho - \rho') - \rho'' = \rho - (\rho' + \rho'')$, and
(2) $(\rho - \rho') - \rho'' = (\rho - \rho'') - \rho'$.

Proof (1) As $\rho' \leq \rho$ and $\rho'' \leq \rho - \rho'$ it holds for each $i, j \in N$ that

$$\rho''(ij) \leq (\rho - \rho')(ij) \leq \rho(ij) - \rho'(ij),$$

thus $\rho'(ij) + \rho''(ij) \leq \rho(ij)$. Therefore both members of the equality are feasible. Now we have

$$\begin{aligned}
[(\rho - \rho') - \rho''](ij) &= [(\rho - \rho') - \rho''](ij) \wedge [(\rho - \rho') - \rho''](ii) \wedge [(\rho - \rho') - \rho''](jj) \\
&= [\rho(ij) - \rho'(ij) - \rho''(ij)] \wedge [\rho(ii) - \rho'(ii) - \rho''(ij)] \\
&\quad \wedge [\rho(jj) - \rho'(jj) - \rho''(ij)] \wedge [\rho(ij) - \rho'(ij) - \rho''(ij)] \\
&\quad \wedge [\rho(ii) - \rho'(ii) - \rho''(ii)] \wedge [\rho(jj) - \rho'(jj) - \rho''(jj)] \\
&= [\rho(ij) - \rho'(ij) - \rho''(ij)] \wedge [\rho(ii) - \rho'(ii) - \rho''(ii)] \\
&\quad \wedge [\rho(jj) - \rho'(jj) - \rho''(jj)] \\
&= [\rho(ij) - (\rho'(ij) + \rho''(ij))] \wedge [\rho(ii) - (\rho'(ii) + \rho''(ii))] \\
&\quad \wedge [\rho(jj) - (\rho'(jj) + \rho''(jj)] = [\rho - (\rho' + \rho'')](ij),
\end{aligned}$$

using that $\rho''(ij) \leq \rho''(ii)$.
(2) Obviously $\rho' + \rho'' = \rho'' + \rho'$. Hence, if we will prove the claim $\rho' \leq \rho - \rho''$ then by (1) we obtain

$$(\rho - \rho') - \rho'' = \rho - (\rho' + \rho'') = \rho - (\rho'' + \rho') = (\rho - \rho'') - \rho'.$$

Let $i, j \in N$. As $\rho''(ij) \leq \rho(ij) - \rho'(ij)$ we get $\rho'(ij) \leq \rho(ij) - \rho''(ij)$. But also $\rho'(ij) \leq \rho(ii) - \rho''(ii)$ because otherwise as $-\rho''(ii) \geq -\rho(ii) + \rho'(ii)$ we will obtain $\rho'(ij) > \rho'(ii)$. □

Now, following the Aubin [2] model we define partitions by levels for fuzzy graphs over N. A partition function chooses a partition for each fuzzy communication structure.

Definition 4.15 Let $\rho \in FG^N$ be a fuzzy communication structure. A partition by levels of ρ is a finite sequence $\{(r_k, s_k)\}_{k=1}^m$ satisfying:

(1) $r_k \in G^N$ and $s_k > 0$, for all $k = 1, \ldots, m$,
(2) $s_1 r_1 \leq \rho$ and for each $k = 2, \ldots, m$

$$s_k r_k \leq \rho - \sum_{p=1}^{k-1} s_p r_p.$$

(3) $\rho - \sum_{k=1}^m s_k r_k = 0$.

A partition function for fuzzy communication structures is a mapping pl determining a partition by levels $pl(\rho)$ of each fuzzy graph $\rho \in FG^N$.

If the reader compares partition by levels for fuzzy graphs and partition by levels for fuzzy coalitions (Definition 2.7) they are slightly different because from the special definition of the subtraction we do not have guaranteed to upset the whole fuzzy communication structure. Let $\rho \in FG^N$. If $\{(r_k, s_k)\}_{k=1}^m$ is a partition by levels of ρ, then for each player $i \in N$, as in (2.6), we have

$$\sum_{\{k: i \in dom\,(r_k)\}} s_k = \rho(ii). \tag{4.8}$$

The above equality is true because the special definition of subtraction does not influence over the vertices, namely $(\rho - \rho')(ii) = \rho(ii) - \rho'(ii)$ if $\rho' \leq \rho$ and $i \in N$.

Now we comment several interesting partition functions for fuzzy communications. These functions are inspired in the proportional (Definition 2.11) and Choquet (Definition 2.12) extensions for fuzzy coalitions. They were introduced in [7].

The proportional by graphs function pg is defined following the proportional behavior in the sense: Players and communications are connected at the same level. Let $\rho \in FG^N$. For each $t \in (0, 1]$ we consider the crisp graph

$$r^\rho[t](ij) = \begin{cases} 1, & \text{if } \rho(ij) = \rho(ii) = \rho(jj) = t \\ 0, & \text{otherwise.} \end{cases} \tag{4.9}$$

For each t this graph selects all the vertices with level t and it establishes a link between two vertices if that link has level t.

Table 4.1 *pg*-algorithm of Example 4.11

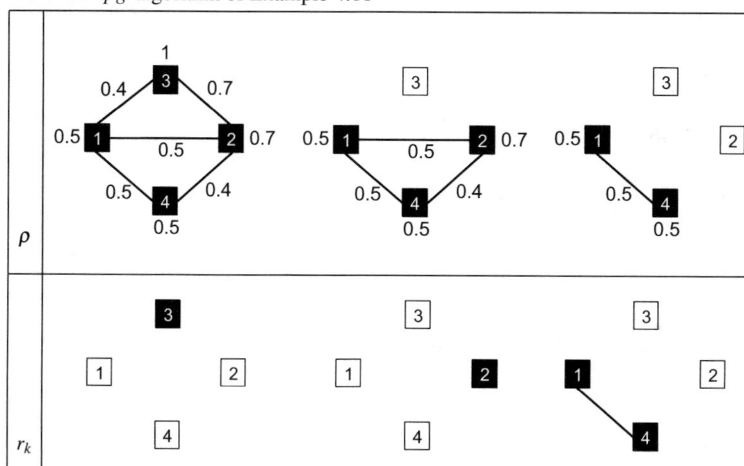

Definition 4.16 The proportional by graphs function pg chooses for each $\rho \in FG^N$ the partition by levels of ρ obtained by the following algorithm (pg-algorithm),

Take $k = 0$, $pg = \emptyset$ and $\rho = \rho$
While $\rho \neq 0$ do
..... $k = k + 1$
..... $s_k = \vee \rho$
..... $r_k = r^{\rho}[s_k]$
..... $pg = pg \cup \{(r_k, s_k)\}$
..... $\rho = \rho - s_k r_k$
$pg(\rho) = pg$

Example 4.12 We can see in Table 4.1 the pg-algorithm applied to the fuzzy graph in Example 4.11. Player 3 is the only element in the graph with level 1, then she plays alone. But if we delete vertex 3 then the links from 3 are deleting too. The same now with player 2. Coalition $\{1, 4\}$ is feasible at level 0.5.

The Choquet by graph function cg is defined following the Choquet behavior in the sense: Players look for connecting using the largest graph.

Table 4.2 *cg*-algorithm of Example 4.11

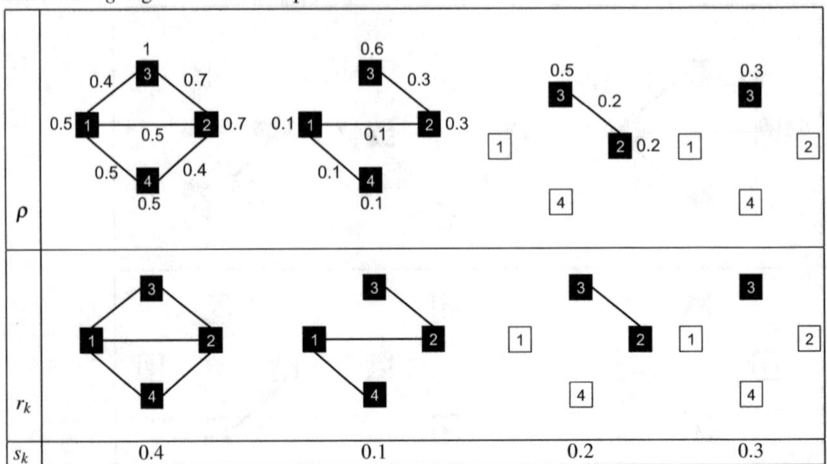

Definition 4.17 The Choquet by graphs function chooses for each $\rho \in FG^N$ the partition by levels of ρ obtained by the following algorithm (cg-algorithm),

Take $k = 0$, $cg = \emptyset$ and $\rho = \rho$
While $\rho \neq 0$ do
..... $k = k + 1$
..... $s_k = \wedge \rho$
..... $r_k = r^\rho$
..... $cg = cg \cup \{(r_k, s_k)\}$
..... $\rho = \rho - s_k r_k$
$cg(\rho) = cg$

Suppose ρ as a fuzzy set in $[0, 1]^{N \times N}$ and $im(\rho) = \{\lambda_1 < \cdots < \lambda_m\}$. Observe that we choose $s_1 = \lambda_1$ and $r_1 = r^\rho = [\rho]_1$. In the next step $\rho(ij) = \rho(ij) - s_1$, the subtraction coincides with the usual difference. Thus $s_2 = \lambda_2 - \lambda_1$. Repeating the process, we prove

$$cg(\rho) = \{(\lambda_k - \lambda_{k-1}, [\rho]_k)\}_{k=1}^m \tag{4.10}$$

Example 4.13 We can see in Table 4.2 the algorithm applied to the fuzzy graph in Example 4.11.

Given a game and a partition function we can define from the crisp graph-worth (Definition 4.5) a "measure" of the profit for fuzzy communication structures.

Definition 4.18 Let pl be a partition function for fuzzy communication structures. If $v \in \mathcal{G}^N$ and $\rho \in FG^N$ then the pl-worth is

$$\gamma^{pl}(v, \rho) = \sum_{k=1}^{m} s_k g(v, r_k),$$

where $pl(\rho) = \{(r_k, s_k)\}_{k=1}^{m}$.

Example 4.14 Let $N = \{1, 2, 3, 4\}$. Suppose the game $v(N) = 10$, $v(S) = 4$ if $|S| = 3$, $v(\{2, 3\}) = v(\{1, 4\}) = 1$ and $v(S) = 0$ otherwise. Using Tables 4.1 and 4.2 we calculate the worth of the fuzzy communication structure in Example 4.11. So, the pg-worth

$$\gamma^{pg}(v, \rho) = g(v, r_1) + 0.7g(v, r_2) + 0.5g(v, r_3) = 0.5.$$

The cg-worth

$$\gamma^{cg}(v, \rho) = 0.4g(v, r_1) + 0.1g(v, r_2) + 0.3g(v, r_3) + 0.3g(v, r_4) = 5.3.$$

But not all the pl-worth are fuzzy graph-worths. Next example shows this fact.

Example 4.15 The *proportional by communications function pc* is defined following the proportional behavior in the sense: Players look for connecting to maximal level although they do not upset their levels. Let $\rho \in FG^N$. For each $t \in (0, 1]$ we consider the crisp graph $r_*^{\rho}[t]$ with $r_*^{\rho}[t](ii) = 1$ if and only if there is $j \neq i$ with $\rho(ij) = t$, and $r_*^{\rho}[t](ij) = 1$ with $i \neq j$ if and only if $\rho(ij) = t$. Particularly $r_*^{\rho}[0] = r^{\rho}$. The partition by levels of ρ is obtained by the following pc-algorithm,

```
Take k = 0, pc = ∅ and ρ = ρ
While ρ ≠ 0 do
..... k = k + 1
..... t = ∨{ρ(ij) : i ≠ j}
..... if t = 0 then
.......... s_k = ∧{ρ(ii) : i ∈ N}
.......... r_k = r^ρ
..... else
.......... s_k = t
.......... r_k = r_*^ρ[s_k]
..... pc = pc ∪ {(r_k, s_k)}
..... ρ = ρ − s_k r_k
pc(ρ) = pc
```

We can see in Table 4.3 the algorithm applied to the fuzzy graph in Example 4.11. Now we consider the game $v \in \mathcal{G}_{sa}^N$ with $v(N) = 5$ and $v(S) = 0$ otherwise. First

Table 4.3 pc-algorithm of ρ in Example 4.11

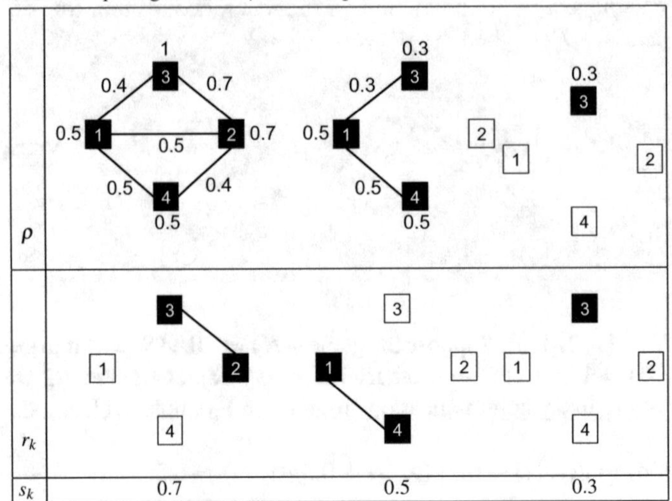

Table 4.4 pc-algorithm of $\rho^{0.2}_{-23}$

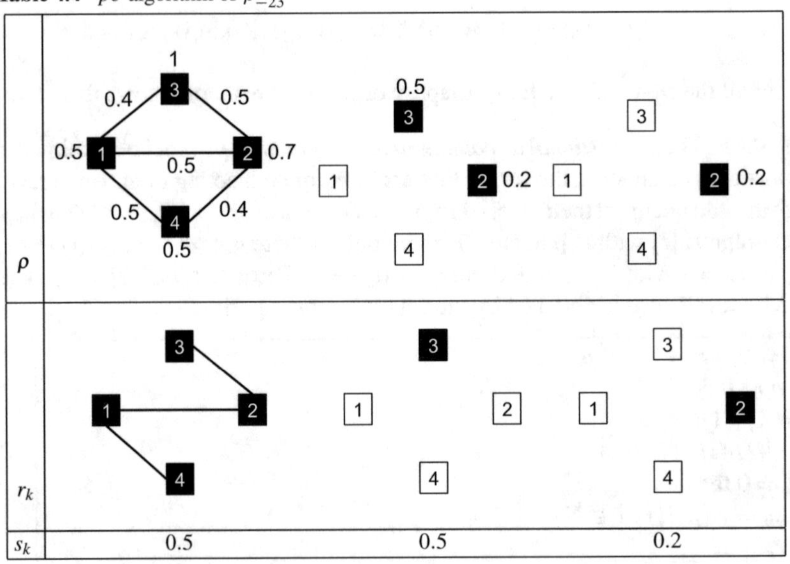

$\gamma^{pc}(v,\rho) = 0$ because N is not connected in the partition. Second we take $\rho^{0.2}_{-23}$. We need to apply the algorithm again to this fuzzy graph, its pc-partition is in Table 4.4. Finally we determine the pc-worth of $\rho^{0.2}_{-23}$, $\gamma^{pc}(v,\rho^{0.2}_{-23}) = 3.5$. Thus

$$\gamma^{pc}(v,\rho^{0.2}_{-23}) > \gamma^{pc}(v,\rho).$$

The pc-worth does not satisfies monotonicity by links.

Although we consider all the pl-worths interesting (their importance should come from their applications) we introduce the following concept over the set of partition functions.

Definition 4.19 A partition function pl for fuzzy communication structures is admissible if the pl-worth is also a fuzzy graph-worth.

Extension inheritable condition (Definition 3.7) was introduced in Chap. 3 for partitions by levels of fuzzy coalitions, now we extend the idea to fuzzy communication structures.

Definition 4.20 A partition function pl for fuzzy communication structures is an inherited extension if

(1) $pl(r) = \{(r, 1)\}$ for every $r \in G^N$ and,
(2) for all $\rho \in FG^N$ and $S \subseteq N$ with $\rho_S \neq 0$ it holds

$$pl(\rho_S) = \left\{ \left(r', \sum_{\{(r,s)\in pl(\rho):r_S=r'\}} s \right) : r' \neq 0, \exists (r, s) \in pl(\rho) \text{ with } r_S = r' \right\}.$$

We will prove that the partition functions in Definitions 4.2 and 4.3 are inherited.

Proposition 4.17 *The partition functions proportional by graphs and Choquet by graphs are inherited extensions.*

Proof It is trivial to get $pg(r) = cg(r) = \{(r, 1)\}$ for all $r \in G^N$.
Let $\rho \in FG^N$ and $S \subseteq N$. Suppose the proportional case. Each player plays once in the pg-algorithm. Also each link ij is used only once if $\rho(ij) = \rho(ii) = \rho(jj)$, otherwise it is no used. Moreover if $\tau \in [0, 1]^N$ with $\tau(i) = \rho(ii)$ then the levels in $pl(\rho)$ go round $im(\tau)$ but in decreasing order, and then the levels in $pl(\rho_S)$ go round $im(\tau \times e^S)$. So wc obtain

$$pg(\rho_S) = \{(r_S, s) : (r, s) \in pg(\rho)\}.$$

Now consider the Choquet case. Let $cg(\rho) = \{(r_k, s_k)\}_{k=1}^m$ and $im\,(\rho) = \{\lambda_1 < \cdots < \lambda_m\}$. We take the first $k_1 \in 1, \ldots, m$ such that $\wedge \rho_S = \lambda_{k_1}$. For all $k = 1, \ldots, k_1$ we have $(r_k)_S = ([\rho]_1)_S$. Thus the first chosen graph in the algorithm for ρ_S is $([\rho]_1)_S$. We also get by construction (4.10) that

$$\sum_{k=1}^{k_1} s_k = \lambda_{k_1} = \wedge \rho_S.$$

We can repeat the reasoning with the next level in ρ_S. □

Inherited extensions are not in general admissible because they do not satisfy monotonicity by links in general, although they verify the other conditions.

Lemma 4.1 *If pl is an inherited extension then the pl-worth satisfies extension and additivity by components.*

Proof Let $v \in \mathcal{G}^N$. If $r \in G^N$ then $pl(r) = \{(r, 1)\}$, thus $\gamma^{pl}(v, r) = g(v, r)$. We have

$$\begin{aligned}\gamma^{pl}(v, \rho) &= \sum_{k=1}^{m} s_k g(v, r_k) = \sum_{k=1}^{m} s_k \sum_{S \in N/r_k} g(v, (r_k)_S) \\ &= \sum_{T \in N/\rho} \sum_{k=1}^{m} s_k \sum_{S \in N/(r_k)_T} g(v, (r_k)_S) = \sum_{T \in N/\rho} \sum_{k=1}^{m} s_k g(v, (r_k)_T) \\ &= \sum_{T \in N/\rho} \gamma^{pl}(v, \rho_T).\end{aligned}$$

□

Next we prove that the proportional and Choquet by graph functions are admissible.

Proposition 4.18 *The inherited extensions proportional by graphs and Choquet by graphs are admissible.*

Proof Since the above proposition we only need to prove that the extensions are monotone by links. Let $ij \in L(\rho)$ and $t \in (0, \rho(ij)]$.

First consider the pg-algorithm applied to ρ and ρ^t_{-ij}. If player i (or j) satisfies $\rho(ij) < \rho(ii)$ then the algorithm obtains the same result for both fuzzy graphs because the link is not used. If $\rho(ij) = \rho(ii) = \rho(jj)$ then the link is used for ρ but not for ρ^t_{-ij}. As the level of the vertices do not change, then there is only on element in the partition which is different, we denote these elements (r', s') and (r'', s'') for ρ and ρ^t_{-ij}. We have $s' = s''$ and $r'' = r'_{-ij}$. So, as the crisp graph-worth is monotone by links $g(v, r') \geq g(v, r'')$.

Finally we study the cg-algorithm applied to ρ and ρ^t_{-ij}. Let $cg(\rho) = \{(r_p, s_p)\}_p$ and $cg(\rho^t_{-ij}) = \{(r'_q, s'_q)\}_q$. We consider two different situations.

(a) Suppose $t = \rho(ij)$. While $\wedge\rho < \rho(ij)$ we get $(r'_p, s'_p) = ((r_p)_{-ij}, s_p)$. Let k the step where $\wedge\rho = \rho(ij)$. We have two options. If $\wedge\rho^t_{-ij} = \rho(ij)$ too then $(r'_k, s'_k) = ((r_k)_{-ij}, s_k)$ and later $\rho = \rho^t_{-ij}$ in the next steps. As $g(v, r_k) \geq g(v, r'_k)$ we get the required inequality. If $\wedge\rho^t_{-ij} \geq \rho(ij)$ then $s_k < s'_k$ and $r'_k = (r_k)_{-ij}$. Level s'_k is just the next minimal level to s_k in ρ. Unfortunately $\rho \neq \rho^t_{-ij}$ in the next step $k+1$, but $s_{k+1} = s'_k - s_k$ and $r_{k+1} = r'_k$. Hence we compare two steps of the algorithm for ρ with only one for ρ^t_{-ij}. The fuzzy graphs are the same from that step. We obtain

$$s'_k g(v, r'_k) \leq s_k g(v, r_k) + s_{k+1} g(v, r_{k+1}),$$

and $g(v, r'_p) \leq g(v, r_p)$ for the before steps. Therefore $\gamma^{cg}(v, \rho^t_{-ij}) \leq \gamma^{cg}(v, \rho)$.
(b) Suppose $t < \rho(ij)$. Let k be the first step such that the algorithm is different. We denote $\alpha = \rho(ij)$ in the step k. Following in this step, $\wedge\rho^t_{-ij} = \alpha - t$. We get $\alpha - t \leq s_k \leq \alpha$ and $s'_k = \alpha - t$. Moreover $r'_k = r_k$. We begin the next step with

$$\rho^t_{-ij} = \rho^t_{-ij} - (\alpha - t) r_k.$$

Now the algorithm chooses $s'_{k+1} = s_k - \alpha + t$ and

$$r'_{k+1} = \begin{cases} (r_k)_{-ij}, & \text{if } s_k < \alpha \\ r_k, & \text{otherwise.} \end{cases}$$

Since $r'_{k+1} \leq r_k$ (if we use the same vertices) we get

$$s'_k g(v, r'_k) + s'_{k+1} g(v, r'_{k+1}) \leq s_k g^v(r_k).$$

Consider the fuzzy graphs ρ in the step $k+1$ and ρ^t_{-ij} in the step $k+2$. They are in situation (a) because $t = \rho(ij)$. Thus $\gamma^{cg}(v, \rho^t_{-ij}) \leq \gamma^{cg}(v, \rho)$. □

Jiménez-Losada et al. [6] showed an example of interesting fuzzy graph-worth bases also in a partition function depending on the game but it not inherited. A connected acyclic graph is named a *tree*. Given a set of vertices $S \subset N$, a *spanning tree* for S is a tree connecting all the players in S, namely $r \in G^N$ connected acyclic and with $N^r = S$. Next figure shows, with $N = \{1, 2, 3, 4\}$, a spanning tree for coalition $\{1, 2, 3\}$ (Fig. 4.13)

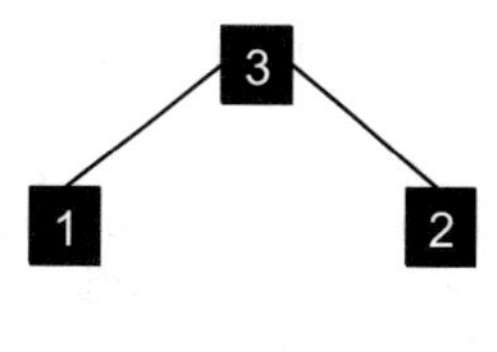

Fig. 4.13 Spanning tree for coalition $\{1, 2, 3\}$

Definition 4.21 Let ρ a fuzzy communication structure. The connection level of ρ is the number

$$c(\rho) = \vee\{t \in (0.1] : [\rho]_t \text{ connected}\}.$$

For instance, the connection level of ρ in Example 4.11 is $c(\rho) = 0.5$ because $[\rho]_{0.5}$ is connected and for any $t > 0.5$ players 1 and 4 are not connected to the others.

Definition 4.22 Let $\rho \in FG^N$. A Choquet by vertices (cv) partition of ρ is a partition by levels cv obtained by the following algorithm

Take $k = 0$, $cv = \emptyset$ and $\rho = \rho$
While $\rho \neq 0$ do
..... $k = k + 1$
..... choose $S \in N/\rho$
..... $s_k = c(\rho_S)$
..... choose r_k a spanning tree for S with $r_k \leq [\rho]_{s_k}$
..... $cv = cv \cup \{(r_k, s_k)\}$
..... $\rho = \rho - s_k r_k$
$cv(\rho) = cv$

Obviously there exist in general several cv-partitions but also the number of feasible cv-partitions is finite. We set $CV(\rho)$ the family of cv-partitions of the fuzzy graph ρ.

Example 4.16 Suppose the fuzzy graph ρ in Fig. 4.14. The connection level of ρ is $c(\rho) = 0.4$, but there are three feasible spanning trees to choose (they are in Fig. 4.14). Depending on the chosen tree the algorithm continues with a different fuzzy graph but the second step in this example obtain only one option for each tree. In Fig. 4.15 we can see the new fuzzy graph at the beginning of the second step and the new tree for the three options in the order given in the above figure. In the second step the connection level of all the components in the new graph is 0.3. So, $CV(\rho)$ has three feasible partitions.

Given a fuzzy graph ρ and taken a Choquet by vertices partition $cv(\rho) \in CV(\rho)$ we can determine a worth by a game v following Definition 4.18, $\gamma^{cv}(v, \rho)$. Each one of these partitions gets a cv-worth. We look for the best option, i.e. we take a Choquet by vertices partition obtaining the maximal cv-worth.

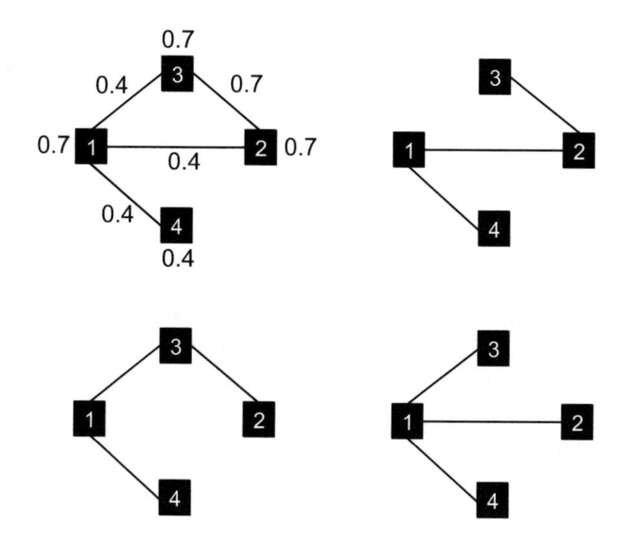

Fig. 4.14 First step in cv-algorithm

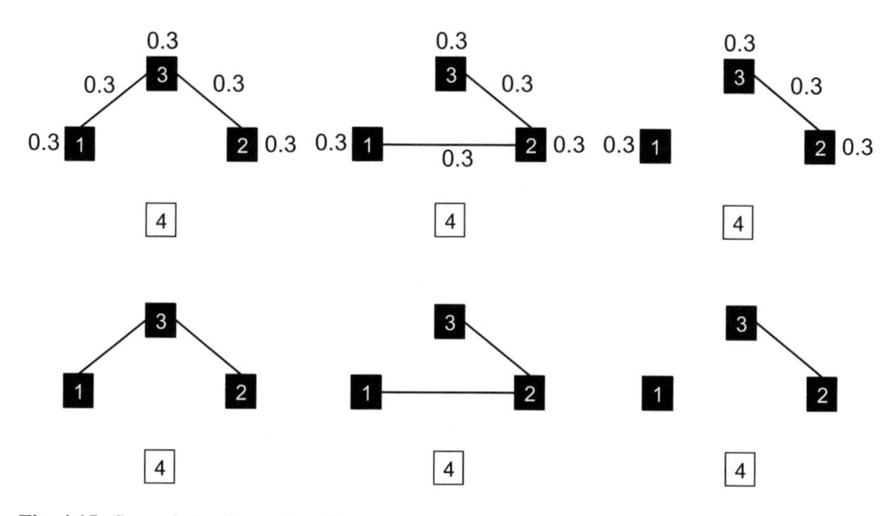

Fig. 4.15 Second step in cv-algorithm

Definition 4.23 Let $v \in \mathscr{G}^N$. The CV-worth is defined for each fuzzy communication structure ρ as the maximal cv-worth with $cv \in CV(\rho)$, namely

$$\gamma^{CV}(v, \rho) = \bigvee_{cv \in CV(\rho)} \gamma^{cv}(v, \rho).$$

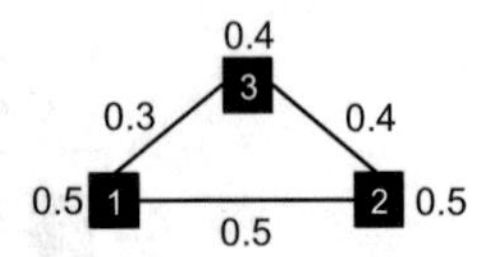

Fig. 4.16 Fuzzy graph in Example 4.18

Choquet by vertices functions are those partition functions that choose any Choquet by vertices partition obtaining the CV-worth for each fuzzy graph.

Example 4.17 Consider the anonymous game $v(S) = |S| - 1$ for all non-empty coalition $S \subseteq N = \{1, 2, 3, 4\}$. We calculate the CV-worth of the fuzzy communication structure ρ given in Example 4.16. The first two partitions in $CV(\rho)$ obtain the same worth, that is $0.4v(N) + 0.3v(\{1, 2, 3\} = 1.8$. But the third one has a different worth, $0.4v(N) + 0.3v(\{2, 3\}) + 0.3v(\{1\}) = 1.5$. Thus $\gamma^{CV}(v, \rho) = 1.8$.

Next example shows that a Choquet by vertices function is neither an extension nor inherited in general.

Example 4.18 Suppose the communication structure with matrix

$$r = \begin{bmatrix} 111 \\ 111 \\ 111 \end{bmatrix}.$$

Any Choquet by vertices function must choose between this three options

$$\begin{bmatrix} 111 \\ 110 \\ 101 \end{bmatrix}, \begin{bmatrix} 110 \\ 111 \\ 011 \end{bmatrix}, \begin{bmatrix} 101 \\ 011 \\ 111 \end{bmatrix}.$$

Hence, it is not an extension. Now take the fuzzy graph in Fig. 4.16. The algorithm has only one option for this example, the reader can see it in Table 4.5. But if we take $S = \{1, 3\}$ the partition uses link 13. The function is not inherited.

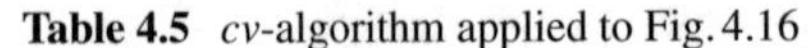

Table 4.5 *cv*-algorithm applied to Fig. 4.16

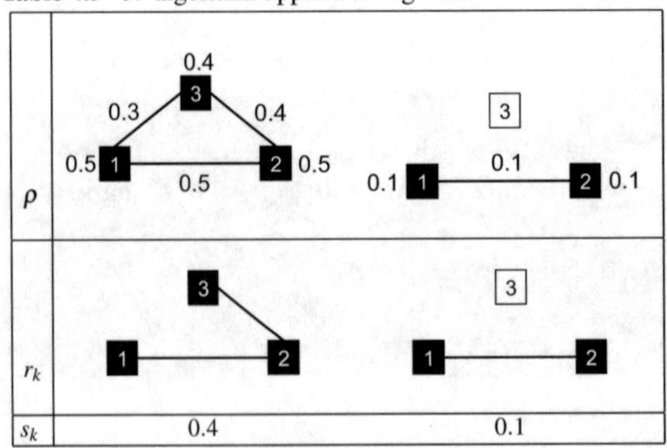

But, the method CV obtains also a fuzzy graph-worth.

Proposition 4.19 *The CV-worth is a fuzzy graph-worth.*

Proof Given a communication structure, although the chosen partition by levels is not $\{(r, 1)\}$ the worth is the same because r is a spanning tree, namely $\gamma^{CV}(v, r) = g(v, r)$.

Suppose $v \in \mathcal{G}_{sa}^N$. Let ρ_{-ij}^t with $\rho \in FG^N$ and $t \in (0, \rho(ij)]$. Suppose $cv(\rho_{-ij}^t) \in CV(\rho')$ with $\gamma^{CV}(v, \rho_{-ij}^t) = \gamma^{cv}(v, \rho_{-ij}^t)$. For each step and component in $N^{\rho_{-ij}^t}$ there is always a path between any pair of players in ρ with greater level, thus by superadditivity and the fact that the Myerson graph-worth we get a Choquet by vertices partition $cv(\rho)$ with

$$\gamma^{CV}(v, \rho_{-ij}^t) \leq \gamma^{cv}(v, \rho) \leq \gamma^{CV}(v, \rho).$$

Hence we get link monotonicity.

As the algorithm is working by components, the fuzzy graph-worth satisfies additivity by components. □

4.5 Fuzzy Myerson Values

We will follow Myerson [9] using any fuzzy graph-worth to describe models to study fuzzy communication structures. A new game is defined for each partition by levels introducing the information about the fuzzy communication.

Definition 4.24 Let $v \in \mathcal{G}^N$ be a game and pl be a partition function for fuzzy communication structures. For each $\rho \in FG^N$ the pl-vertex game is $(v/\rho)^{pl} \in \mathcal{G}^N$ where for any coalition S it holds

$$(v/\rho)^{pl}(S) = \gamma^{pl}(v, \rho_S)$$

Example 4.19 Suppose the fuzzy communication structure ρ in Fig. 4.17. We calculate the cg-vertex game taking the anonymous game $v \in \mathcal{G}_a^N$ with $v(S) = |S| - 1$ for all non-empty coalition. The cg-partition by level of this graph was obtained in Table 4.2. Remember that, as cg is inherited, to get the partitions by levels of the coalitions we only have to intersect the partition in Table 4.2 with each coalition. The worth of the great coalition N was actually calculated in Example 4.14,

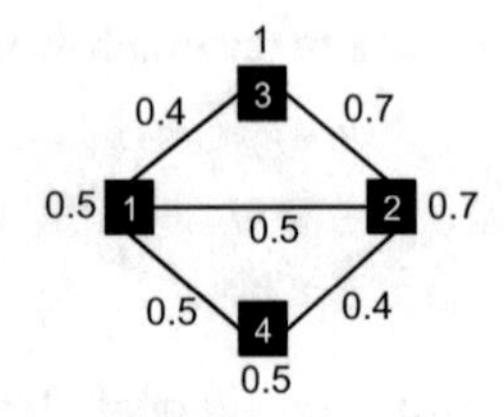

Fig. 4.17 Fuzzy communication structure ρ in Example 4.19

Table 4.6 Game $(v/\rho)^{cg}$

S	$\{1\}$	$\{2\}$	$\{3\}$	$\{4\}$	$\{1,2\}$	$\{1,3\}$	$\{1,4\}$	$\{2,3\}$
$(v/\rho)^{cg}(S)$	0	0	0	0	0.5	0.4	0.5	0.7
S	$\{2,4\}$	$\{3,4\}$	$\{1,2,3\}$	$\{1,2,4\}$	$\{1,3,4\}$	$\{2,3,4\}$	N	
$(v/\rho)^{cg}(S)$	0.4	0	1.2	1	0.9	1.1	5.3	

$$(v/\rho)^{cg}(N) = \gamma^{cg}(v,\rho) = 5.3.$$

Consider $S = \{1, 2, 4\}$. The worth of this coalition for the new game is

$$(v/\rho)^{cg}(S) = 0.5v(\{1,2,4\} + 0.2v(\{2\}) = 1.$$

Table 4.6 shows the worths of all the coalitions for the new characteristic function.

Proposition 4.20 *Let $v \in \mathcal{G}^N$, $\rho \in FG^N$ and let pl be a partition function.*

(1) If v is additive then $(v/\rho)^{pl}$ is additive, moreover $(v/\rho)^{pl} = (\rho(ii)v_i)_{i\in N}$.
(2) $((av+bw)/\rho)^{pl} = a(v/\rho)^{pl} + b(w/\rho)^{pl}$ for all $a, b \in \mathbb{R}$. Hence $(-v/\rho)^{pl} = -(v/\rho)^{pl}$.
(3) If v, w are strategically equivalent then so are $(v/\rho)^{pl}, (w/\rho)^{pl}$.
(4) $(v^{svg}/\rho)^{pl} = ((v/\rho)^{pl})^{svg}$.

Proof (1) Remember that the components form a partition of the domain of a relation. Let $v \in \mathbb{R}^N$. We take for each coalition S its partition $pl(\rho_S) = \{(r_k, s_k)\}_{k=1}^m$. Following Definition 4.24, Proposition 4.7 and (4.8) we get

$$(v/\rho)^{pl}(S) = \sum_{k=1}^m s_k g^v(r_k) = \sum_{k=1}^m s_k (v/r_k)(S) = \sum_{k=1}^m s_k \sum_{i\in dom\,(r_k)} v_i$$
$$= \sum_{i\in S} \sum_{\{k: i\in dom\,(r_k)\}} s_k v_i = \sum_{i\in S} \rho(ii) v_i.$$

(2), (3) They are trivial from Proposition 4.7.
(4) We determine the saving game, for each coalition. If $i \in N$ then $(v/\rho)^{pl}(\{i\}) = \rho(ii)v(\{i\})$. Let $S \subseteq N$ with $pl(\rho_S) = \{(r_k, s_k)\}_{k=1}^m$, using Proposition 4.7 and (4.8)

$$\begin{aligned}
(v^{svg}/\rho)^{pl}(S) &= \gamma^{pl}(v^{svg}, \rho_S) = \sum_{k=1}^{m} s_k g(v^{svg}, r_k) = \sum_{k=1}^{m} s_k (v^{svg}/r_k)(S) \\
&= \sum_{k=1}^{m} s_k (v/r_k)^{svg}(S) = \sum_{k=1}^{m} s_k \left[\sum_{i \in S} v/r_k(\{i\}) - v/r_k(S) \right] \\
&= \sum_{i \in S} \left[\sum_{\{k: i \in N^{r_k}\}} s_k \right] v(\{i\}) - \sum_{k=1}^{m} s_k v/r_k(S) \\
&= \sum_{i \in S} \rho(ii) v(\{i\}) - \sum_{k=1}^{m} s_k g(v, r_k) = \sum_{i \in S} (v/\rho)^{pl}(\{i\}) - (v/\rho)^{pl}(S) \\
&= ((v/\rho)^{pl})^{svg}(S).
\end{aligned}$$

□

The above result revised which properties are satisfies by the new game for all the partition functions. Now we consider inherited extensions to get more properties of the pl-vertex game. In this case it is possible to describe the pl-vertex game by the vertex games of the graphs in the partition.

Proposition 4.21 *Let pl be a inherited extension for fuzzy communication struture. If $v \in \mathcal{G}^N$ and $\rho \in FG^N$ with $pl(\rho) = \{(r_k, s_k)\}_{k=1}^m$ then*

$$(v/\rho)^{pl} = \sum_{k=1}^{m} s_k v/r_k.$$

Proof Suppose $S \subseteq N$ non-empty. As the chosen partition function is inherited,

$$pl(\rho_S) = \left\{ \left(r', \sum_{\{k:(r_k)_S = r'\}} s_k \right) : r' \neq 0, \exists\, k \text{ with } (r_k)_S = r' \right\}.$$

By Definitions 4.24 and 4.5, we have taking into account that $g(v, 0) = 0$

$$
\begin{aligned}
(v/\rho)^{pl}(S) = \gamma^{pl}(v, \rho_S) &= \sum_{(r',s')\in pl(\rho_S)} s' g(v, r') \\
&= \sum_{(r',s')\in pl(\rho_S)} \sum_{\{k:(r_k)_S = r'\}} s_k g(v, r') \\
&= \sum_{k=1}^{m} s_k g(v, (r_k)_S) = \sum_{k=1}^{m} s_k v/r_k(S).
\end{aligned}
$$

□

Examples 4.5 and 4.6 showed in the crisp case that monotonicity and convexity are not inheritable in general and, obviously, nor does the fuzzy options. Superadditivity needs the inherited condition for the partition. In order to get the convexity we introduce the following concept.

Definition 4.25 Let pl be a partition function for fuzzy communication structures. A fuzzy graph $\rho \in FG^N$ is pl-cycle-complete if r is cycle-complete for all $(r, s) \in pl(\rho)$.

Proposition 4.22 *Let pl be an inherited extension.*

(1) If $v \in \mathcal{G}_{sa}^N$ *then* $(v/\rho)^{pl} \in \mathcal{G}_{sa}^N$ *for all* $\rho \in FG^N$.
(2) If $v \in \mathcal{G}_{sa}^N$ *and* $v \geq 0$ *then* $(v/\rho)^{pl} \in \mathcal{G}_m^N$ *for all* $\rho \in FG^N$.
(3) If $v \in \mathcal{G}_c^N$ *then* $(v/\rho)^{pl} \in \mathcal{G}_c^N$ *for all* $\rho \in FG^N$ *which is pl-cycle-complete.*

Proof Superadditive (monotone, convex) games form positive cones in the sense that a linear combination with positive coefficients of superadditive (monotone, convex) games is also a superadditive (monotone, convex) game. All the steps are true from Propositions 4.21, 4.7 and 4.9. □

Definition 4.26 Let γ be a fuzzy graph-worth. The γ-vertex game for a game v and a fuzzy communication structure ρ is defined as

$$
(v/\rho)^{\gamma}(S) = \gamma(v, \rho_S)
$$

for all coalition S.

Observe that all the pl-vertex games are γ^{pl}-vertex games when pl is admissible. It is difficult for studying property transmission of the general concept of fuzzy graph-worth, therefore it must be studied each worth in particular. Superadditivity and monotonicity are obtained for all the fuzzy graph-worths.

Proposition 4.23 *Let $v \in \mathcal{G}^N$ and $\rho \in FG^N$. For each fuzzy graph-worth γ it holds:*

(1) If $v \in \mathcal{G}^N_{sa}$ then $(v/\rho)^\gamma \in \mathcal{G}^N_{sa}$ for all $\rho \in FG^N$.
(2) If $v \in \mathcal{G}^N_{sa}$ and $v \geq 0$ then $(v/\rho)^\gamma \in \mathcal{G}^N_m$ for all $\rho \in FG^N$.

Proof (1) If $S \cap T = \emptyset$ then $\rho_S + \rho_T \leq \rho_{S \cup T}$. Moreover ρ_S and ρ_T form different components in $\rho_S + \rho_T$. Applying link monotonicity several times we have

$$\gamma(v, \rho_{S \cup T}) \geq \gamma(v, \rho_S + \rho_T).$$

As γ is additive by components then

$$(v/\rho)^\gamma(S \cup T) \geq \gamma(v, \rho_S + \rho_T) = \gamma(v, \rho_S) + \gamma(v, \rho_T).$$

(2) Using (1) we get $(v/\rho)^\gamma \in \mathcal{G}^N_{sa}$ and obviously $(v/\rho)^\gamma \geq 0$. □

We analyze the case of the CV-worth, namely the CV-vertex game $(v/\rho)^{CV}$ for all $\rho \in FG^N$. But the concept of cycle-complete is not easy to extend it to the CV-worth.

Proposition 4.24 *Let $v \in \mathcal{G}^N$ and $\rho \in FG^N$.*

(1) If v is additive then $(v/\rho)^{CV}$ is additive, moreover $(v/\rho)^{CV} = (\rho(ii)v_i)_{i \in N}$.
(2) $((av + bw)/\rho)^{CV} = a(v/\rho)^{CV} + b(w/r)^{CV}$ for all $a, b \in \mathbb{R}$. Hence $(-v/\rho)^{CV} = -(v/\rho)^{CV}$.
(3) If v, w are strategically equivalent then so are $(v/\rho)^{CV}, (w/\rho)^{CV}$.
(4) $((v^{svg})/\rho)^{CV} = ((v/\rho)^{CV})^{svg}$.

Proof (1)–(4) For each coalition $S \subseteq N$ we consider a partition by levels $cv \in CV(\rho_S)$ with $\gamma^{CV}(v, \rho_S) = \gamma^{cv}(v, \rho_S)$. We have following Proposition 4.20 that $(v/\rho)^{cv}(S) = \gamma^{cv}(v, \rho_S)$ satisfies the four properties. Thus $(v/\rho)^{CV}$ verifies them too. □

A value for games with fuzzy communication structure is

$$f : \mathcal{G}^N \times FG^N \to \mathbb{R}^N.$$

We can define several of values in the way of Myerson using fuzzy graph-worths an the fuzzy versions of the vertex game. Now we define Myerson values for fuzzy graphs.

Definition 4.27 Let γ a fuzzy graph-worth. The γ-Myerson value is a value for games over N with fuzzy communication structure defined for each game $v \in \mathcal{G}^N$ and $\rho \in FG^N$ as

$$\mu^\gamma(v, \rho) = \phi((v/\rho)^\gamma).$$

In order to find an axiomatization of the Myerson values we take into account that players are asymmetric in the structure. Let $f : \mathcal{G}^N \times FG^N \to \mathbb{R}^N$ a value function for games over N with fuzzy communication structure.

γ-**Efficiency by components**. Let γ be a fuzzy graph-worth. If $v \in \mathcal{G}^N$ and $\rho \in FG^N$ then $f(v, \rho)(T) = \gamma(v, \rho_T)$ for all $T \in N/\rho$.

Fuzzy fairness. Let $v \in \mathcal{G}^N$ and $\rho \in FG^N$, it holds for all $ij \in L(\rho)$ and $t \in [0, \rho(ij)]$ that

$$f_i(v, \rho) - f_i(v, \rho^t_{-ij}) = f_j(v, \rho) - f_j(v, \rho^t_{-ij}).$$

Carrier. Let $v \in \mathcal{G}^N$ and $\rho \in FG^N$. For all player $i \notin N^\rho$ it holds $f_i(v, \rho) = 0$.

The following axiomatization was given in [7].

Theorem 4.5 *Let γ be a fuzzy graph-worth. The γ-Myerson value is the only value for games over N with fuzzy communication structure satisfying γ-efficiency by components, fuzzy fairness and carrier.*

Proof First we test that the γ-Myerson value verifies the three axioms.
CARRIER. Suppose $i \notin dom\,(\rho)$. For all coalition $S \subseteq N \setminus \{i\}$, player i does not change the components in the coalition, namely $N/\rho_{S\cup\{i\}} = N/\rho_S$. Thus $(v/\rho)^\gamma(S \cup \{i\}) = (v/\rho)^\gamma(S)$ and player i is null for $(v/\rho)^\gamma$. As Shapley value verifies null player property (Proposition 1.11) then $\mu^\gamma_i(v, \rho) = 0$.
γ-EFFICIENCY BY COMPONENTS. Let $T \in N/\rho$ be a component in a fuzzy communication structure ρ. As γ is additive by components we obtain for any coalition $S \subseteq N$

$$(v/\rho)^\gamma(S) = \gamma(v, \rho_S) = \sum_{R\in N/\rho} \gamma(v, (\rho_R)_S) = \sum_{R\in N/\rho} (v/\rho_R)^\gamma(S).$$

Shapley vale is a linear function, therefore for all $i \in N$

$$\mu_i^\gamma(v,\rho) = \phi_i((v/\rho)^\gamma) = \sum_{R\in N/\rho} \phi_i((v/\rho_R)^\gamma) = \sum_{R\in N/\rho} \mu_i(v,\rho_R).$$

But if $i \in T$ then carrier axiom says that $\mu_i(v,\rho_R) = 0$ for $R \in (N/\rho)\setminus\{T\}$. So, we have by efficiency of Shapley value

$$\sum_{i\in T}\mu_i(v,\rho) = \sum_{i\in T}\mu_i(v,\rho_T) = \sum_{i\in T}\phi_i((v/\rho_T)^\gamma) = (v/\rho_T)^\gamma(N) = \gamma(v,\rho_T).$$

FAIRNESS. Let $ij \in L(\rho)$ and $t \in (0, \rho(ij)]$. We repeat the proof in Theorem 4.3 using again the game w^{ij} in (4.6) defined now as

$$w^{ij} = (v/\rho)^\gamma - (v/\rho^t_{-ij})^\gamma.$$

Take now two values f^1, f^2 for games with fuzzy communication structure satisfying the three axioms. We will prove that $f^1 = f^2$. If $i \notin dom(\rho)$ then carrier implies $f_i^1(v,\rho) = f_i^2(v,\rho) = 0$. For the rest of players we work by induction in $|L(\rho)|$. If $|L(\rho)| = 0$ then each player $i \in dom(\rho)$ verifies $\{i\} \in N/\rho$ and by γ-efficiency by components $f_i^1(v,\rho) = f^2(v,\rho) = \rho(ii)v(\{i\})$. Suppose that the equality of both values is true if $|L(\rho)| = k-1$ and consider r with $|L(\rho)| = k$. Let $T \in N/\rho$ and $i \in T$. If $j \in T$ is such $ij \in L(\rho)$ then fuzzy fairness and induction say

$$\begin{aligned} f_i^1(v,\rho) - f_j^1(v,\rho) &= f_i^1\left(v,\rho_{-ij}^{\rho(ij)}\right) - f_j^1\left(v,\rho_{-ij}^{\rho(ij)}\right) \\ &= f_i^2\left(v,\rho_{-ij}^{\rho(ij)}\right) - f_j^2\left(v,\rho_{-ij}^{\rho(ij)}\right) = f_i^2(v,\rho) - f_j^2(v,\rho). \end{aligned}$$

Otherwise, $j \in T$ but $ij \notin L(\rho)$, there is a path between i and j and applying the above reasoning in each link in the path we get also the same equality. Hence, for all $j \in T$, $f_j^1(v,\rho) - f^2(v,\rho) = A$, being $A = f_i^1(v,\rho) - f_i^2(v,\rho)$. So, using γ-efficiency in component T,

$$0 = \sum_{j\in T} f_j^1(v,\rho) - f_j^2(v,\rho) = |T|A.$$

We get $A = 0$ □

Example 4.20 We determine the CV-Myerson value of the game $v(S) = |S| - 1$ with the fuzzy communication structure in Fig. 4.14. First we need to calculate the CV-vertex game. In Example 4.16 the worth of the great coalition was calculated, $(v/\rho)^{CV}(N) = \gamma^{CV}(v,\rho) = 1.8$. As the graph has only one component the CV-Myerson value is efficient on this worth, namely it is an allocation of 1.8 among the players. Game v is 0-normalized and so the vertex game is. The worth of a coalition $\{i, j\}$ is easy to determine, $(v/\rho)^{CV}(\{i, j\}) = \rho(ij)$. The calculus of $(v/\rho)^{CV} = 1.4$ is similar to the calculus of the worth of the great coalition. Finally, the rest of

coalitions with three players has only one option in the algorithm. Next table shows the worths of all the coalitions for the new characteristic function.
So, the CV-Myerson value is

$$\mu^{CV}(v, \rho) = \phi((v/\rho)^{CV}) = (0.5666, 0.5166, 0.5166, 0.2).$$

If $\gamma = \gamma^{pl}$ with pl an admissible partition function for fuzzy communication structures then we denote the γ^{pl}-Myerson value as μ^{pl} instead of $\mu^{\gamma^{pl}}$ and we name it the pl-Myerson value. The pl-Myerson values with pl fuzzy graph-worths derived from inherited extensions have a nice formula of calculus.

Theorem 4.6 *If pl is an admissible inherited extension for fuzzy communication structures then for all game v and fuzzy graph ρ it holds*

$$\mu^{pl}(v, \rho) = \sum_{k=1}^{m} s_k \mu(v, r_k),$$

where $pl(\rho) = \{(r_k, s_k)\}_{k=1}^{m}$.

Proof The proof follows from Proposition 4.21 and the linearity of the Shapley value. □

Particularly, we observe the case of the cg-Myerson value. In this case, we consider the signed capacity $\mu_i(v) : G^N \to \mathbb{R}$ given by $\mu_i(v)(r) = \mu_i(v, r)$ (taking each r as a subset of elements, its support).

Theorem 4.7 *For all $v \in \mathcal{G}^N$ and $\rho \in FG^N$ the cg-Myerson value satisfies*

$$\mu_i^{cg}(v, \rho) = \int_c \rho \, d\mu_i(v).$$

Proof The proof follows from (4.10) and the above theorem. □

Example 4.21 In Example 4.19 we determine the cg-vertex game $(v/\rho)^{cg}$ (see Table 4.7) for the anonymous game $v(S) = |S| - 1$ and the fuzzy communication structure ρ in Fig. 4.17. So, now we calculate the cg-Myerson value

$$\mu^{cg}(v, \rho) = \phi((v/\rho)^{cg}) = (1.333, 1.433, 1.316, 1.216).$$

But we can also use the above formula integrated in the cg-algorithm.

Table 4.7 Game $(v/\rho)^{CV}$

S	$\{1\}$	$\{2\}$	$\{3\}$	$\{4\}$	$\{1,2\}$	$\{1,3\}$	$\{1,4\}$	$\{2,3\}$
$(v/\rho)^{CV}(S)$	0	0	0	0	0.4	0.4	0.4	0.7
S	$\{2,4\}$	$\{3,4\}$	$\{1,2,3\}$	$\{1,2,4\}$	$\{1,3,4\}$	$\{2,3,4\}$	N	
$(v/\rho)^{CV}(S)$	0	0	1.4	0.8	0.8	0.7	1.8	

Table 4.8 cg-algorithm of Example 4.20 with Myerson values

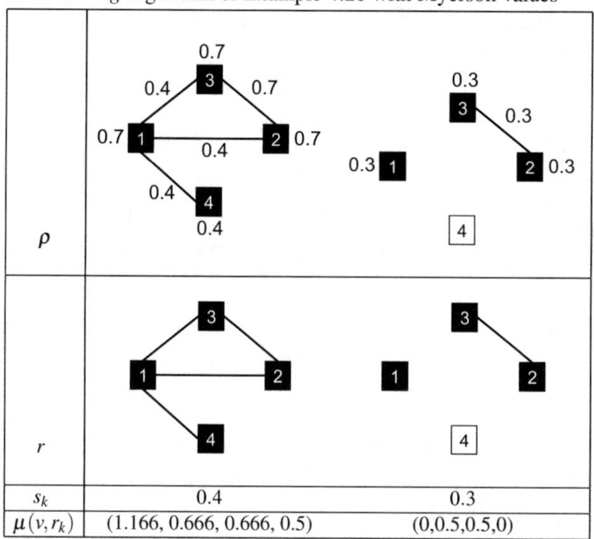

s_k	0.4	0.3
$\mu(v, r_k)$	(1.166, 0.666, 0.666, 0.5)	(0,0.5,0.5,0)

Example 4.22 We use the fuzzy graph in Example 4.14 to compare the cg-Myerson value with the CV-Myerson value of v calculated in Example 4.20. Table 4.8 includes a new line in the cg-algorithm determining the Myerson value of each graph in the partition. The combination of these Myerson values following Theorem 4.6 obtains

$$\mu^{cg}(v, \rho) = (0.466, 0.416, 0.416, 0.2).$$

Remember that

$$\mu^{CV}(v, \rho) = (0.5666, 0.5166, 0.5166, 0.2).$$

The CV-Myerson and cg-Myerson values verify different efficiencies because we get $\gamma^{CV}(v, \rho) = 1.8$ and $\gamma^{cg}(v, \rho) = 1.5$.

Given $\rho \in FG^N$ a communication structure and $\theta \in \Theta^N$ a permutation over N we introduce the new relation $\theta\rho$ is such $\theta\rho(\theta(i), \theta(j)) = \rho(i, j)$. Next we see properties of the pl-Myerson values with an inherited admissible partition pl. All of them are true from Theorem 4.6 and Proposition 4.11.

Proposition 4.25 *Let pl be an inherited admissible partition function for fuzzy communication structures. The pl-Myerson value satisfies the following properties for a fuzzy graph ρ and a game v.*

(1) If $\theta \in \Theta^N$ then $\mu^{pl}_{\theta(i)}(\theta v, \theta\rho) = \mu^{pl}_i(v, \rho)$ for all $i \in N$.
(2) If $v, w \in \mathcal{G}^N$ then $\mu^{pl}(av + bw, \rho) = a\mu^{pl}(v, \rho) + b\mu^{pl}(w, \rho)$.
(3) If $i \in dom(\rho)$ is a necessary player for $v \in \mathcal{G}^N_{sa}$ and also $v \geq 0$ then $\mu^{pl}_j(v, \rho) \leq \mu^{pl}_i(v, \rho)$ for any another player j.
(4) If $i \in dom(\rho)$ and $v \in \mathcal{G}^N_{sa}$ then $\mu^{pl}_i(v, \rho) \geq \rho(ii)v(\{i\})$.
(5) If $S \subseteq N$, ρ is pl-cycle-complete, and $v \in \mathcal{G}^N_c$ then $\mu^{pl}(v, \rho)(S) \geq (v/\rho)^{pl}(S)$.
(6) $\mu^{pl}(-v, \rho) = -\mu^{pl}(v, \rho)$.
(7) For all $i \in dom(\rho)$ and $v \in \mathcal{G}^N$ it holds

$$\mu^{pl}_i(v, \rho) = \rho(ii)v(\{i\}) - \mu^{pl}_i(v^{svg}, \rho).$$

Two of the specific properties in Proposition 4.12 of the Myerson value can be extended for all the fuzzy Myerson values following the same proofs.

Proposition 4.26 *Let γ be a fuzzy graph-worth. The γ-Myerson value satisfies the following properties for a fuzzy communication structure ρ.*

(1) If $ij \in L(\rho)$ and $v \in \mathcal{G}^N_{sa}$ then $\mu^\gamma_i(v, \rho) \geq \mu^\gamma_i(v, \rho^t_{-ij})$.
(2) If $ij \in L(\rho)$ and $v \in \mathcal{G}^N$ satisfies $(v/\rho)^\gamma = v/\rho^t_{-ij}$ then $\mu^\gamma(v, \rho) = \mu^\gamma(v, \rho^t_{-ij})$.

4.6 Transitive Fuzzy Communication Structures

If we take a transitive fuzzy communication structure then we get a "situation" similar to the Aumann-Dreze model.

Definition 4.28 A transitive fuzzy communication structure is a fuzzy relation satisfying: weakly reflexivity, symmetry and transitivity.

A particular case of transitive fuzzy graph is ρ^τ with $\tau \in [0, 1]^N$ (see Example 3.5). Let $i, j, k \in N$. Suppose $\tau(i) \geq \tau(j)$, we have

$$\rho^\tau(ij) = \tau(i) \wedge \tau(j) \geq \rho^\tau(ik) \wedge \rho^\tau(kj).$$

Transitivity in the crisp version implies the independence of two players in a component to communicate between them with reward to the rest of the players. Moreover, the Myerson value (Definition 4.9) for these communication structures coincides with the coalitional value (Definition 4.3). In the fuzzy case, transitivity guarantees for two players greater communication level straightly than using other agents. Depending on the chosen fuzzy graph-worth, the above crisp condition is satisfied or not in the fuzzy version (for instance the CV-worth can use the straight link between the players and also the path by another one). Obviously we can consider a transitive fuzzy communication structure as any fuzzy graph and then we can apply all the models (Definition 4.27) in the above sections without problems. But if we want to keep the crisp spirit of the coalitional structures explained before, we have some troubles. We should look for an internal model using always transitive elements. Hence the general concept of fuzzy graph-worth is not good, because it is not controllable and the third condition in Definition 4.13 uses a fuzzy graph which is not always transitive. If we use partition functions, the crisp graph in each step should be transitive too. The axiomatization of the γ-Myerson values is also a problem because fuzzy fairness is not internal for transitive structures, therefore we should try to get an axiomatization in the way of the coalitional value.

Definition 4.29 A partition function pl for fuzzy communication structures is transitive if for all transitive fuzzy graph ρ with $pl(\rho) = \{(r_k, s_k)\}_{k=1}^m$ it holds that r_k is a transitive communication structure for all $k = 1, \ldots, m$.

Fortunately both partition functions defined in the above sections (Definitions 4.16 and 4.17), the cg-partition and the pg-partition, are transitive.

Proposition 4.27 *The cg-partition function and the pg-partition function are transitive.*

Proof Let ρ be a transitive fuzzy communication structure.
CHOQUET BY GRAPHS. Since (4.10) we know that $r_k = [\rho]_k$ where $im_0(\rho) = \{\lambda_0 < \lambda_1 < \cdots < \lambda_m\}$. Denote $cg(\rho) = \{(r_k, s_k)\}_{k=1}^m$. We will prove that for all $k = 1, \ldots, m$ the graph r_k is transitive. Suppose three different players $i, i', i'' \in N$ such that $r_k(ii') = r_k(i'i'') = 1$. Hence $\rho(ii'), \rho(i'i'') \geq \lambda_k$. As ρ is transitive,

$$\rho(ii'') \geq \rho(ii') \wedge \rho(i'i'') \geq \lambda_k,$$

and then we get $r_k(ii'') = 1$.
PROPORTIONAL BY GRAPHS. We will prove again that for all $k = 1, \ldots, m$ the graph r_k is transitive. Suppose three different players $i, i', i'' \in N$ such that $r_k(ii') = r_k(i'i'') = 1$. Following the pg-algorithm, $\rho(ii'), \rho(i'i'') = s_k$. Also $\rho(ii) = \rho(i'i')$ $= \rho(i'', i'') = s_k$. As ρ is transitive,

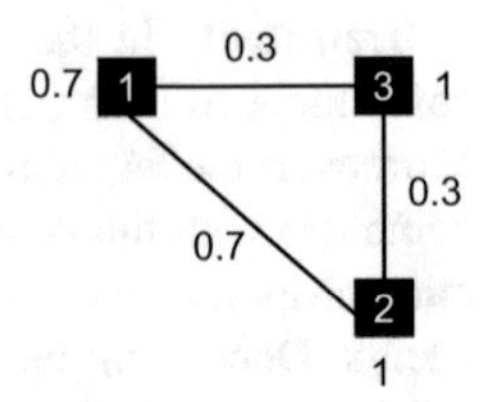

Fig. 4.18 Transitive fuzzy communication structure

$$s_k = \rho(ii) \wedge \rho(i''i'') \geq \rho(ii'') \geq \rho(ii') \wedge \rho(i'i'') = s_k,$$

and then we get $r_k(ii'') = 1$. □

The pl-Myerson values (with pl an inherited transitive partition function) applied to transitive fuzzy communication structures can be considered as particular extensions of the model proposed by Aumann and Dreze.

Proposition 4.28 *Let pl be an transitive inherited extension for fuzzy communication structures. If $\rho \in FG^N$ is transitive and $pl(\rho) = \{(r_k, s_k)\}_{k=1}^m$ then for all game v, $(v/\rho)^{pl} = \sum_{k=1}^m s_k v^{N/r_k}$ and the pl-Myerson value satisfies*

$$\mu^{pl}(v, \rho) = \sum_{k=1}^{m} s_k \phi(v, N/r_k).$$

Proof Theorem 4.6 says that if $pl(\rho) = \{(r_k, s_k)\}_{k=1}^m$ with ρ transitive then

$$\mu^{pl}(v, \rho) = \sum_{k=1}^{m} s_k \mu(v, r_k).$$

As pl is transitive and also ρ each r_k is a coalition structure. Hence $\mu(v, r_k) = \phi(v, N/r_k)$. □

Example 4.23 Suppose the anonymous game $v(S) = |S|^2$ over $N = \{1, 2, 3\}$. Figure 4.18 shows a transitive fuzzy communication structure ρ among the players. We calculate the cg-Myerson value $\mu^{cg}(v, \rho)$, that we can consider as a fuzzy coalitional solution of the game. All the players are symmetric in the only component of the fuzzy graph because the game is anonymous but the solution is

$$\mu^{cg}(v, \rho) = 0.3(3, 3, 3) + 0.4(2, 2, 1) + 0.3(0, 1, 1) = (1.7, 2, 1.4).$$

Remark 4.2 Transitive fuzzy communication structures represents actually a special class of fuzzy coalition structures. The generalization of coalition structure (Definition 4.1) in a fuzzy way is the following: a family of disjoint (two and two)

Table 4.9 cg-Algorithm of Example 4.23 with coalitional values

ρ	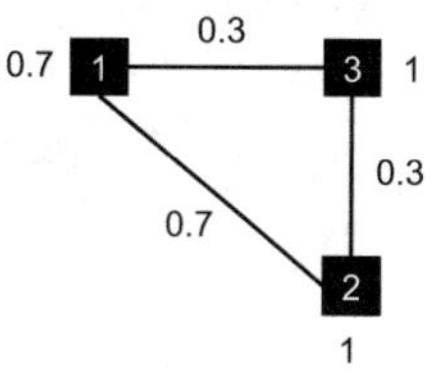	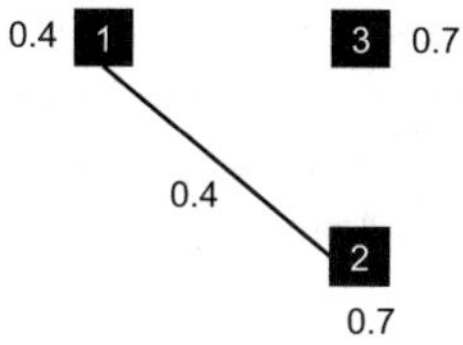	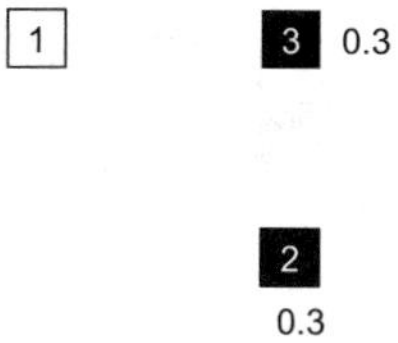
r_k	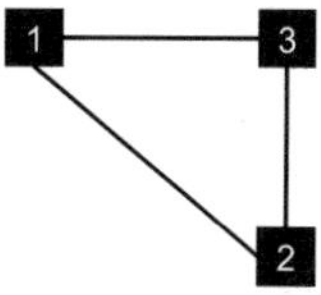	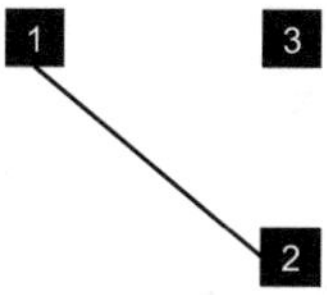	
N/r_k	$\{\{1,2,3\}\}$	$\{\{1,2\},\{3\}\}$	$\{\{2\},\{3\}\}$
s_k	0.3	0.4	0.3
$\phi(v,N/r_k)$	(3, 3, 3)	(2,2,1)	(0,1,1)

fuzzy coalitions with union into the great coalition. Hence, this concept depends on the used T-norm to do the intersection and the union. The more general concept is the following. A *fuzzy coalitional structure* over N is a finite family $\mathscr{B} = \{\beta_1, \ldots, \beta_m\}$ with $\beta_k \in [0, 1]^N$ for any k and $\sum_{k=1}^{m} \beta_k \leq e^N$. But there is no way to represent these structures as fuzzy binary relations keeping all the information.

We propose now an axiomatization for Myerson values over the transitive fuzzy communication structures. Let f be a value for games with transitive fuzzy communication structure and pl be a transitive partition function for fuzzy graphs (Table 4.9).

pl-Efficiency by components. For all $v \in \mathscr{G}^N$ and $\rho \in FG^N$ transitive, $f(v, \rho)(T) = \gamma^{pl}(v, \rho_T)$ for every $T \in N/\rho$.

Restricted null player. Let $i \in N$ be a null player in the game v, it holds $f_i(v, \rho) = 0$ if $i \in N^\rho$.

Necessary player. If $i \in N^\rho$ is a necessary player for a game v and $\rho \in FG^N$ transitive then $f_i(v, \rho) \geq f_j(v, \rho)$.

Linearity. For all $v, w \in \mathscr{G}^N$, $a, b \in \mathbb{R}$ and a transitive fuzzy graph ρ it holds

$$f(av + bw, \rho) = af(v, \rho) + bf(w, \rho).$$

Carrier. Let $v \in \mathscr{G}^N$ and $\rho \in FG^N$ transitive. For all player $i \notin dom\,(\rho)$ it holds $f_i(v, \rho) = 0$.

Theorem 4.8 *Let pl be an transitive inherited extension for fuzzy communication structures. The pl-Myerson value is the only value for games with transitive fuzzy communication structure satisfying pl-efficiency by components, restricted null player, necessary player, linearity and carrier.*

Proof First we prove that our value satisfies all the axioms. Proposition 4.28 says that if $pl(\rho) = \{(r_k, s_k)\}_{k=1}^{m}$ with ρ transitive then

$$\mu^{pl}(v, \rho) = \sum_{k=1}^{m} s_k \mu(v, r_k) = \sum_{k=1}^{m} s_k \phi(v, N/r_k).$$

The coalitional value verifies efficiency by components, restricted null player, additivity, carrier (Theorem 4.1) and also necessary player (Proposition 4.3).
pl- EFFICIENCY BY COMPONENTS. It was proved in Theorem 4.5.
RESTRICTED NULL PLAYER. Let i be a null player for v and $i \in N^{\rho}$. We have $\phi_i(v, N/r_k) = 0$ for each k from restricted null player and carrier axioms of the coalitional value. So, $\mu_i^{pl}(v, \rho) = 0$.
NECESSARY PLAYER. Suppose $i \in N^{\rho}$ necessary player for $v \in \mathcal{G}_m^N$ and $j \in N \setminus \{i\}$. If k satisfies that $i \notin N^{r_k}$ then $v^{N/r_k} = 0$. Hence,

$$\mu_i^{pl}(v, \rho) = \sum_{k=1}^{m} s_k \phi_i(v, N/r_k) \geq \sum_{k=1}^{m} s_k \phi_j(v, N/r_k) = \mu_j^{pl}(v, \rho).$$

LINEARITY. Proposition 4.20 says that $((av + bw)/\rho)^{pl} = a(v/\rho)^{pl} + b(w/\rho)^{pl}$. Linearity of the Shapley value implies the axiom.
CARRIER. It was proved in Theorem 4.5.

Now, considering linearity, we take an unanimity game u_T. Players are divided in three groups in order to determine their payoffs: out of the domain the payoff is zero by carrier, out of T the payoff is also zero by restricted null player. Let $i \in T \cap N^{\rho}$, the payoff must be the same by necessary player. Efficiency by components implies the uniqueness. □

The coalitional value satisfies symmetry in components, but now symmetric players in the same component can have different payoffs (see Example 4.23).

Another interesting difference between coalition structures and communication structures in general is that convexity is an inheritable condition (see Proposition 4.1) for the first one and not for the second one. We obtain the same result for convex games in the fuzzy case.

Proposition 4.29 *Let pl be an inherited transitive partition function. If* $v \in \mathcal{G}_c^N$ *then* $(v/\rho)^{pl} \in \mathcal{G}_c^N$ *for all* ρ *transitive fuzzy communication structure. Moreover*

$$\mu^{pl}(v, \rho)(S) \geq (v/\rho)^{pl}(S).$$

Proof Convex games form a non-negative cone, therefore the linear combination of convex games with non-negative coefficients is another convex game. Let $pl(\rho) = \{(r_k, s_k)\}_{k=1}^m$ for a transitive fuzzy graph. From Proposition 4.1 we know that v^{N/r_k} is convex when $v \in \mathcal{G}_c^N$ for each k. Proposition 4.28 implies that $(v/\rho)^{pl}$ is convex. Using Proposition 4.3 and the fact that pl is inherited we get the required inequality. □

Let pl be an inherited transitive partition function for fuzzy communication structures. In Example 3.5 we introduced a fuzzy relation ρ^τ from a given fuzzy coalition τ. For each $ij \in N \times N$ we put $\rho^\tau(ij) = \tau(i) \wedge \tau(j)$. We define the *induced partition function* for fuzzy coalitions from pl as pl^* where for any $\tau \in [0, 1]^N$,

$$pl^*(\tau) = \{(N^{r_k}, s_k)\}_{k=1}^m \tag{4.11}$$

with $pl(\rho^\tau) = \{(r_k, s_k)\}_{k=1}^m$. Remember that we have seen before that ρ^τ is transitive for all fuzzy coalition τ. For instance, we will prove that $cg^* = ch$. Let $cg(\rho^\tau) = \{(r_k, s_k)\}_{k=1}^m$. From (4.10) we have that $im_0(\tau) = \{\lambda_0 < \lambda_1 < \cdots < \lambda_m\} = im(\rho^\tau)$ with $s_k = \lambda_k - \lambda_{k-1}$ and $r_k = [\rho^\tau]_k$ for any k. Observe that if $i \in N^{r_k}$ then $\tau(i) \geq \lambda_k$. If $i, j \in N^{r_k}$ then $\rho^\tau(ij) = \tau(i) \wedge \tau(j) \geq \lambda_k$. So, $N/r_k = \{[\tau]_k\}$ and then $cg^*(\tau) = ch(\tau)$. The reader can test that $pg^* = pr$.

Proposition 4.30 *Let pl be an transitive inherited extension for fuzzy communication structures. For all* $\tau \in [0, 1]^N$ *and* $v \in \mathcal{G}^N$ *it holds*

$$\mu^{pl}(v, \rho^\tau) = \phi^{pl^*}(v, \tau).$$

Proof Let $\tau \in [0, 1]^N$. If $pl(\rho^\tau) = \{(r_k, s_k)\}_{k=1}^m$ then $N/r_k = \{N^{r_k}\}$ for all k because r_k is transitive. Thus $\phi(v, N/r_k) = \phi(v, N^{r_k})$ for all k. Hence, Proposition 4.28 implies

$$\mu^{pl}(v, \rho^\tau) = \sum_{k=1}^m s_k \phi(v, N/r_k) = \sum_{k=1}^m s_k \phi(v, N^{r_k}) = \phi^{pl^*}(v, \tau).$$

□

4.7 Probabilistic Communication Structures

Calvo et al. [4] introduced probabilistic communication situations before fuzzy communication structures. Their model can be include in the fuzzy model described in the book. But we have to adapt slightly the concept of fuzzy graph.

Definition 4.30 A probabilistic communication structure over N is a (undirected) probabilistic graph ρ, namely a fuzzy bilateral relation over N which satisfies:

(1) ρ is reflexive, $\rho(ii) = 1$ for all players $i \in N$,
(2) ρ is symmetric, $\rho(i, j) = \rho(j, i) = \rho(ij)$ for all $i \neq j$.

Thus the difference between probabilistic graph and fuzzy graph is the definition of reflexivity. Observe that a probabilistic graph is a proximity relation (see Sect. 3.3) but perhaps this name will be more successful. Now the interpretation of $\rho(ij)$ is the probability of communication between players i and j, and then it is supposed that each player i is always availability. All the crisp reflexive graphs are probabilistic graphs. All the probabilistic graphs are fuzzy graphs but, not all the fuzzy graphs are probabilistic too.

Example 4.24 Let $N = \{1, 2, 3, 4, 5, 6, 7, 8\}$. Fuzzy graph ρ in Fig. 4.10 is not a probabilistic because $\rho(ii) < 1$ for all $i \geq 4$. In Fig. 4.19 we show a probabilistic communication structure over N.

Obviously all the develop of games with fuzzy communication structure given in Sects. 4.4 and 4.5 is applicable to probabilistic communication structures because they are also fuzzy graphs. For this special family of fuzzy communication structure Calvo et al. [4] introduced a Myerson value which is really a pl-Myerson value using a partition function based in the multilinear extension of Owen [11] (see Definition 2.10).

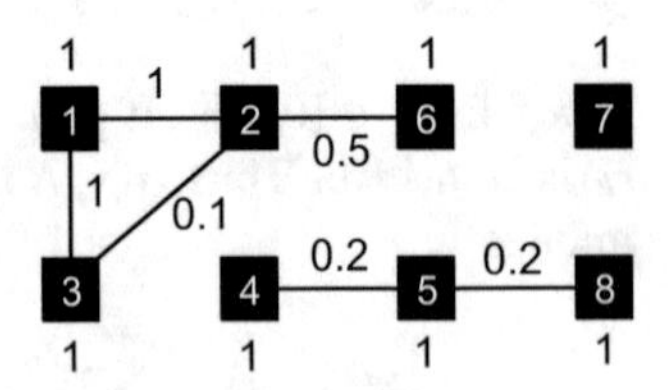

Fig. 4.19 Probabilistic communication structure

Definition 4.31 The probabilistic function pb is defined for each ρ probabilistic communication structure as

$$pb(\rho) = \left\{ \left(r, \prod_{ij \in L(r)} \rho(ij) \prod_{ij \in L(\rho) \setminus L(r)} (1 - \rho(ij)) \right) \right\}_{\{r \in G^N : [\rho]_1 \leq r \leq r^\rho\}}.$$

The pb function for a probabilistic graph takes all the possible communication structures into ρ (those with probability non-zero) and their probability of happening. We will prove that pb is an admissible inherited extension.

Proposition 4.31 *The mapping pb is an admissible inherited extension for probabilistic communication structures.*

Proof First we prove that pb is a partition function. Let ρ be a probabilistic graph. As $[\rho]_1 \leq r \leq r^\rho$ then $\rho(ij) > 0$ for all $ij \in L(r)$ and $\rho(ij) < 1$ for all $ij \notin L(r)$, thus

$$\prod_{ij \in L(r)} \rho(ij) \prod_{ij \in L(\rho) \setminus L(r)} (1 - \rho(ij)) > 0.$$

We consider any ordering for the elements in $\{r \in G^N : [\rho]_1 \leq r \leq r^\rho\}$ and let r the first of them in that ordering. For each $i_0 j_0 \in L(r)$ we rewrite, in order to apply Lemma 2.1, the sum of all the levels in the partition with graphs containing this element in the following sense

$$\begin{aligned}
&\sum_{\{r \in G^N : [\rho]_1 \leq r \leq r^\rho, r(i_0 j_0) = 1\}} \prod_{ij \in L(r)} \rho(ij) \prod_{ij \in L(\rho) \setminus L(r)} (1 - \rho(ij)) \\
&= \rho(i_0 j_0) \sum_{R \subseteq L(N) \setminus \{i_0 j_0\}} \prod_{ij \in R} \rho(ij) \prod_{L(N) \setminus (R \cup \{i_0 j_0\})} (1 - \rho(ij)) \\
&= \rho(i_0 j_0).
\end{aligned}$$

Hence $\rho = \sum_{(r,s) \in pb(\rho)} sr = 0$ and for each $(r_0, s_0) \in pb(\rho)$ it holds

$$s_0 r_0 \leq \rho - \sum_{(r,s) \in pb(\rho) \setminus \{(r_0, s_0)\}} sr.$$

The proof to test that pb is inherited is equal to the proof of Proposition 3.9 where we proved that the multilinear extension is inherited. As we did before, we must deal with ρ as a fuzzy set of $L(N)$.

Finally we prove that pb is admissible. As pb is inherited we only have to prove link monotonicity from Lemma 4.1.[4] Let $v \in \mathcal{G}_{sa}^N$. If ρ is a probabilistic graph then $\rho^t_{-i_0 j_0}$ is also a probabilistic graph for all $i_0 j_0 \in L(\rho)$ and $t \in (0, \rho(i_0 j_0)]$. We denote as $R = \{r \in G^N : dom(r) = N\}$. From Definition 4.18 we can use

$$\gamma^{pb}(v, \rho) = \sum_{\{r \in G^N : [\rho]_1 \leq r \leq r^\rho\}} \prod_{ij \in L(r)} \rho(ij) \prod_{ij \in L(\rho) \setminus L(r)} (1 - \rho(ij)) g(v, r),$$

$$\gamma^{pb}(v, \rho) = \sum_{r \in R} \prod_{ij \in L(r)} \rho(ij) \prod_{ij \in L(N) \setminus L(r)} (1 - \rho(ij)) g(v, r), \tag{4.12}$$

and also

$$\gamma^{pb}(v, \rho^t_{-i_0 j_0}) = \sum_{r \in R} \prod_{ij \in L(r)} \rho^t_{-i_0 j_0}(ij) \prod_{ij \in L(N) \setminus L(r)} (1 - \rho^t_{-i_0 j_0}(ij)) g(v, r), \tag{4.13}$$

For each $r \in R$ with $i_0 j_0 \in L(r)$ we find $r' \in R$ with $L(r') = L(r) \setminus \{i_0 j_0\}$. We obtain for ρ in (4.12)

$$\prod_{ij \in L(r)} \rho(ij) \prod_{ij \in L(N) \setminus L(r)} (1 - \rho(ij)) g(v, r) + \prod_{ij \in L(r')} \rho(ij) \prod_{ij \in L(N) \setminus L(r')} (1 - \rho(ij)) g(v, r')$$

$$= \prod_{ij \in L(r')} \rho(ij) \prod_{ij \in L(N) \setminus L(r)} (1 - \rho(ij)) \left[\rho(i_0 j_0) g(v, r) + (1 - \rho(i_0 j_0)) g(v, r') \right]. \tag{4.14}$$

Also in (4.13), using that $\rho^t_{-i_0 j_0}(ij) = \rho(ij)$ for all $ij \in L(N) \setminus \{i_0 j_0\}$, we get

$$\prod_{ij \in L(r')} \rho(ij) \prod_{ij \in L(N) \setminus L(r)} (1 - \rho(ij)) \left[(\rho(i_0 j_0) - t) g(v, r) + (1 - \rho(i_0 j_0) + t) g(v, r') \right]. \tag{4.15}$$

We do (4.14) and (4.15) obtaining

$$t \prod_{ij \in L(r')} \rho(ij) \prod_{ij \in L(N) \setminus L(r)} (1 - \rho(ij)) \left[g(v, r) - g(v, r') \right].$$

But, as g satisfies link monotonicity we have $g(v, r) \geq g(v, r')$ and so

$$\gamma^{pb}(v, \rho) - \gamma^{pb}(v, \rho^t_{-i_0 j_0}) \geq 0.$$

□

So, we introduce the probabilistic Myerson value as the pb-Myerson value applied to games with a probabilistic communication structure.

---

[4] Extension property in this case only use reflexive communication structure.

Definition 4.32 The probabilistic Myerson value is a value for games with probabilistic communication structure defined for each game v and each probabilistic graph ρ as

$$\mu^{pb}(v,\rho) = \phi((v/\rho)^{pb}),$$

with $(v/\rho)^{pb}(S) = \gamma^{pb}(v,\rho_S)$ for all coalition S.

We test that our concept of probabilistic Myerson value coincides with that one in [4] directly from Theorem 4.6.

Proposition 4.32 *The probabilistic Myerson value satisfies for all game v and probabilistic graph ρ that*

$$\mu^{pb}(v,\rho) = \sum_{\{r\in G^N:[\rho]_1\leq r\leq r^\rho\}} \prod_{ij\in L(r)} \rho(ij) \prod_{ij\in L(\rho)\setminus L(r)} (1-\rho(ij))\mu(v,r).$$

if $pb(\rho) = \left\{\left(r, \prod_{ij\in L(r)} \rho(ij) \prod_{ij\in L(\rho)\setminus L(r)} (1-\rho(ij))\right)\right\}_{\{r\in G^N:[\rho]_1\leq r\leq r^\rho\}}.$

Axiomatization in Theorem 4.5 is applicable for the probabilistic Myerson value, which is slight different to that in [4], and also all those properties given in Sects. 4.4 and 4.5.

Transitivity condition must be changed in probabilistic situations.

Definition 4.33 A transitive probabilistic communication structure is a probabilistic graph ρ satisfying semi-transitivity.

Theorem 4.8 and Proposition 4.29 are valid for the probabilistic Myerson value over transitive probabilistic communication structures.

Remark 4.3 Probabilistic communication structures can be extended in the following sense. A *probabilistic graph* is a fuzzy relation ρ verifying semi-reflexivity ($\rho(ij) \leq \rho(ii)\rho(jj)$) and symmetry. In the case $\rho(ij)$ is also the probability of communication between i and j, and $\rho(ii)$ the probability to be active player i. For each graph $r \in G^N$ the probability by ρ of obtaining r is

$$\prod_{ij\in L(r)} \rho(ij) \prod_{\{ij\notin L(r):i,j\in N^r\}} \rho(ii)\rho(jj)(1-\rho(ij))$$
$$\prod_{\{ij\notin L(r):i\in N^r,j\notin N^r\}} \rho(ii)(1-\rho(jj))(1-\rho(ij))$$
$$\prod_{\{ij\notin L(r):ij\notin N^r\}} (1-\rho(ii))(1-\rho(jj))(1-\rho(ij)).$$

Results and proofs are similar to the studied case, but the expression are very long and complicate.

References

1. Aumann, R.J., Dreze, J.H.: Cooperative games with coalition structures. Int. J. Game Theory **3**(4), 217–237 (1974)
2. Aubin, J.P.: Cooperative fuzzy games. Math. Oper. Res. **6**(1), 1–13 (1981)
3. Barberá, S., Gerber, A.: On coalition formation: durable coalition structure. Math. Soc. Sci. **45**(2), 185–203 (2003)
4. Calvo, E., Lasaga, J., van den Nouweland, A.: Values for games with probabilistic graphs. Math. Soc. Sci. **37**, 70–95 (1999)
5. Fernández, J.R.: Complexity and algorithms in cooperative games. Ph.D. thesis. University of Seville. Spain (2000)
6. Jiménez-Losada, A., Fernández, J.R., Ordóñez, M., Grabrisch, M.: Games on fuzzy communication structures with Choquet players. Eur. J. Oper. Res. **207**, 836–847 (2010)
7. Jiménez-Losada, A., Fernández, J.R., Ordóñez, M.: Myerson values for games with fuzzy communication structure. Fuzzy Sets Syst. **213**, 74–90 (2013)
8. Mordeson, J.N., Nair, P.S.: Fuzzy graphs and fuzzy hypergraphs. Studies in Fuzzines and Soft Computing. Springer, Berlin (2000)
9. Myerson, R.B.: Graphs and cooperation in games. Math. Oper. Res. **2**(3), 225–229 (1977)
10. Myerson R.B: Conference structures and fair allocation rules. Int. J. Game Theory **9** (3), 169–182 (1980)
11. Owen, G.: Multilinear extensions of games. Manag. Sci. **18**(5), 64–79 (1972)
12. Owen, G.: Values of graph-restricted games. SIAM J. Algebr. Discret. Methods **7**(2), 210–220 (1986)
13. Slikker, M., van den Nouweland, A.: Social and Economic Networks in Cooperative Game Theory. Theory and Decision Library. Series C: Game Theory, Mathematical Programming and Operations Research. Kluwer Academic Publishers, Boston (2001)
14. van den Nouweland, A., Borm, P.: On the convexity of communication games. Int. J. Game Theory **19**(4), 421–430 (1991)
15. van den Nouweland, A.: Cost allocation and communication. Nav. Res. Logist. **40**(5), 733–744 (1993)

Chapter 5
A Priori Fuzzy Unions

5.1 Introduction

The model explained in the book studies the value of a cooperative game with an additional information about the relations among the players. This information is showed as a mathematical structure in order to introduce it in the characteristic function of the game. But the same mathematical structure can be interpreted by different situations and so, the way to modify the game is also different.

Aumann and Dreze [2] considered that a partition of the set of players is formed. Later Myerson [12] supposed that also the links used to form these groups are known (the communications). These systems are represented by graphs. Owen [13] proposed a different model to deal with a coalition structure. The Aumann and Dreze model (see Sect. 4.2) considered the coalition structure as a final distribution of the players in unions, and then there is not side-payment between these unions. Hence each final coalition obtains its worth for the characteristic function and then the players in it bargain, separately from the rest, a payoff vector with reward to the restricted game over the coalition (Definition 3.5). On the other hand, Owen saw the coalition structure as a priori union system. Now, $\mathscr{B}$ is a "starting point for further negotiations", as said Sánchez-Soriano and Pulido [15]. So, each union represents a group of players with similar interests, ideas or conditions. The unions must be considered in the game when players use them to bargain. Suppose a situation where players have to establish the terms of an agreement about certain subjects upon which the payoffs of the agents depend. Players are distributed in unions with similar positions respect to the terms of the agreement. Obviously, the position of the unions in the bargaining should be the focus to determine the payoffs. Hart and Kurz analyzed deeply a priory unions in [9]. Casajus [6] followed to Myerson in the way of the a priori unions, namely he proposed to take into account in each a priori union the links producing the group.

We studied in Chap. 4 extensions to fuzzy communication structures based on Jiménez-Losada et al. [10]. In this chapter we analyze fuzzy extensions of the Owen

A. Jiménez-Losada, *Models for Cooperative Games with Fuzzy Relations among the Agents*, Studies in Fuzziness and Soft Computing 355,
DOI 10.1007/978-3-319-56472-2_5

model [13] and the Casajus variant [6]. Fernández et al. [7] use proximity relations to study fuzzy relations a priori. These relations allow us to describe more realistic situations, because the closeness of ideas between agents has different degrees. The crisp model will be seen as a simplification of the problem.

5.2 A Priori Unions. The Owen Model

Let $\mathscr{B} = \{B_1, \ldots, B_m\}$ be a coalition structure (Definition 4.1) over the finite set of players N in a game $v \in \mathscr{G}^N$. We take $\mathscr{B}$ in the classical way of Aumann and Dreze [2], namely $\mathscr{B}$ is a partition of N, and we named the elements in $\mathscr{B}$ as unions. Owen [13] proposed a different model to deal with a coalition structure. The Aumann and Dreze model (see Sect. 4.2) considered the coalition structure as a final distribution of the players in unions, and then there is not side-payment between these unions. Hence each union $B \in \mathscr{B}$ obtains $v(B)$ and then the players in B bargain, separately from the rest, a payoff vector with reward to the restricted game over B (Definition 3.5). On the other hand, Owen saw the coalition structure $\mathscr{B}$ as a priori union system, a starting point for further negotiations. So, each union $B \in \mathscr{B}$ represents a group of players with similar interests, ideas or conditions. The unions must be considered in the game when players use them to bargain. Suppose a situation where players have to establish the terms of an agreement about certain subjects upon which the payoffs of the agents depend. Players are distributed in unions with similar positions respect to the terms of the agreement. Obviously, the position of the unions in the bargaining should be the focus to determine the payoffs.

Definition 5.1 A priori union system for N is a non-empty family of coalitions $\mathscr{B} = \{B_1, \ldots, B_m\}$ such that $B_p \cap B_q = \emptyset$ for all $p \neq q$ and $\bigcup_{p=1}^{m} B_p = N$. The set of a priori union systems over N is denoted as P^N. Each element of $\mathscr{B}$ is named a union.

Each a priori union system $\mathscr{B}$ is associated to an equivalence relation (reflexive, symmetric and transitive) $r_{\mathscr{B}}$ where $N^{r_{\mathscr{B}}} = N$ and the equivalence classes are the unions, i.e. $\mathscr{B} = N/r_{\mathscr{B}}$. As we said before, a union is formed a priori in the sense that its players have similar interests and not thinking about the communication level among them. Hence the relation must be reflexive and not weakly reflexive, because any player has the same ideas as herself.

Example 5.1 Figure 5.1 shows a priori union system over $N = \{1, 2, 3, 4, 5, 6, 7, 8\}$, the set of seller in Example 4.1. In this case sellers also organize themselves by the city where they sell, $\mathscr{B} = \{\{1, 2, 3, 4\}, \{5, 6, 7\}, \{8\}\}$. But they would like in this case to establish an agreement in order to avoid competition in the same city (perhaps some of them want to earn more money and visit the other cities). Hence, they put their profits in common and they must decide how to allocate this amount. Matrix $r_{\mathscr{B}}$ represents the system.

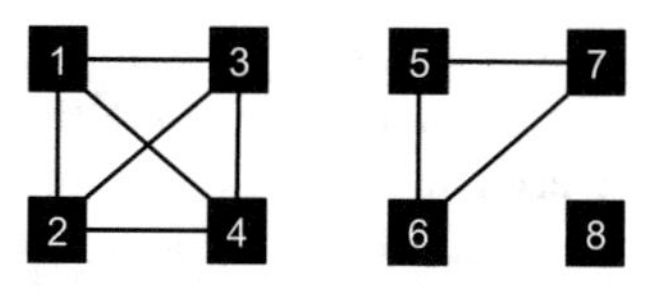

Fig. 5.1 A priori union system

$$r_{\mathscr{B}} = \begin{bmatrix} 1\,1\,1\,1\,0\,0\,0\,0 \\ 1\,1\,1\,1\,0\,0\,0\,0 \\ 1\,1\,1\,1\,0\,0\,0\,0 \\ 1\,1\,1\,1\,0\,0\,0\,0 \\ 0\,0\,0\,0\,1\,1\,1\,0 \\ 0\,0\,0\,0\,1\,1\,1\,0 \\ 0\,0\,0\,0\,1\,1\,1\,0 \\ 0\,0\,0\,0\,0\,0\,0\,1 \end{bmatrix}$$

Reflexive coalition structures and a priori union systems are the same mathematical objects, thus the difference of interpretation must appear in the definition of the instrumental game incorporating the information to an original game. Owen [13] proposed un way in two steps to get the new game, first the unions determine how much each one receives and later the agents in each union fix how these receipts are to be divided.

Definition 5.2 Let $v \in \mathscr{G}^N$ be a game and $\mathscr{B} = \{B_1, \ldots, B_m\} \in P^N$ be a priori union system. Let $M = \{1, \ldots, m\}$. The quotient game is the game $w^{v,\mathscr{B}} \in \mathscr{G}^M$ defined for all $Q \subseteq M$ as

$$w^{v,\mathscr{B}}(Q) = v\left(\bigcup_{q \in Q} B_q\right).$$

The quotient game is still not our instrumental game, it is only the tool for the first step. We will apply the Shapley value for games over M in order to obtain a payoff for each union, $\phi^M(w^{\mathscr{B}})$. But each player, and also each coalition, into a union aspires to get a payoff according to their possibilities in the whole game and not only within the union. Otherwise they would break out the union. Owen [13], to take into account this fact, proposed to define a quotient game for each coalition in a union. Let $\mathscr{B} = \{B_1, \ldots, B_m\}$ be a priori union system. Suppose a union $B_p \in \mathscr{B}$ and $S \subseteq B_p$. The *restricted a priori union system* of $\mathscr{B}$ to S is

$$\mathscr{B}_S = \{B_1, \ldots, B_{p-1}, S, B_{p+1}, \ldots, B_m\} \in P^{N \setminus (B_p \setminus S)}. \tag{5.1}$$

Hence, the above system explains the same situation but without the action of rest of players in the union. Next we define the restricted game for this model as a game in each union.

Definition 5.3 Let $\mathscr{B} = \{B_1, \ldots, B_m\} \in P^N$, $M = \{1, \ldots, m\}$ and $p \in M$. The p-union game is $v_p^{\mathscr{B}} \in \mathscr{G}^{B_p}$ defined for each $S \subseteq B_p$ as

$$v_p^{\mathscr{B}}(S) = \phi_p^M\left(w^{v,\mathscr{B}_S}\right).$$

Example 5.2 Let $N = \{1, 2, 3, 4\}$. Suppose the game in Table 5.1. The a priori union system $\mathscr{B} = \{B_1 = \{1, 2, 3\}, B_2 = \{4\}\}$ (Fig. 5.2) represents for instance the fact that players 1, 2 and 3 are relatives. In this case $M = \{1, 2\}$ and then we have two union games. First we determine $v_1^{\mathscr{B}}$. If we take coalition $S = \{1\} \subset B_1$ then $\mathscr{B}_S = \{\{1\}, \{4\}\}$. We construct the quotient game $w^{v,\mathscr{B}_S}$ over M,

$$w^{v,\mathscr{B}_S}(\{1\}) = v(\{1\}) = 0,\ w^{v,\mathscr{B}_S}(\{2\}) = v(B_2) = 0,$$

$$w^{v,\mathscr{B}_S}(\{1, 2\}) = v(\{1, 4\}) = 5.$$

So, $\phi^M(w^{v,\mathscr{B}_S}) = (2.5, 2.5)$ and the worth $v_1^{\mathscr{B}}(\{1\}) = 2.5$. In the same way,

$$v_1^{\mathscr{B}}(\{2\}) = v_1^{\mathscr{B}}(\{3\}) = v_1^{\mathscr{B}}(\{1, 3\}) = v_1^{\mathscr{B}}(\{2, 3\}) = 2.5.$$

If $S = \{1, 2\}$ then $\mathscr{B}_S = \{\{1, 2\}, \{4\}\}$ and

$$w^{v,\mathscr{B}_S}(\{1\}) = v(\{1, 2\}) = 1,\ w^{v,\mathscr{B}_S}(\{2\}) = v(B_2) = 0,$$

$$w^{v,\mathscr{B}_S}(\{1, 2\}) = v(\{1, 2, 4\}) = 5.$$

Table 5.1 Game v in Example 5.2

S	$\{1\}$	$\{2\}$	$\{3\}$	$\{4\}$	$\{1,2\}$	$\{1,3\}$	$\{1,4\}$	$\{2,3\}$
v	0	0	0	0	1	0	5	0
S	$\{2,4\}$	$\{3,4\}$	$\{1,2,3\}$	$\{1,2,4\}$	$\{1,3,4\}$	$\{2,3,4\}$	N	
v	5	5	2	5	5	5	8	

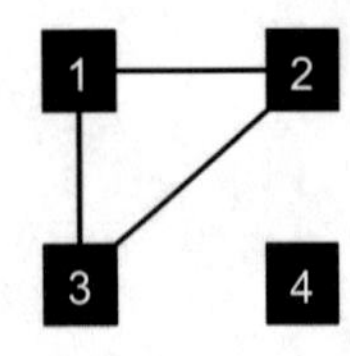

Fig. 5.2 A priori union system in Example 5.2

We get now $\phi^M(w^{v,\mathscr{B}_S}) = (3, 2)$ and the worth $v_1^{\mathscr{B}}(\{1, 2\}) = 3$. If $S = B_1$ then $\mathscr{B}_S = \mathscr{B}$ and

$$w^{v,\mathscr{B}_S}(\{1\}) = v(B_1) = 2, \ w^{v,\mathscr{B}_S}(\{2\}) = v(B_2) = 0,$$

$$w^{v,\mathscr{B}_S}(\{1, 2\}) = v(N) = 8.$$

The new Shapley value is $\phi^M(w^{v,\mathscr{B}_S}) = (5, 3)$ and so $v_1^{\mathscr{B}}(B_1) = 5$. Observe that we also get $v_2^{\mathscr{B}}(B_2) = 3$ from the before game.

We analyze in the next proposition several properties of the union games. Given an a priori union system $\mathscr{B} = \{B_1, \ldots, B_m\}$ we will use for each non-empty $Q \subseteq M = \{1, \ldots, m\}$

$$B_Q = \bigcup_{q \in Q} B_q. \tag{5.2}$$

Proposition 5.1 *Let $\mathscr{B} = \{B_1, \ldots, B_m\} \in P^N$, $M = \{1, \ldots, m\}$ and $p \in M$. For all $v \in \mathscr{G}^N$ the p-union game satisfies*

(1) If v is additive then $v_p^{\mathscr{B}}$ is additive, moreover $v_p^{\mathscr{B}} = v_{B_p}$.
(2) If v is convex then so is $v_p^{\mathscr{B}}$.
(3) If v is monotone then $v_p^{\mathscr{B}}$ is monotone.
(4) $(av + bw)_p^{\mathscr{B}} = av_p^{\mathscr{B}} + bw_p^{\mathscr{B}}$ for all $a, b \in \mathbb{R}$. Hence $(-v)_p^{\mathscr{B}} = -(v_p^{\mathscr{B}})$.
(5) If v, w are strategically equivalent then so are $v_p^{\mathscr{B}}, w_p^{\mathscr{B}}$.
(6) $(v^{svg})_p^{\mathscr{B}} = (v_p^{\mathscr{B}})^{svg}$.
(7) $(v^{dual})_p^{\mathscr{B}} = (v_p^{\mathscr{B}})^{dual}$.

Proof (1) Let $v \in \mathbb{R}^N$. For each $S \subseteq B_p$ we define the vector $w^S \in \mathbb{R}^M$ with

$$w_q^S = \begin{cases} \sum_{i \in B_q} v_i, & \text{if } q \neq p \\ \sum_{i \in S} v_i, & \text{if } q = p. \end{cases}$$

If $Q \subseteq M$ with $p \notin Q$ then

$$w^{v,\mathscr{B}_S}(Q) = v\left(B_Q\right) = \sum_{i \in B_Q} v_i = \sum_{q \in Q} \sum_{i \in B_q} v_i = \sum_{q \in Q} w_q^S.$$

Following the same way, if $p \in Q$ then

$$w^{v,\mathscr{B}_S}(Q) = v\left(B_{Q \setminus \{p\}} \cup S\right) = \sum_{q \in Q \setminus \{p\}} \sum_{i \in B_q} v_i + \sum_{i \in S} v_i = \sum_{q \in Q} w_q^S.$$

Hence, we obtain that $w^{v,\mathscr{B}_S} = w^S$. Example 1.16 showed that the Shapley value of a vector is the own vector, namely

$$v_p^{\mathscr{B}}(S) = \phi_p^M(w^{v,\mathscr{B}_S}) = w_p^{\mathscr{B}_S} = \sum_{i \in S} v_i.$$

(2) Let $S, T \subseteq B_p$. Consider $Q \subseteq M \setminus \{p\}$. Convexity of v implies

$$v\left(B_Q \cup (S \cup T)\right) + v\left(B_Q \cup (S \cap T)\right) \geq v\left(B_Q \cup S\right) + v\left(B_Q \cup T\right).$$

So, from Definition 5.2 we have

$$\left(w^{v,\mathscr{B}_{S\cup T}} + w^{v,\mathscr{B}_{S\cap T}}\right)(Q \cup \{p\}) \geq \left(w^{v,\mathscr{B}_S} + w^{v,\mathscr{B}_T}\right)(Q \cup \{p\}).$$

Furthermore

$$\left(w^{v,\mathscr{B}_{S\cup T}} + w^{v,\mathscr{B}_{S\cap T}}\right)(Q) = \left(w^{v,\mathscr{B}_S} + w^{v,\mathscr{B}_T}\right)(Q),$$

thus we obtain

$$\left(w^{v,\mathscr{B}_{S\cup T}} + w^{v,\mathscr{B}_{S\cap T}}\right)(Q \cup \{p\}) - \left(w^{v,\mathscr{B}_{S\cup T}} + w^{v,\mathscr{B}_{S\cap T}}\right)(Q) \geq$$
$$\left(w^{v,\mathscr{B}_S} + w^{v,\mathscr{B}_T}\right)(Q \cup \{p\}) - \left(w^{v,\mathscr{B}_S} + w^{v,\mathscr{B}_T}\right)(Q).$$

We apply that the Shapley value satisfies marginality (Proposition 1.12), and then

$$\phi_p^M\left(w^{v,\mathscr{B}_{S\cup T}} + w^{v,\mathscr{B}_{S\cap T}}\right) \geq \phi_p^M\left(w^{v,\mathscr{B}_S} + w^{v,\mathscr{B}_T}\right).$$

Linearity of the Shapley value (Proposition 1.8) and Definition 5.3 conclude that

$$v_p^{\mathscr{B}}(S \cup T) + v_p^{\mathscr{B}}(S \cap T) \geq v_p^{\mathscr{B}}(S) + v_p^{\mathscr{B}}(T).$$

(3) Let $S \subseteq T \subseteq B_p$. Consider $Q \subseteq M \setminus \{p\}$ and notation B_Q again. Monotonicity of v implies

$$v\left(B_Q \cup T\right) \geq v\left(B_Q \cup S\right).$$

We obtain then

$$w^{v,\mathscr{B}_T}(Q \cup \{p\}) - w^{v,\mathscr{B}_T}(Q) \geq w^{v,\mathscr{B}_S}(Q \cup \{p\}) - w^{v,\mathscr{B}_S}(Q).$$

Marginality of the Shapley value (Proposition 1.12) say that

$$v_p^{\mathscr{B}}(T) = \phi_p^M\left(w^{v,\mathscr{B}_T}\right) \geq \phi_p^M\left(w^{v,\mathscr{B}_S}\right) = v_p^{\mathscr{B}}(S).$$

(4) For each coalition $S \subseteq N$ and $Q \subseteq M$

$$w^{av+bw,\mathscr{B}_S}(Q) = \left\{ \begin{array}{ll} (av+bw)(B_{Q\setminus\{p\}} \cup S) & \text{if } p \in Q \\ (av+bw)(B_Q) & \text{if } p \notin Q \end{array} \right\} = aw^{v,\mathscr{B}_S}(Q) + bw^{w,\mathscr{B}_S}(Q)$$

Linearity of Shapley value implies again

$$(av+bw)_p^{\mathscr{B}}(S) = \phi_p^M\left(w^{av+bw,\mathscr{B}_S}\right) = av_p^{\mathscr{B}}(S) + bw_p^{\mathscr{B}}(S).$$

(5) It is a consequence of steps (4) and (1).
(6) Let $v' \in \mathbb{R}^N$ with $v'_i = v(\{i\})$. We have $v^{svg} = v' - v$. For each coalition $S \subseteq N$ and $Q \subseteq M$

$$w^{v^{svg},\mathscr{B}_S}(Q) = \left\{ \begin{array}{ll} v^{svg}(B_{Q\setminus\{p\}} \cup S) & \text{if } p \in Q \\ v^{svg}(B_Q) & \text{if } p \notin Q \end{array} \right\} = w^{v'-v,\mathscr{B}_S}(Q).$$

So, by steps (4) and (1)

$$(v^{svg})_p^{\mathscr{B}}(S) = \phi_p(w^{v^{svg},\mathscr{B}_S}) = (v'-v)_p^{\mathscr{B}}(S) = \sum_{i\in S} v(\{i\}) - v_p^{\mathscr{B}}(S) = (v_p^{\mathscr{B}})^{svg}.$$

(7) First we calculate $w^{v^{dual},\mathscr{B}_S}$ for each coalition $S \subseteq B_p$. Let $Q \subseteq M \setminus \{p\}$,

$$w^{v^{dual},\mathscr{B}_S}(Q) = v(N) - v(N \setminus B_Q) = v(N) - v\left(B_{M\setminus(Q\cup\{p\})} \cup B_p\right)$$

$$w^{v^{dual},\mathscr{B}_S}(Q \cup \{p\}) = v(N) - v(N \setminus (B_Q \cup S)) = v(N) - v\left(B_{M\setminus(Q\cup\{p\})} \cup (B_p \setminus S)\right).$$

We denote $Q' = M \setminus (Q \cup \{p\})$, so

$$\begin{aligned} w^{v^{dual},\mathscr{B}_S}(Q \cup \{p\}) - w^{v^{dual},\mathscr{B}_S}(Q) &= v\left(B_{Q'} \cup B_p\right) - v\left(B_{Q'} \cup (B_p \setminus S)\right) \\ &= w^{v,\mathscr{B}}(Q' \cup \{p\}) - w^{v,\mathscr{B}_{B_p\setminus S}}(Q' \cup \{p\}). \end{aligned}$$

Using that for all $R \subseteq B_p$ the number $w^{v,\mathscr{B}_R}(Q')$ is the same we obtain

$$\begin{aligned} w^{v^{dual},\mathscr{B}_S}(Q \cup \{p\}) &- w^{v^{dual},\mathscr{B}_S}(Q) \\ &= \left[w^{v,\mathscr{B}}(Q' \cup \{p\}) - w^{v,\mathscr{B}}(Q')\right] \\ &\quad - \left[w^{v,\mathscr{B}_{B_p\setminus S}}(Q' \cup \{p\}) - w^{v,\mathscr{B}_{B_p\setminus S}}(Q')\right] \end{aligned}$$

Hence, from Theorem 1.1 we get

$$\begin{aligned} (v^{dual})_p^{\mathscr{B}}(S) &= \phi_p^M(w^{v^{dual},\mathscr{B}_S}) = \phi_p^M(w^{v,\mathscr{B}}) - \phi_p^M(w^{v,\mathscr{B}_{B_p\setminus S}}) \\ &= v_p^{\mathscr{B}}(B_p) - v_p^{\mathscr{B}}(B_p \setminus S) = (v_p^{\mathscr{B}})^{dual}(S). \end{aligned}$$

□

But superadditivity is not inherited for a priori union systems as next example given by Owen [14] shows.

Example 5.3 Consider the superadditive simple game given by the voting situation $v = [4; 3, 1, 1, 1]$ (see Example 1.14) with the a priori union system $\mathscr{B} = \{\{1\}, \{2\}, \{3, 4\}\} = \{B_1, B_2, B_3\}$. We will calculate $v_3^{\mathscr{B}} \in \mathscr{G}^{\{3,4\}}$. For coalition $\{3\}$ we have $w^{v,\mathscr{B}_{\{3\}}} = [4; 3, 1, 1]$, namely it is also a voting game with the same quota and the three unions $\mathscr{B}_{\{3\}} = \{\{1\}, \{2\}, \{3\}\}$. Then $v_3^{\mathscr{B}}(\{3\}) = \phi_3(w^{v,\mathscr{B}_{\{3\}}}) = 1/6$. Also $w^{v,\mathscr{B}_{\{4\}}} = [4; 3, 1, 1]$ with $\mathscr{B}_{\{4\}} = \{\{1\}, \{2\}, \{4\}\}$ and therefore $v_3^{\mathscr{B}}(\{4\}) = \phi_3(w^{v,\mathscr{B}_{\{4\}}}) = 1/6$. For coalition $\{3, 4\}$ we get $w^{v,\mathscr{B}_{\{3,4\}}} = [4; 3, 1, 2]$, namely it is also a voting game with the same quota and the three unions $\mathscr{B}_{\{3,4\}} = \mathscr{B}$. Then $v_3^{\mathscr{B}}(\{3, 4\}) = \phi_3(w^{v,\mathscr{B}_{\{3,4\}}}) = 1/6$. But then game $v_3^{\mathscr{B}}$ is not superadditive.

For communication situations convexity was a non-inherited property (Example 4.6) and it was proposed a particular kind of communication structures where this condition is inherited, to be cycle-complete. But this option is not feasible here if we observe the simple a priori union system used in the above example. Moreover, except for very particular cases, there exists a superadditive game with some non-superadditive union game.

Union games of simple games are not simple games although they were superadditive (that did not happen with coalition structures). Next example shows this circumstance with an unanimity game.

Example 5.4 Suppose u_R an unanimity game with $R \subseteq N$ and $\mathscr{B} = \{B_1, B_2\}$ such that $R \cap B_1 \neq \emptyset$ and $R \cap B_2 \neq \emptyset$. Let $S \subseteq B_1$, we get $w^{u_R,\mathscr{B}_S}(\{1\}) = u_R(S) = 0$, $w^{u_R,\mathscr{B}_S}(\{2\}) = u_R(B_2) = 0$ and

$$w^{u_R,\mathscr{B}_S}(\{1, 2\}) = u_R(S \cup B_2) = \begin{cases} 1, & \text{if } B_1 \cap R \subseteq S \\ 0, & \text{otherwise.} \end{cases}$$

Hence, we obtain

$$(u_R)_1^{\mathscr{B}}(S) = \phi_1(w^{u_R,\mathscr{B}_S}) = \begin{cases} 1/2, & \text{if } B_1 \cap R \subseteq S \\ 0, & \text{otherwise.} \end{cases}$$

A value function for games with a priori union system is

$$f : \mathscr{G}^N \times P^N \to \mathbb{R}^N.$$

Each pair $(v, \mathscr{B}) \in \mathscr{G}^N \times P^N$ is called a *game over N with a priori union system*. We define a Shapley value for games with an a priori union system using the Shapley value of the modification of each union game.

Definition 5.4 The Owen value is a value for games over N with a priori union system defined for each $v \in \mathcal{G}^N$, $\mathcal{B} \in P^N$ and $i \in N$ as

$$\omega_i(v, \mathcal{B}) = \phi_i(v_p^{\mathcal{B}}),$$

where $i \in B_p$ and $\mathcal{B} = \{B_1, \ldots, B_m\}$.

Example 5.5 Following from Example 5.2 we need to calculate the Shapley value of both p-union games obtained there. So, as all the players are symmetric in each game,

$$\phi(v_1^{\mathcal{B}}) = (1.75, 1.75, 1.5), \quad \phi(v_2^{\mathcal{B}}) = 3.$$

The Owen value is

$$\omega(v, \mathcal{B}) = (1.75, 1.75, 1.75, 3).$$

We can observe that players 1, 2 and 3 increase the profit with their a priori union, because the Shapley value for game v is $\phi(v) = (1.416, 1.416, 1.25, 3.91)$. But actually, this fact may not always be the case, remember that the unions are established a priori for reasons out of characteristic function. In [3] the reader can find how to use a priori unions in a context of coalition formation.

The Owen value has been modified using the same two-steps model, as the reader can see in [11] or [4].

The classical situation in this case it happens when $\mathcal{B} = \{\{i\} : i \in N\}$ because there is not any a priori relation, but also $\mathcal{B} = \{N\}$ can be interpreted in this sense. Next proposition shows that the Owen value coincides with the Shapley value in both cases.

Proposition 5.2 *Let $v \in \mathcal{G}^N$. If $\mathcal{B}_0 = \{\{i\} : i \in N\}$ and $\mathcal{B}_1 = \{N\}$ then the Owen value satisfies*

$$\omega(v, \mathcal{B}_0) = \omega(v, \mathcal{B}_1) = \phi(v).$$

Proof If we take $\mathcal{B}_0$ then each union is identified with a player. We get $(\mathcal{B}_0)_S = \mathcal{B}_0$ for all non-empty coalition S, hence $w^{v,\mathcal{B}_0} = v$. So $v_i^{\mathcal{B}_0} \in \mathcal{G}^{\{i\}}$ with $v_i^{\mathcal{B}_0}(\{i\}) = \phi_i(v)$. If we take $\mathcal{B}_1$ then there is only one union N, so $(\mathcal{B}_1)_S = \{S\}$ for all non-empty coalition. Game $w^{v,(\mathcal{B}_1)_S} \in \mathcal{G}^{\{1\}}$ and $w^{v,(\mathcal{B}_1)_S}(1) = v(S)$. Now $v_1^{\mathcal{B}_1} = v$, thus $\omega(v, \mathcal{B}_1) = \phi(v)$. □

Owen [13] axiomatized his value using one axiom more than the Shapley value. Other axiomatizations were given in [8, 9] or [1]. Let $f : \mathcal{G}^N \times P^N \to \mathbb{R}^N$ a value for games over N with a priori union system.

In an a priori union system the great coalition is formed because the coalition structure is only a priori.

Efficiency. For all $v \in \mathcal{G}^N$ and $\mathcal{B} \in P^N$, $f(v, \mathcal{B})(N) = v(N)$.

Null player. Let $i \in N$ be a null player in the game v, it holds $f_i(v, \mathcal{B}) = 0$ for all $\mathcal{B} \in P^N$.

Symmetry in unions. Let $\mathcal{B} \in P^N$ and $B \in \mathcal{B}$. If $i, j \in B$ with are symmetric for a game v then $f_i(v, \mathcal{B}) = f_j(v, \mathcal{B})$.

We have modified the next axiom from the original of Owen in order to be described without using the quotient game. Let $(v, \mathcal{B}) \in \mathcal{G}^N \times P^N$. Two unions $B, B' \in \mathcal{B}$ are *symmetric* for the game v if for all $S \subseteq N \setminus (B \cup B')$ we have

$$v(S \cup B) = v(S \cup B').$$

Symmetry in the quotient. Let v be a game and $\mathcal{B}$ be an a priori union system. If $B, B' \in \mathcal{B}$ are symmetric for v then

$$f(v, \mathcal{B})(B) = f(v, \mathcal{B})(B').$$

Linearity. For all $v, w \in \mathcal{G}^N$, $a, b \in \mathbb{R}$ and coalition T it holds

$$f(av + bw, T) = af(v, T) + bf(w, T).$$

We follow the proof of Owen [13] but adapting it to the new axiom.

Theorem 5.1 *The Owen value is the only value for games over N with a priori union system satisfying efficiency, null player, symmetry in unions, symmetry in the quotient and linearity.*

Proof We test that the Owen value verifies the axioms. Let $\mathcal{B} = \{B_1, \ldots, B_m\} \in P^N$ and $M = \{1, \ldots, m\}$. Let also $v \in \mathcal{G}^N$.
EFFICIENCY. Efficiency of the Shapley value (Proposition 1.9) implies

$$\sum_{i \in N} \omega_i(v, \mathcal{B}) = \sum_{p=1}^{m} \sum_{i \in B_p} \phi_i(v_p^{\mathcal{B}}) = \sum_{p=1}^{m} v_p^{\mathcal{B}}(B_p) = \sum_{p=1}^{m} \phi_p\left(w^{v,\mathcal{B}}\right)$$

$$= w^{v,\mathcal{B}}(M) = v\left(\bigcup_{p \in M} B_p\right) = v(N).$$

NULL PLAYER. Let $i \in N$ be a null player for a game v. We consider $i \in B_p$. Player i is also a null player of $v_p^{\mathscr{B}}$. If $S \subseteq B_p \setminus \{i\}$ and $Q \subseteq M \setminus \{p\}$ then as i is null for v we obtain

$$w^{v,\mathscr{B}_{S\cup\{i\}}}(Q \cup p) = v\left(B_Q \cup S \cup \{i\}\right) = v\left(B_Q \cup S\right) = w^{v,\mathscr{B}_S}(Q \cup p).$$

Always it is true that $w^{v,\mathscr{B}_{S\cup\{i\}}}(Q) = w^{v,\mathscr{B}_S}(Q)$. If all the marginal contributions of player p are the same for both games $w^{v,\mathscr{B}_{S\cup\{i\}}}$, $w^{v,\mathscr{B}_S}$ then Theorem 1.1 says that the Shapley values are also the same, thus

$$v_p^{\mathscr{B}}(S \cup \{i\}) = \phi_p\left(w^{v,\mathscr{B}_{S\cup\{i\}}}\right) = \phi_p\left(w^{v,\mathscr{B}_S}\right) = v_p^{\mathscr{B}}(S).$$

By Definition 5.4 and Proposition 1.11 it holds $\omega_i(v, \mathscr{B}) = \phi_i(v_p^{\mathscr{B}}) = 0$.
SYMMETRY IN UNIONS. Let $i, j \in B_p$ be symmetric players with $p \in M$. We will prove the claim i, j are symmetric for game $v_p^{\mathscr{B}}$. We take $Q \subseteq M \setminus \{p\}$, which verifies $w^{v,\mathscr{B}_{S\cup\{i\}}}(Q) = w^{v,\mathscr{B}_{S\cup\{j\}}}(Q)$ for all $S \subseteq N \setminus \{i, j\}$, and as i, j are symmetric for v

$$w^{v,\mathscr{B}_{S\cup\{i\}}}(Q \cup \{p\}) = v(B_Q \cup S \cup \{j\}) = v(B_Q \cup S \cup \{i\}) = w^{v,\mathscr{B}_{S\cup\{j\}}}(Q).$$

Theorem 1.1 again implies that

$$v_p^{\mathscr{B}}(S \cup \{i\}) = \phi_p\left(w^{v,\mathscr{B}_{S\cup\{i\}}}\right) = \phi_p\left(w^{v,\mathscr{B}_{S\cup\{j\}}}\right) = v_p^{\mathscr{B}}(S \cup \{j\}).$$

As Shapley value satisfies symmetry (Proposition 1.10) it holds $\omega_i(v, \mathscr{B}) = \omega_j(v, \mathscr{B})$.
SYMMETRY IN THE QUOTIENT. Suppose $p \neq q \in M$ with B_p, B_q symmetric for v. We test that p, q are symmetric for game $w^{v,\mathscr{B}}$. If $Q \subseteq M \setminus \{p, q\}$ then

$$w^{v,\mathscr{B}}(Q \cup \{p\}) = v(B_Q \cup B_p) = v(B_Q \cup B_q) = w^{v,\mathscr{B}}(Q \cup \{q\}).$$

Therefore from efficiency and symmetry of Shapley value,

$$\begin{aligned}\sum_{i\in B_p} \omega_i(v, \mathscr{B}) &= \sum_{i\in B_p} \phi_i(v_p^{\mathscr{B}}) = v_p^{\mathscr{B}}(B_p) = \phi_p(w^{v,\mathscr{B}}) \\ &= \phi_q(w^{v,\mathscr{B}}) = v_p^{\mathscr{B}}(B_q) = \sum_{i\in B_p} \phi_i(v_p^{\mathscr{B}}) = \sum_{i\in B_p} \omega_i(v, \mathscr{B}).\end{aligned}$$

LINEARITY. Proposition 5.1 implies the result.

Consider f a value for games with a priori union structure satisfying the axioms. From linearity we only have to obtain the uniqueness for the unanimity games. Let $T \subseteq N$ be a non-empty coalition. Players out of T are null for u_T thus their payoffs are zero. We set $C = \{p \in M : B_p \cap T \neq \emptyset\}$. For all $p, q \in C$ we have that B_p, B_q

are symmetric for u_T because both are required to complete T. Symmetry for the quotient (and also null player axiom) says that there exists $A \in \mathbb{R}$ with

$$\sum_{i \in B_p \cap T} f_i(u_T, \mathcal{B}) = A$$

for all $p \in C$. Efficiency implies

$$\sum_{i \in N} f_i(u_T, \mathcal{B}) = \sum_{p \in C} \sum_{i \in B_p \cap T} f_i(u_T, \mathcal{B}) = |C|A = 1.$$

Hence $A = 1/|C|$. Two players $i, j \in B_p \cap T$ with $p \in C$ are symmetric for v because both are necessary to complete T, therefore there is $K_p \in \mathbb{R}$ with $f_i(u_T, \mathcal{B}) = K_p$ for all $i \in B_p$. So

$$\sum_{i \in B_p \cap T} f_i(u_T, \mathcal{B}) = |B_p \cap T| K_p = A = \frac{1}{|C|},$$

and

$$K_p = \frac{1}{|C||B_p \cap T|}.$$

□

We have needed five axioms to characterize the Owen value, we must analyze the independence of the axioms.

Remark 5.1 We find values different to the Owen value verifying all the axioms except one of them.

- Consider value f^1 defined for each $v \in \mathcal{G}^N$ and $\mathcal{B} \in P^N$ as $f^1(v, \mathcal{B}) = a\omega(v, \mathcal{B})$ with $a \in \mathbb{R} \setminus \{1\}$. This value satisfies all the axioms except efficiency.
- Let $\theta \in \Theta^N$ and suppose $n \geq 2$. In Remark 1.4 we proved that the marginal function m^θ is a value for games satisfying efficiency, null player and linearity. Now we follow the two steps model to generate a new value. Let $\mathcal{B} = \{B_1, \ldots, B_m\}$, $M = \{1, \ldots, m\}$ and $p \in M$. We denote θ_p the induced permutation in B_p by θ ($i <_{\theta_p} j$ if and only if $i <_\theta j$ for all $i, j \in B_p$). For each player $i \in B_p$, we take

$$f_i^2(v, \mathcal{B}) = m_i^{\theta_p}(v_p^{\mathcal{B}}).$$

  Following the proof of Theorem 5.1, as ϕ, m^{θ_p} verifies efficiency, null player and additivity then f^2 so is. Value f^2 satisfies symmetry in the quotient because ϕ verifies symmetry and m^{θ_p} is efficient (see again the proof of Theorem 5.1). As m^{θ_p} does not verify symmetry then f^2 does not verify symmetry in the unions.
- Similar to the egalitarian value (see Remark 1.4) is defined the egalitarian value into the unions as

$$f_i^3(v, \mathcal{B}) = \frac{v(N)}{mb_p},$$

where m is the number of unions and $i \in B_p$ with $|B_p| = b_p$. This value satisfies all the axioms except the null player property.

- Following Remark 1.4 again we take $Null(v) = \{i \in N : i \text{ null player in } v\}$. Let $m' = |\{B \in \mathcal{B} : B \cap (N \setminus Null(v)) \neq \emptyset\}|$. If $i \in B_p$, then $b'_p = |\{B_p \setminus Null(v)\}|$. It defines

$$f_i^4(v) = \begin{cases} \dfrac{v(N)}{m'b'_p}, & \text{if } i \notin Null(v) \\ 0, & \text{if } i \in Null(v). \end{cases}$$

This value satisfies all the axioms except additivity.

- Finally if we consider the Shapley value, $f^5(v, \mathcal{B}) = \phi(v)$, this value satisfies all the axioms except symmetry in the quotient.

The Owen value admits similar formulas than the Shapley value. A permutation $\theta \in \Theta^N$ is *compatible* with an a priori union system $\mathcal{B} = \{B_1, \ldots, B_m\}$, $M = \{1, \ldots, m\}$ if $\theta = (\theta^{\pi(1)}, \ldots, \theta^{\pi(m)})$ with $\theta^{\pi(p)} \in \Theta^{B_{\pi(p)}}$ and $\pi \in \Theta^M$, namely those permutations θ such that for each $p \in M$ and for all $i, i' \in B_p$ there is no $j \notin B_p$ with $i <_\theta j <_\theta i'$. The family of compatible permutations with $\mathcal{B}$ is $\Theta^N_{\mathcal{B}}$.

Theorem 5.2 *Let $\mathcal{B} = \{B_1, \ldots, B_m\} \in P^N$, $M = \{1, \ldots, m\}$ and $v \in \mathcal{G}^N$. It holds:*

(1) $\omega(v, \mathcal{B}) = \dfrac{1}{|\Theta^N_{\mathcal{B}}|} \displaystyle\sum_{\theta \in \Theta^N_{\mathcal{B}}} m^\theta(v).$

(2) For each $i \in B_p$,

$$\omega_i(v, \mathcal{B}) = \sum_{Q \subseteq M \setminus \{p\}} \sum_{S \subseteq B_p \setminus \{i\}} c^m_{|Q|} c^{B_p}_{|S|} [v(B_Q \cup S \cup \{i\}) - v(B_Q \cup S)].$$

Proof (1) If we denote $b_q = |B_q|$ for all $q \in M$ then the number of compatible permutations for $\mathcal{B}$ is

$$|\Theta^N_{\mathcal{B}}| = \prod_{q \in M} b_q!.$$

Let $i \in B_p \in \mathcal{B}$, by Definitions 1.9, 1.10, 5.3 and 5.4 we obtain

$$\omega_i(v, \mathscr{B}) = \phi_i^{B_p}(v_p^{\mathscr{B}}) = \frac{1}{b_p!}\sum_{\theta^p \in \Theta^{B_p}} m_i^{\theta^p}(v_p^{\mathscr{B}})$$
$$= \frac{1}{b_p!}\sum_{\theta^p \in \Theta^{B_p}} \left[v_p^{\mathscr{B}}(S_{\theta^p}^i \cup \{i\}) - v_p^{\mathscr{B}}(S_{\theta^p}^i)\right]$$
$$= \frac{1}{b_p!}\sum_{\theta^p \in \Theta^{B_p}} \left[\phi_p^M\left(w^{v,\mathscr{B}_{S_\theta^i \cup \{i\}}}\right) - \phi_p^M\left(w^{v,\mathscr{B}_{S_\theta^i}}\right)\right]$$
$$= \frac{1}{b_p!}\sum_{\theta^p \in \Theta^{B_p}} \frac{1}{m!}\sum_{\pi \in \Theta^M}\left[m_p^\pi\left(w^{v,\mathscr{B}_{S_\theta^i \cup \{i\}}}\right) - m_p^\pi\left(w^{v,\mathscr{B}_{S_\theta^i}}\right)\right].$$

The difference between games $w^{v,\mathscr{B}_{S_\theta^i \cup \{i\}}}$ and $w^{v,\mathscr{B}_{S_\theta^i}}$ is their application over sets containing p, therefore they obtain the same worths for Q. So,

$$m_p^\pi\left(w^{v,\mathscr{B}_{S_\theta^i \cup \{i\}}}\right) - m_p^\pi\left(w^{v,\mathscr{B}_{S_\theta^i}}\right) = w^{v,\mathscr{B}_{S_\theta^i \cup \{i\}}}(Q_\pi^p \cup \{p\}) - w^{v,\mathscr{B}_{S_\theta^i \cup \{i\}}}(Q_\pi^p)$$
$$-w^{v,\mathscr{B}_{S_\theta^i}}(Q_\pi^p \cup \{p\}) + w^{v,\mathscr{B}_{S_\theta^i}}(Q_\pi^p)$$
$$= v\left(B_{Q_\pi^p} \cup S_\theta^i \cup \{i\}\right) - v\left(B_{Q_\pi^p} \cup S_\theta^i\right).$$

For each θ^p and π fixed, we consider all the compatible permutations θ for $\mathscr{B}$ ordering M by π and using θ_p into B_p. We get $\prod_{q \in M\setminus\{p\}} b_q!$ of these permutations such that

$$m_i^\theta(v) = v\left(B_{Q_\pi^p} \cup S_\theta^i \cup \{i\}\right) - v\left(B_{Q_\pi^p} \cup S_\theta^i\right).$$

Thus we have,

$$\omega_i(v, \mathscr{B}) = \frac{1}{m!\prod_{q \in M} b_q!}\sum_{\theta^p \in \Theta^{B_p}}\left(\prod_{q \in M\setminus\{p\}} b_q!\right)\sum_{\pi \in \Theta^M} m_i^\theta(v)$$
$$= \frac{1}{|\Theta_{\mathscr{B}}^N|}\sum_{\theta \in \Theta_{\mathscr{B}}^N} m_i^\theta(v).$$

(2) Following the above proof, we have

$$\omega_i(v, \mathscr{B}) = \frac{1}{m!}\frac{1}{b_p!}\sum_{\theta^p \in \Theta^{B_p}}\sum_{\pi \in \Theta^M} v\left(B_{Q_\pi^p} \cup S_\theta^i \cup \{i\}\right) - v\left(B_{Q_\pi^p} \cup S_\theta^i\right)$$

As in the proof of Theorem 1.1, for each $Q \subseteq M \setminus \{p\}$ with cardinality $|Q| = q$ and $S \subseteq B_p \setminus \{i\}$ with cardinality $|S| = s$ we have $q!(m - q - 1)!$ permutations of M and $s!(b_p - s - 1)!$ of B_p such that the combination of which obtain always the same marginal contribution, $v\left(B_Q \cup S \cup \{i\}\right) - v\left(B_Q \cup S\right)$. Hence, using again Theorem 1.1

$$\omega_i(v,\mathscr{B}) = \sum_{Q\subseteq M\setminus\{p\}} \sum_{S\subseteq B_p\setminus\{i\}} \frac{q!(m-q-1)!}{m!}\frac{s!(b_p-s-1)!}{b_p!} v\left(B_Q\cup S\cup\{i\}\right) - v\left(B_Q\cup S\right)$$
$$= \sum_{Q\subseteq M\setminus\{p\}} \sum_{S\subseteq B_p\setminus\{i\}} c_q^m c_s^{b_p} v\left(B_Q\cup S\cup\{i\}\right) - v\left(B_Q\cup S\right).$$

□

Next we summarize other properties of the Owen value.

Proposition 5.3 *The Owen value satisfies the following properties for an a priori union system $\mathscr{B}=\{B_1,\ldots,B_m\}$. Let $M=\{1,\ldots,m\}$ and $p\in M$.*

(1) If $i\in B_p$ is a necessary player for $v\in\mathscr{G}_m^N$ then $\omega_j(v,\mathscr{B})\le\omega_i(v,\mathscr{B})$ for all $j\in B_p$.
(2) Let $v\in\mathscr{G}_c^N$. If $S\subseteq B_p$ then $\omega(v,\mathscr{B})(S)\ge v(S)$.
(3) If $v,v'\in\mathscr{G}^N$ and $a,b\in\mathbb{R}$ then $\omega(av+bv',\mathscr{B})=a\omega(v,\mathscr{B})+b\omega(v',\mathscr{B})$.
(4) For all $i\in B_p$ and $v\in\mathscr{G}^N$ it holds $\omega_i(v,\mathscr{B})=v_p^{\mathscr{B}}(\{i\})-\omega_i(v^{svg},\mathscr{B})$.
(5) $\omega(v^{dual},\mathscr{B})=\omega(v,\mathscr{B})$.

Proof (1) We see that i is a necessary player for $v_p^{\mathscr{B}}$. If $S\subseteq B_p$ with $i\notin S$ then $w^{v,\mathscr{B}_S}=0$. Hence, the p-union game verifies

$$v_p^{\mathscr{B}}(S)=\phi_p^M(w^{v,\mathscr{B}_S})=0.$$

Proposition 5.1 implies that $v_p^{\mathscr{B}}\in\mathscr{G}_m^{B_p}$. As Shapley value satisfies necessary player property (Proposition 1.11) we get the result.
(2) Suppose $S\subseteq B_p$. Since Proposition 5.1 we know that $v_p^{\mathscr{B}}$ is convex, thus Proposition 1.14 says

$$\omega(v,\mathscr{B})(S)=\phi(v_p^{\mathscr{B}})(S)\ge v_p^{\mathscr{B}}(S)=\phi_p^M(w^{v,\mathscr{B}_S}).$$

We will prove that $w^{v,\mathscr{B}_S}$ is superadditive using $v\in\mathscr{G}_c^N$. Let $Q,Q'\subseteq M$ with $Q\cap Q'=\emptyset$. If $p\notin Q\cup Q'$

$$w^{v,\mathscr{B}_S}(Q\cup Q')=v(B_Q\cup B_{Q'})\ge v(B_Q)+v(B_{Q'})=w^{v,\mathscr{B}_S}(Q)+w^{v,\mathscr{B}_S}(Q').$$

If $p\in Q$ (or Q') then

$$w^{v,\mathscr{B}_S}(Q\cup Q')=v(B_{Q\setminus\{p\}}\cup B_{Q'}\cup S)\ge v(B_{Q\setminus\{p\}}\cup S)+v(B_{Q'})$$
$$=w^{v,\mathscr{B}_S}(Q)+w^{v,\mathscr{B}_S}(Q').$$

So, Proposition 1.13

$$\omega(v, \mathscr{B})(S) \geq w^{v,\mathscr{B}_S}(\{p\}) = v(S).$$

(3) It is trivial from the linearity of the Shapley value.
(4) Propositions 5.1 and 1.15 imply

$$\begin{aligned}\omega_i(v^{svg}, \mathscr{B}) &= \phi_i^{B_p}((v^{svg})_p^{\mathscr{B}}) = \phi_i^{B_p}((v_p^{\mathscr{B}})^{svg}) \\ &= v_p^{\mathscr{B}}(\{i\}) - \phi_i^{B_p}(v_p^{\mathscr{B}}) = v_p^{\mathscr{B}}(\{i\}) - \omega_i(v, \mathscr{B}).\end{aligned}$$

(5) It is trivial from Propositions 5.1 and 1.15. □

Example 5.3 showed that the union games are not always superadditive. Necessary player property cannot extended to the whole set of players as we see in the next example.

Example 5.6 Consider $N = \{1, 2, 3\}$ and game v with $v(N) = 4$ and $v(S) = 0$ for the other coalitions S. All the players are necessary. We take $\mathscr{B} = \{\{1\}, \{2, 3\}\}$. If the Owen value satisfies necessary player property then all the payoffs of the players will be the same. But, we get $\omega(v, \mathscr{B}) = (2, 1, 1)$.

Next proposition is removed from the proof of the uniqueness in Theorem 5.1.

Proposition 5.4 *Let $R \subseteq N$, $R \neq \emptyset$. Let $\mathscr{B} = \{B_1, \ldots, B_m\} \in P^N$, $M = \{1, \ldots, m\}$ and $Q_R = \{q \in M : B_q \cap R \neq \emptyset\}$. For each $i \in B_p$ with $p \in M$ it holds*

$$\omega_i(u_R, \mathscr{B}) = \begin{cases} \dfrac{1}{|Q_R||B_p \cap R|}, & \text{if } i \in B_p \cap R \\ 0, & \text{otherwise.} \end{cases}$$

We can describe the Owen value from the dividends using the above proposition.

Theorem 5.3 *For each $v \in \mathscr{G}^N$ and a priori union system $\mathscr{B} = \{B_1, \ldots, B_m\}$ the Owen value of a player $i \in B_p$ with $p \in \{1, \ldots, m\}$ is*

$$\omega_i(v, \mathscr{B}) = \sum_{i \in S \subseteq N} \frac{\Delta_S^v}{|S \cap B_p||Q_S|},$$

where $Q_S = \{q \in M : B_q \cap S \neq \emptyset\}$.

Proof From Proposition 1.1 and linearity of the Owen value (Proposition 5.3) we get

Table 5.2 Dividends of the game v

S	$\{1\}$	$\{2\}$	$\{3\}$	$\{1,2\}$	$\{1,3\}$	$\{2,3\}$	N
Δ_S^v	0	0	5000	10000	25000	20000	−10000

$$\omega_i(v,\mathscr{B}) = \sum_{i\in S\subseteq N} \Delta_S^v \omega_i(u_S,\mathscr{B}) = \sum_{i\in S\subseteq N} \Delta_S^v \frac{1}{|Q_S||B_p \cap S|}.$$

□

Example 5.7 Suppose the bankruptcy game in Example 1.15 with three creditors. The capital of the firm is $Q = 50000$ €and the demand vector of the creditors is $q = (25000, 20000, 40000)$. Table 1.3 represented the worths of the game. We got the dividends of the coalitions in Table 1.7, that we recover as Table 5.2, and also the Shapley value $\phi(v) = (14166.6, 13333.3, 22500)$. Consider now that minor creditors (1 and 2) are actually two business divisions of the same firm. They must work as a union, hence we have $\mathscr{B} = \{\{1,2\},\{3\}\}$. We use formula in Theorem 5.3 to get the Owen value. So,

$$\omega_1(v,\mathscr{B}) = \frac{\Delta_{\{1\}}^v}{1} + \frac{\Delta_{\{1,2\}}^v}{2} + \frac{\Delta_{\{1,3\}}^v}{2} + \frac{\Delta_N^v}{4} = 5000 + 12500 - 2500 = 15000.$$

$$\omega_2(v,\mathscr{B}) = \frac{\Delta_{\{2\}}^v}{1} + \frac{\Delta_{\{1,2\}}^v}{2} + \frac{\Delta_{\{2,3\}}^v}{2} + \frac{\Delta_N^v}{4} = 5000 + 10000 - 2500 = 12500.$$

$$\omega_3(v,\mathscr{B}) = \frac{\Delta_{\{3\}}^v}{1} + \frac{\Delta_{\{1,3\}}^v}{2} + \frac{\Delta_{\{2,3\}}^v}{2} + \frac{\Delta_N^v}{4} = 5000 + 12500 + 10000 - 5000 = 22500.$$

The solution is $\omega(v,\mathscr{B}) = (15000, 12500, 22500)$. In this case, the a priori union only means a different allocation in the union.

5.3 The Graph Variant

Several papers relate communication structures and a priori unions systems, for instance [17] or [16]. We will use that one proposed by Casajus [6]. We take a graph to represent the relations among the players. The components of the graph represent the a priori unions of the players as in the Owen model but now we have also information about how this groups are formed. For instance, me, my sister and my brother in low form a priori familiar union, but without my sister not.

Definition 5.5 A cooperation structure over N is a bilateral relation r satisfying two conditions:

(1) r is reflexive, $r(ii) = 1$ for all $i \in N$,
(2) r is symmetric, $r(i, j) = r(j, i) = r(ij)$ if $i, j \in N$ with $i \neq j$.

The family of cooperation structures is denoted as C^N

Thus cooperation structures are undirected graphs (communication structures) which are reflexive over N, namely the domain is the whole set of players. The difference as between coalition structures and a priori union systems lies in the interpretation: final organization of the players in coalitions or a priori.

The quotient game is used again for analyzing the bargaining among unions. But in the second step we take the Myerson value (Definition 4.9), considering the asymmetric position of the players in each union. So, we take the union games (Definition 5.3) to obtain one payoff for each coalition in a union. Remember that if r is a reflexive graph then N/r is a partition of N, namely an a priori union system. A value for *games with cooperation structure* over N is a mapping $f : \mathcal{G}^N \times C^N \rightarrow \mathbb{R}^N$ obtaining payoff vectors.

Definition 5.6 The Myerson–Owen value is a value for games over N with cooperation structure defined for each $v \in \mathcal{G}^N$, $r \in C^N$ and $i \in N$ as

$$\omega_i(v, r) = \mu_i\left(v_p^{N/r}, r_{B_p}\right),$$

where $i \in B_p$ and $N/r = \{B_1, \ldots, B_m\}$.

Although the Myerson value can be written by components (Proposition 4.10), in this case it is not possible to join the solution in only one expression because the union games are different for each component.

Example 5.8 Let $N = \{1, 2, 3, 4\}$. Suppose the game in Table 5.1. The a priori union system is the same, $\{\{1, 2, 3\}, \{4\}\}$ because 1, 2 and 3 are relatives. But now player 1 is the husband of player 2 and player 3 the cousin of player 1, thus we use the cooperation structure r in Fig. 5.3. Observe that if player 1 decide not to be active the relativity between players 2 and 3 disappears. The union games are the same,

$$v_1^{N/r}(\{1\}) = v_1^{N/r}(\{2\}) = v_1^{N/r}(\{3\}) = v_1^{N/r}(\{1, 3\}) = v_1^{N/r}(\{2, 3\}) = 2.5,$$
$$v_1^{N/r}(\{1, 2\}) = 3, = v_1^{N/r}(\{1, 2, 3\}) = 5. \quad v_2^{N/r}(\{4\}) = 3.$$

We need to calculate the Myerson value of both p-union games. So, the Myerson–Owen value is

$$\omega(v, \mathcal{B}) = (0.916, 2.166, 1.916, 3).$$

Fig. 5.3 Cooperation structure in Example 5.8

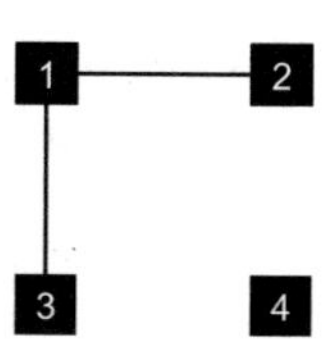

Observe that player 1 is harmed despite her position because game $v_1^{N/r}$ is not superadditive.

Following the definition of the Myerson value, the new solution is also a Shapley value, so if $v \in \mathscr{G}^N$ and $r \in C^N$ we have for each $i \in B_p$ with $N/r = \{B_1, \ldots, B_m\}$,

$$\omega_i(v, r) = \phi_i^{B_p}\left(v_p^{N/r}/r_{B_p}\right).$$

The last game, that we apply the Shapley value over, inherits all the properties and problems satisfied by both union game and vertice game, so we can guarantee neither superadditivity nor convexity. The combination of both tool games satisfies the following properties.

Proposition 5.5 *Let $v \in \mathscr{G}^N$ and $r \in C^N$ with $N/r = \{B_1, \ldots, B_m\}$ and $M = \{1, \ldots, m\}$. For each $p \in M$ it holds:*

(1) If v is additive then $v_p^{N/r}/r_{B_p} = v_{B_p}$.
(2) $(av + bw)_p^{N/r}/r_{B_p} = av_p^{N/r}/r_{B_p} + bw_p^{N/r}/r_{B_p}$ for all $a, b \in \mathbb{R}$. Hence $(-v)_p^{N/r}/r_{B_p} = -(v_p^{N/r}/r_{B_p})$.
(3) If v, w are strategically equivalent then so are $v_p^{N/r}/r_{B_p}, w_p^{N/r}/r_{B_p}$.
(4) $(v^{svg})_p^{N/r}/r_{B_p} = (v_p^{N/r}/r_{B_p})^{svg}$.
(5) $(v^{rdual})_p^{N/r}/r_{B_p} = (v_p^{N/r}/r_{B_p})^{dual}$.
(6) If $v \in \mathscr{G}_c^N$ and r is cycle-complete then $v_p^{N/r}/r_{B_p} \in \mathscr{G}_c^N$.

Proof (1) As $v \in \mathbb{R}^N$ we have from Proposition 5.1 that $v_p^{N/r} = v_{B_p}$. Relation r is reflexive therefore $N^{r_{B_p}} = B_p$, and then using Proposition 4.7 $v_p^{N/r}/r_{B_p} = v_{B_p}$.
(2), (3) and (4) follow directly from Propositions 4.7 and 5.1.
5) If we repeat the proof with v^{rdual} (1.5) in Proposition 5.1 (7) then we obtain that $(v^{rdual})_p^{N/r} = (v_p^{N/r})^{rdual}$. Proposition 4.13 implies that

$$\left((v_p^{N/r})^{rdual}\right)/r_{B_p} = (v_p^{N/r}/r_{B_p})^{dual}.$$

(6) Proposition 5.1 says that $v_p^{N/r}$ is convex if $v \in \mathcal{G}_c^N$. As r is cycle-complete then Proposition 4.9 implies that $v_p^{N/r}/r_{B_p} \in \mathcal{G}_c^N$. □

There are two axiomatizations of the Myerson–Owen value, Casajus [6] and Fernández et al. [7]. The second one allows us to compare better the graph variant with Owen value. Let f be a value for games with cooperation structure over N.

Efficiency. For all $v \in \mathcal{G}^N$ and $r \in C^N$, $f(v,r)(N) = v(N)$.

Null players can obtain profits in a graph from their positions because there is asymmetry, but if all the players connected to them are null then they cannot get benefits. A *null coalition* in a game v is a non-empty coalition $S \subseteq N$ such that if $i \in S$ then i is a null player in v. In a cooperation structure r we denote the set of null unions for a game v as

$$null^v(r) = \{B \in N/r : B \text{ is null for } v\}. \tag{5.3}$$

Null union. Let $r \in C^N$. If $B \in null^v(r)$ for $v \in \mathcal{G}^N$ then $f_i(v,r) = 0$ for all $i \in B$.

Remember that two unions $B, B' \in N/r$ are *symmetric* for the game v if for all $S \subseteq N \setminus (B \cup B')$ we have

$$v(S \cup B) = v(S \cup B').$$

Symmetry in the quotient. Let v be a game and $\mathcal{B}$ be an a priori union system. If $B, B' \in \mathcal{B}$ are symmetric for v then

$$f(v,\mathcal{B})(B) = f(v,\mathcal{B})(B').$$

Linearity. For all $v, w \in \mathcal{G}^N$, $a, b \in \mathbb{R}$ and cooperation structure r it holds $f(av + bw, r) = af(v,r) + bf(w,r)$.

But symmetry in unions is not feasible for cooperation structures because players depend on the position in the subgraph of each union. Nor can we use fairness (in the sense of Myerson, see Sect. 4.3) because if we delete a link then the set of components changes and also the union games. Casajus [6] proposed a modification of the fairness to solve this problem. Let $(v,r) \in \mathcal{G}^N \times C^N$ and $ij \in L(r)$. If $B \in N/r$ with $i, j \in B$ then we denote as B^i, B^j the components containing respectively i, j in r_{-ij} (see Definition 4.6), namely $i \in B^i \in N/r_{-ij}$ and $j \in B^j \in N/r_{-ij}$. Let

$$N_{ij}^i = (N \setminus B) \cup B^i \text{ and } N_{ij}^j = (N \setminus B) \cup B^j. \tag{5.4}$$

Fig. 5.4 Modified fairness

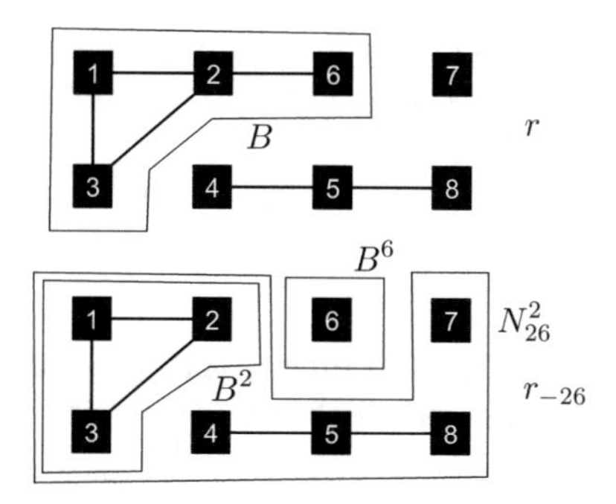

Set N^i_{ij} consists in deleting also the component containing j when the link ij is deleted, if this action disconnects i and j. We need to revise the next axiom for suiting in our context.[1] These changes affect the proof of the next theorem too.

Modified fairness. Let $r \in C^N$ and $B \in N/r$. If $ij \in L(r_B)$ then for any game v it holds

$$f_i(v, r) - f_i(v_{N^i_{ij}}, r_{-ij}) = f_j(v, r) - f_i(v_{N^j_{ij}}, r_{-ij}).$$

Observe that it can be $B^i = B$ because deleting ij we do not disconnect B and then $N^i_{ij} = N$, hence modified fairness coincides with fairness in that case. In order to be consistent with the rest of the book we keep the set of players fixed, and then we will use the restricted game instead the subgame, so we put players in $B \setminus B^i$ as null players. Next example shows the idea.

Example 5.9 Let $N = \{1, 2, 3, 4, 5, 6, 7, 8\}$ and $i = 2$, $j = 6$. Suppose the cooperation structure r in Fig. 5.4. The component containing link 26 is $B = \{1, 2, 3, 6\}$. In the same figure the reader can see the graph r_{-26}. Now, players 2 and 6 are in different components $B^2 = \{1, 2, 3\}$ and $B^6 = \{6\}$. In this graph player 6 will be null because we restrict the game to N^2_{26}. But if we take $ij = 12$ then $B^1 = B^2 = B$ and $N^1_{12} = N^2_{12} = N$.

We will use the following property of the Shapley value. Null players can be removing as players, obtaining the same payoffs for the rest of players.

Lemma 5.1 *Let $v \in \mathcal{G}^N$. If $j \in N$ is a null player in v then for all $i \in N \setminus \{j\}$*

$$\phi^N_i(v) = \phi^{N\setminus\{j\}}_i(v).$$

[1]Modified fairness in [6, 7] uses subgames but we need to use appropriated restricted games in our context.

Proof Observe that if j is null player in v then j is also null player in $v|_{N\setminus\{i\}}$. If $i \neq j$ then by Proposition 1.16 we apply balanced contributions property to the Shapley value

$$\phi_i^N(v) - \phi_i^{N\setminus\{j\}}(v) = \phi_j^N(v) - \phi_j^{N\setminus\{i\}}(v) = 0.$$

□

The proof of the next characterization theorem is slightly different to that in [7] because we use the revised version of modified fairness.

Theorem 5.4 *The Myerson–Owen value is the only value for games over N with cooperation structure satisfying efficiency, null union, symmetry in the quotient, linearity and modified fairness.*

Proof We test that the Myerson–Owen value verifies the axioms. Let $v \in \mathcal{G}^N$, $r \in C^N$, $N/r = \{B_1, \ldots, B_m\}$ and $M = \{1, \ldots, m\}$.
EFFICIENCY. We use that Myerson value satisfies efficiency by components (Theorem 4.3) and the Shapley value satisfies efficiency (Proposition 1.9) to obtain

$$\sum_{i \in N} \omega_i(v, r) = \sum_{p=1}^{m} \sum_{i \in B_p} \mu_i\left(v_p^{N/r}\right) = \sum_{p=1}^{m} v_p^{N/r}(B_p) = \sum_{p=1}^{m} \phi_p\left(w^{v,(N/r)_{B_p}}\right)$$

But for all p we get the same quotient game $w^{v,(N/r)_{B_p}} = w^{v,N/r}$, and so,

$$\sum_{i \in N} \omega_i(v, r) = \sum_{p=1}^{m} \phi_p\left(w^{v,N/r}\right) = w^{v,N/r}(M) = v\left(\cup_{p=1}^{m} B_p\right) = v(N).$$

LINEARITY. Let $p \in M$. We have from Proposition 5.5 that

$$(av + bw)_p^{N/r}/r_{B_p} = av_p^{N/r}/r_{B_p} + bw_p^{N/r}/r_{B_p}.$$

As Shapley value verifies linearity then for each player $i \in B_p$,

$$\begin{aligned}\omega_i(av + bw, r) &= \phi_i\left((av + bw)_p^{N/r}/r_{B_p}\right) = a\phi_i\left(v_p^{N/r}/r_{B_p}\right) + b\phi_i\left(w_p^{N/r}/r_{B_p}\right) \\ &= a\omega_i(v, r) + b\omega_i(w, r).\end{aligned}$$

NULL UNION. Suppose $p \in M$ with $B_p \in null^v(r)$ null union. Let $Q \subseteq M \setminus \{p\}$. For each $S = \{i_1, \ldots, i_l\} \subseteq B_p$ we denote $S_k = \{i_1, \ldots, i_k\}$ with $k = 1, \ldots, l$ and $S_0 = \emptyset$. We get from (5.2)

$$w^{v,(N/r)_S}(Q \cup \{p\}) - w^{v,(N/r)_S}(Q) = v(B_Q \cup S) - v(B_Q)$$
$$= \sum_{k=1}^{l} v(N_Q \cup S_k) - v(N_Q \cup S_{k-1}),$$

As each i_k is a null player $v(N_Q \cup S_k) - v(N_Q \cup S_{k-1}) = 0$ and then

$$w^{v,(N/r)_S}(Q \cup \{p\}) - w^{v,(N/r)_S}(Q) = 0.$$

Hence p is a null player in $w^{v,(N/r)_S}$ and $v_p^{N/r}(S) = \phi_p(w^{v,(N/r)_S}) = 0$ because Shapley value satisfies the null player property. As $v_p^{N/r} = 0$ then $\phi^{B_p}(v_p^{N/r}) = 0$ and $\omega_i(v, r) = 0$ for all $i \in B_p$.

SYMMETRY IN THE QUOTIENT. Consider B_p, B_q symmetry unions in v. If $Q \subseteq M \setminus \{p, q\}$ then

$$w^{v,N/r}(Q \cup \{p\}) = v(B_Q \cup B_p) = v(B_Q \cup B_q) = w^{v,N/r}(Q \cup \{q\}),$$

namely p, q are symmetric in the quotient game. But then, as Shapley value satisfies symmetry (Proposition 1.10),

$$v_p^{N/r}(B_p) = \phi_p(w^{v,N/r}) = \phi_q(w^{v,N/r}) = v_q^{N/r}(B_q).$$

Now we use that Myerson value is efficient by components

$$\omega(v, r)(B_p) = \mu\left(v_p^{N/r}, r_{B_p}\right)(B_p) = v_p^{N/r}(B_p) = v_q^{N/r}(B_q) = \omega(v, r)(B_q).$$

MODIFIED FAIRNESS. Let $ij \in L(r)$ and, without losing generality, take $i, j \in B_m$. We use the notation in (5.4). First we suppose that $N_{ij}^i = N$, namely component B_m is still connected in r_{-ij}. In that case $N/r_{-ij} = N/r$ and then $v_m^{N/r_{-ij}} = v_m^{N/r}$. Theorem 4.3 proved that Myerson value satisfies fairness, thus

$$\omega_i(v, r) - \omega_i(v, r_{-ij}) = \mu_i\left(v_m^{N/r}, r_{B_m}\right) - \mu_i\left(v_m^{N/r}, (r_{B_m})_{-ij}\right)$$
$$= \mu_j\left(v_m^{N/r}, r_{B_m}\right) - \mu_j\left(v_m^{N/r}, (r_{B_m})_{-ij}\right)$$
$$= \omega_j(v, r) - \omega_j(v, r_{-ij}).$$

Now suppose that deleting link ij we get two components from B_m,

$$N/r_{-ij} = N/r \setminus \{B_m\} \cup \{B_m^i, B_{m+1}^j\},$$

with $i \in B_m^i$ and $j \in B_{m+1}^j$. We will prove the claim

$$v_m^{N/r} = \left(v_{N_{ij}^i}\right)_m^{N/r_{-ij}} \quad \text{in } B_m^i.$$

Let $S \subseteq B^i_m$. As all the players in B^j_{m+1} are null in the restricted game $v_{N^i_{ij}}$ (Definition 3.5) then B^j_{m+1} is a null union in that game. Next we see that union $m+1$ is a null player in $w^{v_{N^i_{ij}},(N/r_{-ij})_S}$. Let $B^j_{m+1} = \{j_1, \ldots, j_l\}$. If $Q \subseteq M$ then we $B'_Q = B_Q$ if $m \notin Q$ and $B'_Q = B_{Q\setminus\{m\}} \cup S$ if $m \in Q$. Hence, taking $T_k = \{j_1, \ldots, j_k\}$ for each $k = 1, \ldots, l$

$$w^{v_{N^i_{ij}},(N/r_{-ij})_S}(Q \cup \{m+1\}) - w^{v_{N^i_{ij}},(N/r_{-ij})_S}(Q) = v_{N^i_{ij}}(B'_Q \cup B^j_{m+1}) - v_{N^i_{ij}}(B'_Q)$$
$$= \sum_{k=1}^{l} v(N_Q \cup T_k) - v(N_Q \cup T_{k-1}) = 0,$$

because each i_k is a null player in $v_{N^i_{ij}}$. Lemma 5.1 implies

$$\left(v_{N^i_{ij}}\right)^{N/r_{-ij}}_m (S) = \phi^{M\cup\{m+1\}}_m \left(w^{v_{N^i_{ij}},(N/r_{-ij})_S}\right) = \phi^M_m \left(w^{v_{N^i_{ij}},(N/r_{-ij})_S}\right).$$

But, as $S \subseteq B^i_m$ then $w^{v_{N^i_{ij}},(N/r_{-ij})_S} = w^{v,(N/r)_S}$, thus we obtain the claim

$$\left(v_{N^i_{ij}}\right)^{N/r_{-ij}}_m (S) = \phi^M_m \left(w^{v,(N/r)_S}\right) = v^{N/r}_m(S).$$

We use the properties of decomposability (Proposition 4.10) and fairness (Theorem 4.3) of the Myerson value together with the above claim to get

$$\begin{aligned}
\omega_i(v, r) - \omega_i(v_{N^i_{-ij}}, r_{-ij}) &= \mu_i\left(v^{N/r}_m, r_{B_m}\right) - \mu_i\left((v_{N^i_{-ij}})^{N/r_{-ij}}_m, (r_{-ij})_{B^i_m}\right) \\
&= \mu_i\left(v^{N/r}_m, r_{B_m}\right) - \mu_i\left(v^{N/r}_m, r_{-ij}\right) \\
&= \mu_j\left(v^{N/r}_m, r_{B_m}\right) - \mu_j\left(v^{N/r}_m, r_{-ij}\right) \\
&= \mu_j\left(v^{N/r}_m, r_{B_m}\right) - \mu_j\left((v_{N^j_{-ij}})^{N/r_{-ij}}_m, (r_{-ij})_{B^j_{m+1}}\right) \\
&= \omega_j(v, r) - \omega_j(v, r_{-ij}).
\end{aligned}$$

Let f^1, f^2 be two values[2] satisfying the axioms. Obviously if we take the null game, $v = 0$, then all the unions in $r \in C^N$ are null unions and therefore $f^1(0, r) = f^2(0, r) = 0$ from the null union property. From linearity it is only necessary to prove the uniqueness for the unanimity games. Suppose $r \in C^N$ with $L(r) = \emptyset$, namely r is a diagonal matrix, then $N/r = \{\{i\} : i \in N\}$. Let u_T with T a non-empty coalition be a unanimity game. Remember that all the players out of T are null in the unanimity game and players in T are symmetric (Proposition 1.7). If $i \notin T$ then $\{i\} \in null^{u_T}(r)$ and by null union condition $f^1_i(u_T, r) = 0$. If $i, j \in T$ then $\{i\}, \{j\}$ are symmetric unions and by symmetry in the quotient $f^1_i(u_T, r) = f^1_i(u_T, r)$. Applying efficiency we have for any $i_0 \in T$

[2] It can be used $f^1 = \omega$.

$$f^1(u_T, r)(N) = f^1(u_T, r)(T) = |T| f^1_{i_0}(u_T, r) = u_T(N) = 1.$$

Thus $f^1_{i_0}(u_T, r) = 1/|T|$. We can repeat the reasoning with f^2 and then $f^1(u_T, r) = f^2(u_T, r)$. Suppose true that if $|L(r)| = l$ then $f^1(u_T, r) = f^2(u_T, r)$ for all the unanimity games u_T (and also for the null game). Let $r \in C^N$ be a cooperation structure with $|L(r)| = l + 1$ and u_T a unanimity game. We set

$$M_T(r) = \{B \in N/r : B \cap T \neq \emptyset\}. \tag{5.5}$$

If $B \in N/r$ but $B \notin M_T(r)$ then $B \in null^{u_T}(r)$ and $f^1_i(u_T, r) = f^2_i(u_T, r) = 0$ for all $i \in B$. If $B \in N/r$ with $|B| > 1$ then there exists $ij \in L(r)$ with $i, j \in B$. We use modified fairness to obtain by induction

$$\begin{aligned} f^1_i(u_T, r) - f^1_j(u_T, r) &= f^1_i\left((u_T)_{N^i_{ij}}, r_{-ij}\right) - f^1_j\left((u_T)_{N^j_{ij}}, r_{-ij}\right) \\ &= f^2_i\left((u_T)_{N^i_{ij}}, r_{-ij}\right) - f^2_j\left((u_T)_{N^i_{ij}}, r_{-ij}\right) \\ &= f^2_i(u_T, r) - f^2_j(u_T, r) \end{aligned}$$

taking into account the sets defined in (5.4). Observe that $(u_T)_R$ for any R is the null game or the same unanimity game since Proposition 3.7. Hence

$$f^1_i(u_T, r) - f^2_i(u_T, r) = f^1_j(u_T, r) - f^2_j(u_T, r).$$

Fix $i_0 \in B$, we will see that both payoffs are the same for her. As B is a connected coalition we can trace a path from any $i \in B$ to i_0 by players in B. Repeating the above process sequentially in the path we get

$$f^1_i(u_T, r) - f^2_i(u_T, r) = f^1_{i_0}(u_T, r) - f^2_{i_0}(u_T, r) = K_B, \ \forall i \in B.$$

Let now $B_0 \in M_T(r)$ fixed. If we have another $B \in M_T(r)$ then they are symmetric unions in u_T because if $S \subseteq N \setminus (B_0 \cup B)$ then $u_T(S \cup B_0) = u_T(S \cup B) = 0$. So, symmetry in the quotient says for any $B \in M_T(r)$

$$\begin{aligned} f^1(u_T, r)(B) &= f^1(u_T, r)(B_0) = H_1 \\ f^2(u_T, r)(B) &= f^2(u_T, r)(B_0) = H_2. \end{aligned}$$

The subtraction of both expressions for any $B \in M_T(r)$ is

$$H_1 - H_2 = \sum_{i \in B} f_i^1(u_T, r) - f_i^2(u_T, r) = |B| K_B.$$

Finally we use that both f^1, f^2 satisfy efficiency, thus the payoffs must add $u_T(N) = 1$ for each of them,

$$0 = f^1(u_T, r)(N) - f^2(u_T, r)(N) = \sum_{B \in M_T(r)} f^1(u_T, r)(B) - f^2(u_T, r)(B)$$
$$= |M_T(r)|(H_1 - H_2).$$

Hence $H_1 = H_2$ and we get for each $B \in M_T(r)$,

$$f^1(u_T, r)(B) - f^2(u_T, r)(B) = |B| K_B = H_1 - H_2 = 0.$$

We conclude that $K_B = 0$ and then $f_{i_0}^1(u_T, r) = f_{i_0}^2(u_T, r)$. □

Other properties of the Myerson–Owen value are organized in two propositions. The proof of the first one follows directly from Proposition 5.5.

Proposition 5.6 *The Myerson–Owen value satisfies the following properties for a cooperation structure r with $N/r = \{B_1, \ldots, B_m\}$ and game v. Let $M = \{1, \ldots, m\}$ and $p \in M$.*

(1) For all $i \in B_p$ it holds $\omega_i(v, r) = v_p^{N/r}(\{i\}) - \omega_i(v^{svg}, r)$.
(2) $\omega(v^{rdual}, r) = \omega(v, r)$.

The next proposition speaks about properties which need the convexity of the game to be guaranteed for games with cooperation structure.

Proposition 5.7 *The Myerson–Owen value satisfies the following properties for a cooperation structure r with $N/r = \{B_1, \ldots, B_m\}$ and game $v \in \mathcal{G}_c^N$. Let $M = \{1, \ldots, m\}$ and $p \in M$.*

(1) If $i \in B_p$ is a necessary player and $v \geq 0$ then $\omega_j(v, r) \leq \omega_i(v, r)$ for all $j \in B_p$.
(2) If $i \in B_p$ then $\omega_i(v, r) \geq v(\{i\})$.
(3) If r is cycle-complete and $S \subseteq B_p$ then $\omega(v, r)(S) \geq v_p^{N/r}/r_{B_p}(S)$.

Proof If v is convex from Proposition 5.1 obtains that $v_p^{N/r}$ is convex.

(1) As $v_p^{N/r}$ is convex and non-negative then it is superadditive and non-negative. Let $i \in B_p$ be a necessary player and $j \in B_p$, Proposition 4.11 says that

$$\omega_i(v, r) = \mu_i(v_p^{N/r}, r_{B_p}) \geq \mu_j(v_p^{N/r}, r_{B_p}) = \omega_j(v, r).$$

(2) If $v_p^{N/r}$ is convex then it is superadditive. Proposition 4.11 implies

$$\omega_i(v, r) = \mu_i(v_p^{N/r}, r_{B_p}) \geq v(\{i\}).$$

(3) We use again Proposition 4.11 to get for each $s \subseteq B_p$,

$$\omega(v, r)(S) = \mu(v_p^{N/r}, r_{B_p})(S) \geq v_p^{N/r}/r_{B_p}(S).$$

□

5.4 Games with a Proximity Relation Among Agents

In order to extend the above models we consider a proximity relation.

Definition 5.7 A proximity relation over N is a fuzzy bilateral relation ρ satisfying

(1) Reflexivity: $\rho(i, i) = \rho(ii) = 1$ for all $i \in N$, and
(2) Symmetry: $\rho(i, j) = \rho(j, i) = \rho(ij)$.

The set of proximity relations is denoted by FC^N

Obviously, the family of cooperation structures C^N is a subset of FC^N. We use notation FC^N because we interpret proximity relations as fuzzy cooperation structures. A proximity relation is actually a reflexive fuzzy graph. The difference with the fuzzy communication structures is the interpretation. If ρ is a proximity then $\rho(ij)$ is interpreted as the closeness level (or also penchant, probability of) between players i and j, obviously each player is closed to herself.

We consider again partitions by levels of fuzzy graphs but it seems natural to keep the reflexivity in the crisp steps, namely each graph in the partition must be a cooperation structure (Definition 5.5). Following Definition 4.15 we introduce the next one.

Definition 5.8 Let $\rho \in FC^N$ be a proximity relation. A reflexive partition by levels of ρ is a finite sequence $\{(r_k, s_k)\}_{k=1}^h$ satisfying:

(1) $r_k \in C^N$ and $s_k > 0$, for all $k = 1, \ldots, h$,
(2) $s_1 r_1 \leq \rho$ and for each $k = 2, \ldots, h$

$$s_k r_k \leq \rho - \sum_{p=1}^{k-1} s_p r_p.$$

(3) $\rho - \sum_{k=1}^{h} s_k r_k = 0$.

A reflexive partition function pl for proximity relations determines a reflexive partition by levels for each proximity relation.

From (4.8) we obtain

$$\sum_{k=1}^{h} s_k = 1. \tag{5.6}$$

Observe that we do not need to look for admissible (Definition 4.19) partition functions, but obviously we consider that they are *extensions* in the sense that $pl(r) = \{(r, 1)\}$ if $r \in C^N$.

The Choquet by graphs extension (Definition 4.17) is an example of reflexive extension. Fernández et al. [7] analyzed this particular case using cg-partitions.

Proposition 5.8 *The Choquet by graphs partition function is a reflexive extension for proximity relations.*

Proof We know since Proposition 4.17 that cg is an extension for fuzzy graphs and then for proximity relations. Now remember from (4.10) that if $(r, s) \in cg(\rho)$ then there exists $t \in im(\rho)$ with $r = [\rho]_t$. But, as $\rho(ii) = 1 > t$ then $r(ii) = 1$ for all $i \in N$. Thus cg is reflexive. □

The question of the interpretation is crucial in this case, and it is complicated to explain the meaning of the partitions in general. So we explain the meaning with the cg-partition, following [7]. Let $\rho \in FC^N$ be a proximity relation. To take $t \in (0, 1]$ means to simplify the problem in the following sense: determining certain degree of closeness to be considered a union, so if we connect a coalition with level at least t they have similar ideas. In other words, we think that if the closeness is less than t then it is negligible. Thus we construct cooperation structures with different levels. The cg-partition obtains a sequence of cooperation structures depending on the level of exigency or requisite. Another interpretation focuses on temporal situations, i.e. the proximity relation determine the duration of the a priori relation.

Fig. 5.5 Proximity structure

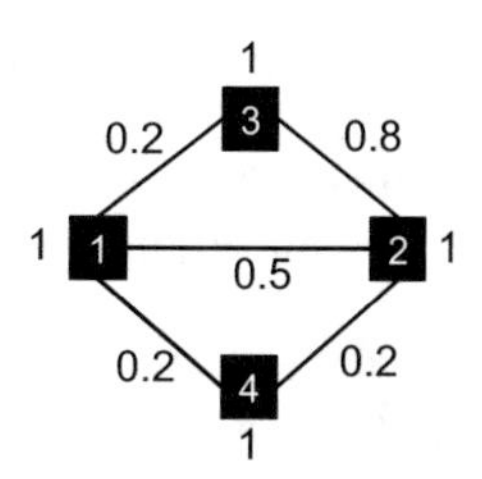

Table 5.3 cg-algorithm applied to Fig. 5.5

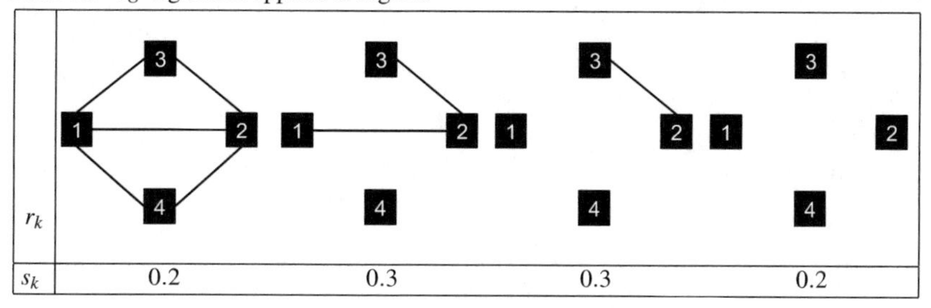

r_k				
s_k	0.2	0.3	0.3	0.2

Example 5.10 Consider the proximity relation ρ in Fig. 5.5 with matrix

$$\rho = \begin{bmatrix} 1 & 0.5 & 0.2 & 0.2 \\ 0.2 & 1 & 0.8 & 0.2 \\ 0.5 & 0.8 & 1 & 0 \\ 0.2 & 0.2 & 0 & 1 \end{bmatrix}.$$

Table 5.3 shows the cg-partition of ρ. First we think that two players are in the same union if and if their closeness is less or equal than 0.2. Obviously "to be in the same union" is transitive. With transitivity here we do not say that two players has certain level of closeness if there is a path with this level between them, but they are in the same group in the bargaining although it can be for different circumstances. Namely we simplify the problem to a crisp one using r_1. In the example all the players are in the same union but this union is formed by a particular communication structure. If we suppose level 0.5 as the minimum closeness to be a union then they use cooperation structure r_2 with two components. If our requirement of closeness is 0.8 then we use a priori union system r_3. Finally, if we consider that players in a union must have exactly the same criteria we obtain the classical situation without a priori unions r_4.

The proportional by graphs extension (Definition 4.16) is also reflexive but it only obtains one graph for all proximity relation because the algorithm takes all the vertices and the links with level 1 in the first step. Next definition introduces a modification of the algorithm for proximity relations similar to the proportional by communication extension (Example 4.15). Let $\rho \in FC^N$. For each $t \in (0, 1]$ consider the crisp graph $\overline{r}^\rho[t]$ with $\overline{r}^\rho[t](ij) = 1$ if and only $i = j$ or $\rho(ij) = t$. Particularly $\overline{r}^\rho[0] = r^\rho$.

Definition 5.9 The prox-proportional extension pp takes the reflexive partition by levels obtained by the following algorithm (pp-algorithm),

Take $k = 0$, $pp = \emptyset$ and $\rho = \rho$
While $\rho \neq 0$ do
..... $k = k + 1$
..... $t = \vee\{\rho(ij) : i \neq j\}$
..... if $t = 0$ then
.... $s_k = \rho(ii)$
.... $r_k = r^\rho$
..... else
.... $s_k = t$
.... $r_k = \overline{r}^\rho[s_k]$
..... $pp = pp \cup \{(r_k, s_k)\}$
..... $\rho = \rho - s_k r_k$
$pp(\rho) = pp$

The pp-algorithm obtains the same result than pg when there exists at least a link with level 1, but otherwise they are different. The pc-extension is not reflexive because each step only chooses those vertices using links at the maximum level.

Example 5.11 Consider the proximity relation in Fig. 5.5. Table 5.4 shows the application of the pp-algorithm. The pp-algorithm works as the pc-algorithm in the sense that we use all the links with the same level (the maximum level of the links), but now all the players are chosen in all the steps, regardless of the level. In this example,

$$\vee\{\rho(ij) : ij \in L(\rho)\} = 0.8.$$

Hence we take (r_1, s_1) where $s_1 = 0.8$ and r_1 is a crisp graph using all the vertices and the unique link with level 0.8.

Unions are the main elements of bargaining in cooperation structures. We introduced the following concept as a generalization of union in a proximity relation.

Table 5.4 *pp*-algorithm applied to Fig. 5.5

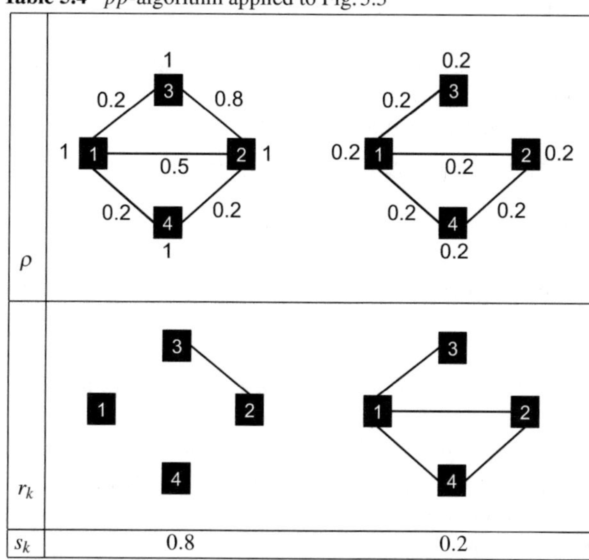

Definition 5.10 Let pl be a reflexive extension for proximity relations and $\rho \in FC^N$. Coalition $B \subseteq N$ is named pl-group in ρ if $B \in N/r$ for some $(r, s) \in pl(\rho)$. The family of all the pl-groups in ρ is denoted by

$$[N/\rho]_{pl} = \bigcup_{(r,s)\in pl(\rho)} N/r.$$

If $B \in [N/\rho]_{pl}$ then the activity set of B is

$$pl(\rho, B) = \{(r, s) \in pl(\rho) : B \in N/r\}.$$

Observe that $[N/\rho]_{pl}$ is not equal to the set of components N/ρ (3.5), but if $\rho \in C^N$ then both coincide.

Example 5.12 The cg-partition of the proximity relation in Fig. 5.5 for $N = \{1, 2, 3, 4\}$ is in Table 5.3. The unions for the different cooperation structures are the following.

$$N/r_1 = \{N\},$$
$$N/r_2 = \{\{1, 2, 3\}, \{4\}\},$$
$$N/r_3 = \{\{1\}, \{2, 3\}, \{4\}\},$$
$$N/r_4 = \{\{1\}, \{2\}, \{3\}, \{4\}\}.$$

The family of groups is

$$[N/\rho]_{cg} = \{N, \{1,2,3\}, \{2,3\}, \{1\}, \{2\}, \{3\}, \{4\}\}.$$

Observe that the components of the fuzzy graph (3.5) form another set, $N/\rho = \{N\}$. For the cg-partition we can write $[N/\rho]_{cg} = \bigcup_{t\in(0,1]} N/[\rho]_t$.

We determine the activity set of several groups:

$$cg(\rho, N) = \{(r_1, 0.2)\},\ cg(\rho, \{2,3\}) = \{(r_3, 0.3)\}$$

$$\text{or } cg(\rho, \{4\}) = \{(r_2, 0.3), (r_3, 0.3), (r_4, 0.2)\}.$$

Using the cg-partition is possible to see the activity set of group B as an activity interval $(t_B, t^B]$, with

$$t_B = \wedge\{t \in (0,1] : B \in N/[\rho]_t\} \text{ and } t^B = \vee\{t \in (0,1] : B \in N/[\rho]_t\}.$$

If $B \in [N/\rho]_{cg}$ then number t_B is an infimum but t^B is a maximum. Group B is contained in a component of $[\rho]_t$ when $t \leq t_B$, it is a component when t is in the activity interval $(t_B, t^B]$ and it is a union of components when $t > t^B$. The activity interval of each group is the following:

$$N \to (0, 0.2]\quad \{1,2,3\} \to (0.2, 0.5]\ \{2,3\} \to (0.5, 0.8]\ \{1\} \to (0.5, 1]$$
$$\{2\} \to (0.8, 1]\ \{3\} \to (0.8, 1]\qquad \{4\} \to (0.2, 1]$$

Fernández et al. [7] introduced several ways of reducing the image of a proximity relation. We introduce in this book a generalization of these process. The scaling of a proximity relation considers insignificant certain levels.

Definition 5.11 Let pl be a reflexive extension for proximity relations. Let $\rho \in FC^N$ with $pl(\rho) = \{(r_k, s_k)\}_{k=1}^h$ and $H = \{1, \ldots, h\}$. For each $K \subseteq H$ the K-scaling of ρ is a new proximity relation ρ_{pl}^K defined as

$$\rho_{pl}^K = \frac{1}{s_K}\sum_{k\in K} s_k r_k,$$

with $s_K = \sum_{k\in K} s_k$.

To test that ρ_{pl}^K is a proximity relation is enough to calculate the level of the vertices. As each r_k is reflexive we get

$$\rho_{pl}^K(ii) = \frac{1}{s_K}\sum_{k\in K} s_k r_k(ii) = \frac{1}{s_K}\sum_{k\in K} s_k = 1.$$

Remark 5.2 (1) Observe that $\rho_{pl}^H = \rho$ is not always true because of the definition of the subtraction of fuzzy graphs (Definition 4.14). For instance we consider the

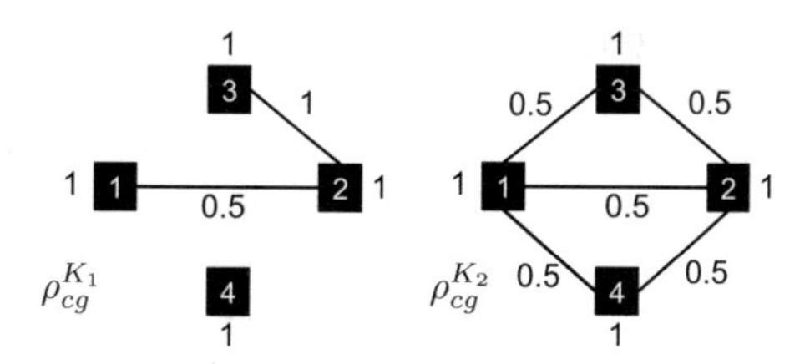

Fig. 5.6 K-scalings of ρ

pp-partition (Definition 5.9). Let $N = \{1, 2, 3\}$ and the proximity relation ρ with matrix

$$\rho = \begin{bmatrix} 1 & 1 & 1 \\ 1 & 1 & 0.5 \\ 1 & 0.5 & 1 \end{bmatrix},$$

the pp-partition is $pp(\rho) = \{(r, 1)\}$ with

$$r = \begin{bmatrix} 1 & 1 & 1 \\ 1 & 1 & 0 \\ 1 & 0 & 1 \end{bmatrix}.$$

So, 23 is not in the partition, and $\rho \neq \rho_{pp}^{H} = r$.
(2) Particularly, we define $\rho_{pl}^{\emptyset} = e^{\{ii:i\in N\}}$ and $s_{\emptyset} = 0$.

Example 5.13 Think of the cg-extension. We follow with the proximity relation ρ in Fig. 5.5 (Example 5.10) and Table 5.3. Let $K_1 = \{(r_2, 0.3), (r_3, 0.3)\}$. Number $s_{K_1} = 0.6$ and the K_1-scaling of ρ is

$$\rho_{cg}^{K_1} = \frac{1}{0.6}[0.3r_2 + 0.3r_3] = \begin{bmatrix} 1 & 0.5 & 0 & 0 \\ 0.5 & 1 & 1 & 0 \\ 0 & 1 & 1 & 0 \\ 0 & 0 & 0 & 1 \end{bmatrix}.$$

Now we consider $K_2 = \{(r_1, 0.2), (r_4, 0.2)\}$. We have

$$\rho_{cg}^{K_2} = \frac{1}{0.4}[0.2r_1 + 0.2r_4] = \begin{bmatrix} 1 & 0.5 & 0.5 & 0.5 \\ 0.5 & 1 & 0.5 & 0.5 \\ 0.5 & 0.5 & 1 & 0 \\ 0.5 & 0.5 & 0 & 1 \end{bmatrix}.$$

It can see both scalings in Fig. 5.6.

Remark 5.3 Fernández et al. [7] introduced the concept of interval scaling[3] in the context of the cg-extension. Let $\rho \in FC^N$. If $a, b \in [0, 1]$ and $a < b$ then

[3]They also defined dual interval scaling of a proximity relation.

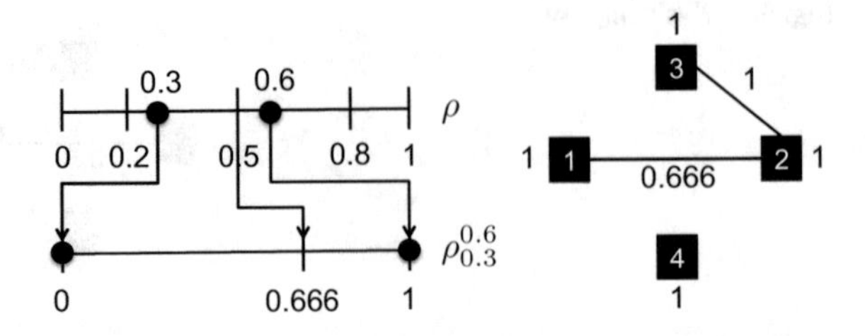

Fig. 5.7 Interval scaling of ρ

$$\rho_a^b(ij) = \begin{cases} 1, & \text{if } \rho(ij) \geq b \\ \dfrac{\rho(ij) - a}{b - a}, & \text{if } \rho(ij) \in (a, b) \\ 0, & \text{if } \rho(ij) \leq a. \end{cases}$$

Figure 5.7 shows the idea of the interval scaling calculating $\rho_{0.3}^{0.6}$ of the proximity relation in Fig. 5.5: to simplify interval [0, 1] to [0.3, 0.6] and rescaling. The interval scaling coincides with a K-scaling when $a, b \in im(\rho)$. Let $im_0(\rho) = \{0 = \lambda_0 < \lambda_1 < \cdots < \lambda_h = 1\}$ and $H = \{1, \ldots, h\}$. Suppose $a, b \in im_0(\rho)$. Hence there exist two numbers $k_a < k_b$ with $\lambda_{k_a} = a$ and $\lambda_{k_b} = b$. We set

$$K_{[a,b]} = \{k \in H : k_a < k \leq k_b\}.$$

The reader can see that $s_{K_{[a,b]}} = b - a$. If $\rho(ij) \geq b$ then $[\rho]_k(ij) = 1$ for all $k \in K_{[a,b]}$, thus $\rho_{cg}^{K_{[a,b]}}(ij) = 1$. If $\rho(ij) \leq a$ then $[\rho]_k(ij) = 0$ for all $k \in K_{[a,b]}$, thus $\rho_{cg}^{K_{[a,b]}}(ij) = 0$. Finally, if $\rho(ij) \in (a, b)$, let $\rho(ij) = \lambda_{k'}$ then $[\rho]_k(ij) = 1$ for all $k_a < k < k'$ (contained in $K_{[a,b]}$), thus

$$\rho_{cg}^{K_{[a,b]}}(ij) = \frac{1}{b - a} \sum_{k=k_a+1}^{k'} (\lambda_k - \lambda_{k-1}) = \frac{\rho(ij) - a}{b - a}.$$

Scaling will allow us to describe axioms for values in this context when the partition function works well with them. We require the extension to be able to obtain the partition by levels of a K-scaling from the partition of the original proximity relation.

Definition 5.12 A reflexive extension pl for proximity relations is inner if for any $\rho \in P^N$ with $pl(\rho) = \{(r_k, s_k)\}_{k=1}^h$, $H = \{1, \ldots, h\}$ and $K \subseteq H$ it holds

$$pl\left(\rho_{cg}^K\right) = \left\{\left(r_k, \frac{s_k}{s_K}\right)\right\}_{k \in K}.$$

Next proposition shows two examples of inner extensions.

Proposition 5.9 *The Choquet by graphs function and the prox-proportional function are inner extensions.*

Proof Let $\rho \in FC^N$. Consider for each extension pl the partition $pl(\rho) = \{(r_k, s_k)\}_{k=1}^{h}$. We denote as $H = \{1, \ldots, h\}$ and take $K \subseteq H$ nonempty (otherwise it is trivial).

CHOQUET BY GRAPHS. We apply the cg algorithm (Definition 4.17) to ρ_{cg}^K. From (4.10) if $im_0(\rho) = \{\lambda_0 < \lambda_1 < \cdots < \lambda_h\}$ then $s_k = \lambda_k - \lambda_{k-1}$ and $r_k = [\rho]_k$. Fixed ij we have

$$\rho_{cg}^K(ij) = \frac{1}{s_K} \sum_{\{k \in K : \rho(ij) \geq \lambda_k\}} (\lambda_k - \lambda_{k-1}).$$

Let $\lambda_p = \bigwedge_{k \in K} \lambda_k$. Pair ij satisfies $\rho(ij) < \lambda_p$ if and only if $\rho_{cg}^K(ij) = 0$. There exists ij with $\rho(ij) = \lambda_p$. So, the cg-algorithm chooses (r_1', s_1') as

$$s_1' = \wedge \rho_{cg}^K = \frac{1}{s_K}(\lambda_p - \lambda_{p-1}) = \frac{s_p}{s_K}.$$

Furthermore $r_1'(ij) = 1$ if and only if $\rho_{cg}^K(ij) \geq s_1'$, namely if and only if $\rho(ij) \geq \lambda_p$. Hence $r_1'(ij) = 1$ if and only if $r_p(ij) = 1$. We get

$$(r_1', s_1') = \left(r_p, \frac{s_p}{s_K}\right).$$

Now, in the second step, we take $\rho_{cg}^K = \rho_{cg}^K - s_1' r_1'$. The cg-algorithm chooses (r_2', s_2'). Let $\lambda_q = \bigwedge_{k \in K \setminus \{p\}} \lambda_k$. If $\rho(ij) < \lambda_q$ then $\rho_{cg}^K(ij) = 0$ in this step. As there is ij with $\rho(ij) = \lambda_q$ then

$$s_2' = \wedge \rho_{cg}^K = \frac{1}{s_K}[(\lambda_q - \lambda_{q-1}) + (\lambda_p - \lambda_{p-1})] - s_1' = \frac{1}{s_K}(\lambda_q - \lambda_{q-1}) = \frac{s_q}{s_K}.$$

We have $r_2'(ij) = 1$ if and only if $\rho_{cg}^K(ij) \geq s_2'$, namely

$$\rho_{cg}^K(ij) = \frac{1}{s_K} \sum_{\{k \in K \setminus \{p\} : \rho(ij) \geq \lambda_k\}} (\lambda_k - \lambda_{k-1}).$$

Therefore $r_2'(ij) = 1$ if and if $\rho(ij) \geq \lambda_q$ if and if $r_q(ij) = 1$. We repeat the process with all the steps.

PROX- PROPORTIONAL. In this case we have $s_1 > \cdots > s_h$. Following the pp-algorithm, we look for the maximum level in ρ_{pp}^K. Each ij, $i \neq j$, appears only once at most in the graphs of the partition (ii appears always), namely or there

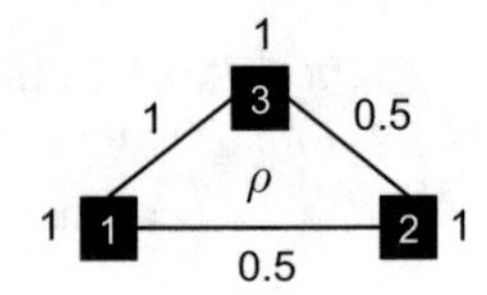

Fig. 5.8 Proximity relation in Example 5.14

Table 5.5 *pb*-partition of Fig. 5.8

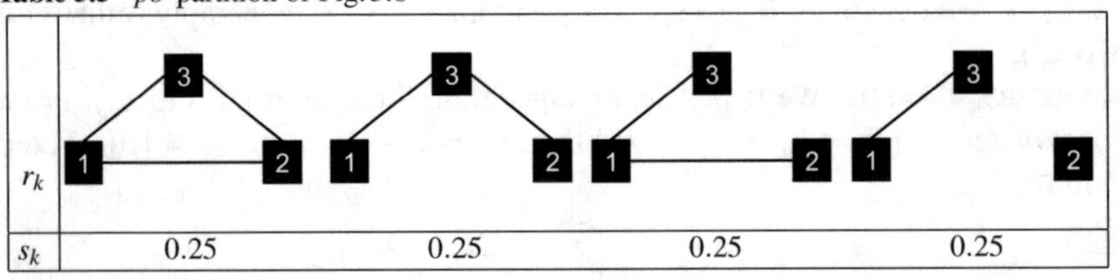

r_k				
s_k	0.25	0.25	0.25	0.25

exists only one $k \in K$ with $r_k(ij) = 1$ or $r_k(ij) = 0$ for all $k \in K$. Let $p \in K$ with $s_p = \bigvee_{k \in K} s_k$. We obtain

$$s_1' = \frac{s_p}{s_K}.$$

Moreover, all the links ij with $r_p(ij) = 1$ are chose by the algorithm and none more, thus $r_1' = r_p$. The process is repeated with the rest of the steps independently because the links of the graphs are disjoint. Observe that although in this algorithm the level of the links can be reduced when they are used (see Example 5.15), this fact is not possible for the scaling ρ^K_{pp}. □

Obviously, the probabilistic extension (Definition 4.31) is also an example of reflexive extension because of vertices are sure events in a proximity relations but it is not inner.

Example 5.14 Consider $N = \{1, 2, 3\}$ and the proximity relation

$$\rho = \begin{bmatrix} 1 & 0.5 & 1 \\ 0.5 & 1 & 0.5 \\ 1 & 0.5 & 1 \end{bmatrix}$$

in Fig. 5.8. The cooperation structures of the *pb*-partition are in Table 5.5. For instance, the probability of obtaining the complete graph r_1 is

$$s_1 = \rho(12)\rho(13)\rho(23) = 0.25.$$

Let $H = \{1, 2, 3, 4\}$. If we take $K = \{2, 3\}$ the scaling ρ^K_{pb}, with $s_K = 0.5$, satisfies

$$\rho^K_{pb} = 0.5 r_2 + 0.5 r_3 = \rho.$$

Hence *pb* is not inner.

Inner extensions satisfy the following lemma which will be used in the next section. Let pl be an extension for proximity relations. Consider $F : FC^N \to \mathbb{R}$. Mapping F is named *consistent* with pl if for all $\rho \in P^N$ with $pl(\rho) = \{(r_k, s_k)\}_{k=1}^h$ happens

$$F(\rho) = \sum_{k=1}^{h} s_k F(r_k). \tag{5.7}$$

Namely, in a consistent mapping we only need to know the worths of the cooperation structures.

Lemma 5.2 *If pl is an inner reflexive extension and F is a consistent mapping with pl over proximity relations then*

$$F(\rho) = \sum_{p=1}^{q} s_{K_p} F\left(\rho_{pl}^{K_p}\right),$$

with $pl(\rho) = \{(r_k, s_k)\}_{k=1}^h$, $H = \{1, \dots, h\}$ and $\{K_p\}_{p=1}^q$ a partition of H.

Proof Suppose $pl(\rho) = \{(r_k, s_k)\}_{k=1}^h$, $H = \{1, \dots, h\}$ and a partition of H given by $\{K_p\}_{p=1}^q$. We obtain for a consistent mapping F with pl

$$\begin{aligned} F(\rho) &= \sum_{k=1}^{h} s_k F(r_k) = \sum_{p=1}^{q} \sum_{k \in K_p} s_k F(r_k) = \sum_{p=1}^{q} s_{K_p} \sum_{k \in K_p} \frac{s_k}{s_{K_p}} F(r_k) \\ &= \sum_{p=1}^{q} s_{K_p} F\left(\rho_{pl}^{K_p}\right), \end{aligned}$$

using that, as pl is inner, $pl\left(\rho_{cg}^{K_p}\right) = \left\{\left(r_k, \frac{s_k}{s_K}\right)\right\}_{k \in K}$. □

5.5 Fuzzy Myerson–Owen Values

The quotient game is used again for analyzing the bargaining among unions. But in the second step we take the Myerson value (Definition 4.9) of the union games for elements in the partition. A *game with a proximity relation* among the agents is a pair (v, ρ) with $v \in \mathcal{G}^N$ and $\rho \in FC^N$. Let ω be the Myerson–Owen value (Definition 5.6).

Definition 5.13 Let pl be an inner extension for proximity relations. The pl-Myerson–Owen value is a value for games over N with proximity relation defined for each $v \in \mathcal{G}^N$, $\rho \in FC^N$ with $pl(\rho) = \{(r_k, s_k)\}_{k=1}^h$ and $i \in N$ as

$$\omega_i^{pl}(v, \rho) = \sum_{k=1}^{h} s_k \omega_i(v, r_k).$$

Although the Myerson value can be written by components (Proposition 4.10), in this case it is not possible to join the solution in only one expression because the union games are different for each component. Observe that, fixed a player $i \in N$ and a game $v \in \mathcal{G}^N$, the pl-Myerson–Owen value is constructed as a consistent mapping (5.7) with pl.

Example 5.15 Consider the unanimity game u_S with $S = \{1, 3, 4\}$ for $N = \{1, 2, 3, 4\}$, and the proximity relation in Fig. 5.5. We use the cooperation structures in Table 5.3. In the first graph r_1, we have only one union. The Myerson–Owen value coincides with the Myerson value, so

$$\omega(u_{\{1,3,4\}}, r_1) = \mu(u_{\{1,3,4\}}, r_1) = (1/3, 0, 1/3, 1/3).$$

In the second graph, r_2, we have two unions $\{1, 2, 3\}$ and $\{4\}$. Observe that both are necessary to get a winning coalition and, into $\{1, 2, 3\}$, player 2 is necessary to connect 1 and 3. So, we get

$$\omega(u_{\{1,3,4\}}, r_2) = (1/6, 1/6, 1/6, 1/2).$$

Graph r_3 is an a priori union system with three unions and then, by Proposition 5.4,

$$\omega(u_{\{1,3,4\}}, r_3) = (1/3, 0, 1/3, 1/3).$$

The last one represents the classical situation an then

$$\omega(u_{\{1,3,4\}}, r_4) = \phi(u_{\{1,3,4\}}) = (1/3, 0, 1/3, 1/3).$$

The cg-Myerson–Owen value is

$$\begin{aligned}\omega^{cg}(u_{\{1,3,4\}}, \rho) &= 0.7(1/3, 0, 1/3, 1/3) + 0.3(1/6, 1/6, 1/6, 1/2)\\ &= (0.283, 0.05, 0.283, 0.383).\end{aligned}$$

We find out a formula for the cg-extension as a Choquet integral. We take for any player $i \in N$ the signed capacity $\omega_i(v)$ over cooperation structures[4] defined for each $r \in C^N$ as

$$\omega_i(v)(r) = \omega_i(v, r),$$

being ω the Myerson–Owen value. The next result follows of the above definition and (4.10).

Theorem 5.5 *Let $\rho \in FC^N$. For each game v and player i it holds*

$$\omega_i^{cg}(v, \rho) = \int_c \rho \, d\omega_i(v).$$

We propose now an axiomatization of the fuzzy Myerson–Owen values inspired by the axioms of the Owen value (Theorem 5.1) and the Myerson–Owen value (Theorem 5.4). This axiomatization is based on that given in [7] for the cg-partition. Let f be a value for games with proximity relation, namely a function which obtains a vector $f(v, \rho) \in \mathbb{R}^N$ for each game with proximity relation (v, ρ). Consider fixed a reflexive extension pl for proximity relations in the following axioms.

Efficiency. For all $v \in \mathcal{G}^N$ and $\rho \in FC^N$ it holds $f(v, \rho)(N) = v(N)$.

Remember that $S \subseteq N$ is a null coalition in a game $v \in \mathcal{G}^N$ if each player $i \in S$ is a null player in v. In a cooperation structure, the agents in a null union have null payoffs. Now, players in a null group do not obtain profit while the coalition was a union. Let ρ be a proximity relation with $pl(\rho) = \{(r_k, s_k)\}_{k=1}^h$ and $H = \{1, \ldots, h\}$. We choose the next set of elements in the partition for a group $B \in [N/\rho]_{pl}$

$$K_B^+ = \left\{k \in H : \exists\, B' \in null^v(r_k) \text{ with } B \subseteq B'\right\},$$

and $K_B^- = H \setminus K_B^+$. Observe that, in $\rho_{pl}^{K_B^+}$, group B is null.

pl-**Null** group. Let $\rho \in FC^N$ and $v \in \mathcal{G}^N$. If $B \in [N/\rho]_{pl}$ is a group then for all $i \in B$ it holds

$$f_i(v, \rho) = s_{K_B^-} f_i\left(v, \rho_{pl}^{K_B^-}\right).$$

Particularly if we consider a cooperation structure r and $B \in N/r$ a null coalition for the game v, the axiom says that $f_i(v, \rho) = 0$ for all $i \in B$, i.e., it coincides with the null union axiom (Sect. 5.3). But pl-null group property also obtains information when B is not null group, because in that case $K_B^- = H$, $s_{K_B^-} = 1$, and then

[4]In this case we add to the family of cooperation structures the graph 0 with worth 0 in order to be a signed capacity.

$$f_i(v,\rho) = f_i(v,\rho_{pl}^H).$$

Two coalitions $S, T \subseteq N$ with $S \cap T = \emptyset$ are symmetry in a game v if $v(R \cup S) = v(R \cup T)$ for all $R \subseteq N \backslash (S \cup T)$. We can suppose that while two groups are symmetry the total payoff for each group is the same, namely if $B, B' \in [N/\rho]_{pl}$ in a proximity relation ρ with $pl(\rho) = \{(r_k, s_k)\}_{k=1}^h$, $H = \{1, \ldots, h\}$ and

$$K_{BB'}^+ = \{k \in H : B, B' \in N/r_k\} \neq \emptyset,$$

then

$$f\left(v, \rho_{pl}^{K_{BB'}^+}\right)(B) = f\left(v, \rho_{pl}^{K_{BB'}^+}\right)(B'). \tag{5.8}$$

But we propose a similar condition using the next axiom (in the same way as in [7]), the part of the payoffs for each group which is not obtained out of the common activity set must be the same. We denote $K_{BB'}^- = H \setminus K_{BB'}^+$.

***pl*-Symmetry in the quotient** Let $\rho \in FC^N$. If $B, B' \in [N/\rho]_{pl}$ with $K_{BB'}^+ \neq \emptyset$ are symmetry in v then

$$f(v,\rho)(B) - s_{K_{BB'}^-} f\left(v, \rho_{pl}^{K_{BB'}^-}\right)(B) = f(v,\rho)(B') - s_{K_{BB'}^-} f\left(v, \rho_{pl}^{K_{BB'}^-}\right)(B').$$

When we take a cooperation structure r the axiom says: if $B, B' \in N/r$ are symmetric in a game v then $f(v,\rho)(B) = f(N, v, \rho)(B')$, i.e., it coincides with symmetry in the quotient, see Sect. 5.3. Observe that, by Lemma 5.2, our pl-Myerson–Owen value satisfies the fuzzy symmetry in the quotient game if and only if it holds (5.8), because it is a consistent mapping with pl.

We extend the modified fairness axiom (see Sect. 5.3) to a fuzzy situation. Let $\rho \in FC^N$ with $pl(\rho) = \{(r_k, s_k)\}_{k=1}^h$ and $H = \{1, \ldots, h\}$. Consider $i, j \in N$ two different players with $\rho(ij) > 0$, we denote as

$$K_{ij} = \{k \in H : ij \in L(r_k)\}.$$

Observe that K_{ij} can be empty.

Modified *pl*-fairness Let ρ be a proximity relation. For each $i, j \in N$, $i \neq j$, with $\rho(ij) > 0$ and $k \in K_{ij}$ it holds

$$\begin{aligned} f_i(v,\rho) - f_j(v,\rho) &= (1 - s_k)\left[f_i\left(v, \rho_{pl}^{H\backslash\{k\}}\right) - f_j\left(v, \rho_{pl}^{H\backslash\{k\}}\right)\right] \\ &+ s_k\left[f_i\left(v_{N_{ij}^i}, (r_k)_{-ij}\right) - f_j\left(v_{N_{ij}^j}, (r_k)_{-ij}\right)\right], \end{aligned}$$

where N_{ij}^i, N_{ij}^j are defined for r_k as in Sect. 5.3.

If we consider a cooperation structure then last axiom coincides with the modified fairness (Sect. 5.3).

Finally, we suppose a common axiom of the Shapley-type values: linearity.

Linearity For all games $v, w \in \mathcal{G}^N$, $a, b \in \mathbb{R}$ and ρ proximity relation over N,

$$f(av + bw, \rho) = af(v, \rho) + bf(w, \rho).$$

Next theorem characterizes the pl-Myerson–Owen value with an inner extension.[5]

Theorem 5.6 *Let pl be an inner extension for proximity relations. The pl-Myerson–Owen value is the only value for games with proximity relation satisfying the following axioms: efficiency, pl-null group, pl-symmetry in the quotient, modified pl-fairness and linearity.*

Proof First we will test each one of the axioms. Suppose $\rho \in FC^N$ and $pl(\rho) = \{(r_k, s_k)\}_{k=1}^h$ with $H = \{1, \ldots, h\}$. Observe that, fixed a game v and a player i, mapping ω_i^{pl} is consistent with pl.

EFFICIENCY. Theorem 5.4 showed that the Myerson–Owen value ω satisfies efficiency. Hence,

$$\sum_{i\in N} \omega_i^{pl}(v, \rho) = \sum_{k=1}^h s_k \sum_{i\in N} \omega_i(v, r_k) = v(N) \sum_{k=1}^h s_k = v(N),$$

because as ρ is a proximity relation then $\sum_{k=1}^h s_k = 1$.

pl-NULL GROUP. Let $B \in [N/\rho]_{pl}$ and $i \in B$.

Suppose $K_B^+ \neq \emptyset$, namely B is a null coalition. Applying Lemma 5.2 to the partition of H, $\{K_B^+, K_B^-\}$.

$$\omega_i^{pl}(v, \rho) = s_{K_B^+}\omega_i^{pl}\left(v, \rho_{pl}^{K_B^+}\right) + s_{K_B^-}\omega_i^{pl}\left(v, \rho_{pl}^{K_B^-}\right).$$

Let $k \in K_B^+$. From the election of K_B^+, there exists $B' \in N/r_k$ with $B \subseteq B'$ and then $i \in B'$. Moreover B' is a union null for v. Since the Myerson–Owen value satisfies null union (Theorem 5.4), we get $\omega_i(v, r_k) = 0$. So, following Definition 5.11

$$\omega_i^{pl}\left(v, \rho_{pl}^{K_B^+}\right) = \sum_{k\in K_B^+} \frac{s_k}{s_{K_B^+}} \omega_i(v, r_k) = 0.$$

[5] An open problem at this moment is to study probabilistic cooperation structures as Calvo et al. [5] analyzed probabilistic communication structures. Remember that pb is not inner.

If B is not a null coalition then $K_B^+ = \emptyset$, thus $K_B^- = H$ and $s_{K_B^-} = 1$. As pl is inner we have $pl(\rho_{pl}^H) = pl(\rho)$. Hence our value satisfies for all player $i \in B$

$$\omega_i^{pl}(v,\rho) = \sum_{k=1} s_k \omega_i(v, r_k) = \omega_i^{pl}(v, \rho_{pl}^H).$$

pl-SYMMETRY IN THE QUOTIENT. Let $B, B' \in [N/\rho]_{pl}$, with $K_{BB'}^+ \neq \emptyset$, symmetric in a game v. Lemma 5.2 over the partition $\{K_{BB'}^+, K_{BB'}^-\}$ implies

$$\sum_{i \in B}\left[\omega_i^{pl}(v,\rho) - s_{K_{BB'}^-}\omega_i^{pl}\left(v, \rho_{pl}^{K_{BB'}^-}\right)\right] = s_{K_{BB'}^+}\sum_{i \in B}\omega_i^{pl}\left(v, \rho_{pl}^{K_{BB'}^+}\right),$$

and also

$$\sum_{j \in B'}\left[\omega_j^{pl}(v,\rho) - s_{K_{BB'}^-}\omega_j^{pl}\left(v, \rho_{pl}^{K_{BB'}^-}\right)\right] = s_{K_{BB'}^+}\sum_{j \in B'}\omega_j^{pl}\left(v, \rho_{pl}^{K_{BB'}^+}\right).$$

If $k \in K_{BB'}^+$ then $B, B' \in N/r_k$. Since Theorem 5.4 the Myerson–Owen value verifies symmetry in the quotient, therefore

$$\sum_{i \in B}\omega_i(v, r_k) = \sum_{j \in B'}\omega_j(v, r_k).$$

Hence,

$$\sum_{i \in B}\omega_i^{pl}\left(v, \rho_{pl}^{K_{BB'}^+}\right) = \sum_{k \in K_{BB'}^+}\frac{s_k}{s_{K_{BB'}^+}}\sum_{i \in B}\omega_i\,(v, r_k) = \sum_{k \in K_{BB'}^+}\frac{s_k}{s_{K_{BB'}^+}}\sum_{j \in B'}\omega_j\,(v, r_k)$$
$$= \sum_{j \in B'}\omega_j^{pl}\left(v, \rho_{pl}^{K_{BB'}^+}\right).$$

MODIFIED pl- FAIRNESS. Suppose ij with $i \neq j$, $\rho(ij) > 0$ and $K_{ij} \neq \emptyset$. Let $k \in K_{ij}$. We use that $\sum_{k \in H} s_k = 1$ to say from Lemma 5.2 again

$$\omega_i^{pl}(v,\rho) = (1 - s_k)\omega_i^{pl}\left(v, \rho_{pl}^{H \setminus \{k\}}\right) + s_k\omega_i(v, r_k),$$

and also

$$\omega_j^{pl}(v,\rho) = (1 - s_k)\omega_j^{pl}\left(v, \rho_{pl}^{H \setminus \{k\}}\right) + s_k\omega_j(v, r_k).$$

The subtraction of both equalities is

$$\omega_i^{pl}(v,\rho) - \omega_j^{pl}(v,\rho) = (1 - s_k)\left[\omega_i^{pl}\left(v, \rho_{pl}^{H \setminus \{k\}}\right) - \omega_j^{pl}\left(v, \rho_{pl}^{H \setminus \{k\}}\right)\right]$$
$$+ s_k[\omega_i(v, r_k) - \omega_j(v, r_k)].$$

Using notation (5.4),

$$\omega_i(v, r_k) - \omega_j(v, r_k) = \omega_i\left(v_{N^i_{ij}}, (r_k)_{ij}\right) - \omega_j\left(v_{N^j_{ij}}, (r_k)_{ij}\right),$$

because of ω satisfies modified fairness since Theorem 5.4.
LINEARITY. We get

$$\begin{aligned}\omega_i^{pl}(av + bw, \rho) &= \sum_{k=1}^{k} s_k\omega_i(av + bw, r_k) \\ &= a\sum_{k=1}^{k} s_k\omega_i(v, r_k) + b\sum_{k=1}^{k} s_k\omega_i(w, r_k) \\ &= a\omega_i^{pl}(v, \rho) + b\omega_i^{pl}(w, \rho).\end{aligned}$$

Suppose f^1, f^2 two values for games with proximity relation satisfying the five axioms. Let $\rho \in FC^N$ with $pl(\rho) = \{(r_k, s_k)\}_{k=1}^h$ and $H = \{1, \ldots, h\}$. We prove the result by induction on the cardinality h.

Let $h = 1$. Of course $pl(\rho) = \{(r, 1)\}$ but perhaps $\rho \neq r$. Hence in this case we will use the uniqueness for the family of cooperation structures (Theorem 5.4). For each $i \in N$ there exists $B \in N/r$ with $i \in B$. We have $K_B^+ = \emptyset$, namely $\rho_{pl}^H = r$, and then pl-null group axiom implies

$$f^1(v, \rho) = f^1(v, r) = f^2(v, r) = f^2(v, \rho).$$

Otherwise B is a null coalition and pl-null group axiom implies

$$f^1(v, \rho) = 0 = f^2(v, \rho),$$

because of $K_B^- = \emptyset$.

We suppose that there is only one value for all the games with a proximity relation ρ with $h < d$.

Consider now a proximity relation ρ with $h = d$. If $f^1 \neq f^2$ linearity implies that there exists a unanimity game u_T satisfying

$$f^1(u_T, \rho) \neq f^2(u_T, \rho).$$

We set $M_T = \{B \in [N/\rho]_{pl} : B \cap T \neq \emptyset\}$. If $B \notin M_T$ then B is a null group for u_T. We apply the pl-null group property for each player $i \in B$,

$$f_i^1(u_T, \rho) = s_{K_B^-} f_i^1\left(u_T, \rho_{pl}^{K_B^-}\right) = s_{K_B^-} f_i^1\left(u_T, \rho_{pl}^{K_B^-}\right) = f_i^2(u_T, \rho),$$

using that $|K_B^-| < d$. Particularly if $|K_B^+| = d$ then $s_{K_B^-} = 0$ and $f_i^1(u_T, \rho) = 0 = f_i^2(u_T, \rho)$. Let $B, B' \in M_T$ with $K_{BB'}^+ \neq \emptyset$. pl-Symmetry in the quotient, as $|K_{BB'}^-| < d$, implies that

$$\begin{aligned} f^1(u_T, \rho)(B) - f^1(u_T, \rho)(B') \\ &= s_{K_{BB'}^-}\left[f^1\left(u_T, \rho_{pl}^{K_{BB'}^-}\right)(B) - f^1\left(u_T, \rho_{pl}^{K_{BB'}^-}\right)(B')\right] \\ &= s_{K_{BB'}^-}\left[f^2\left(u_T, \rho_{pl}^{K_{BB'}^-}\right)(B) - f^2\left(u_T, \rho_{pl}^{K_{BB'}^-}\right)(B')\right] \\ &= f^2(u_T, \rho)(B) - f^2(u_T, \rho)(B'), \end{aligned}$$

Thus $f^1(u_T, \rho)(B) - f^2(u_T, \rho)(B) = f^1(u_T, \rho)(B') - f^2(u_T, \rho)(B')$. Choose any $k \in H$. All $B, B' \in M_T \cap N/r_k$ satisfy $K_{BB'}^+ \neq \emptyset$ and then there exists J_k with

$$f^1(u_T, \rho)(B) - f^2(u_T, \rho)(B) = J_k \quad \forall B \in M_T \cap N/r_k.$$

If $B \in N/r_k$ with $B \notin M_T$ then we got before $f_i^1(u_T, \rho) = f_i^1(u_T, \rho)$ for all $i \in B$. As $N^{r_k} = N$ because ρ is a proximity relation, we obtain using efficiency

$$\begin{aligned} \sum_{i \in N}[f_i^1(u_T, \rho) - f_i^2(u_T, \rho)] &= \sum_{B \in M_T \cap N/r_k} \sum_{i \in B} f_i^1(N, u_T, \rho) - f_i^2(N, u_T, \rho) \\ &= |M_T \cap N/r_k| J_k = 0, \end{aligned}$$

and then $J_k = 0$. Let $i \in N$ such that for all $B \in [N/\rho]_{pl}$ with $i \in B$ it happens that $B \in M_T$. If $\{i\} \in M_T$ then $f_i^1(u_T, \rho) = f_i^2(u_T, \rho)$. For each $i \in B \in M_T \cap N/r_k$ then there exists $j \in B \setminus \{i\}$. If $r_k(ij) = 1$ we apply modified pl-fairness to this link

$$\begin{aligned} f_i^1(u_T, \rho) - f_j^1(u_T, \rho) &= (1 - s_k)\left[f_i^1\left(v, \rho_{pl}^{H\setminus\{k\}}\right) - f_j^1\left(v, \rho_{pl}^{H\setminus\{k\}}\right)\right] \\ &\quad + s_k\left[f_i^1\left(v_{N_{ij}^i}, (r_k)_{-ij}\right) - f_j^1\left(v_{N_{ij}^j}, (r_k)_{-ij}\right)\right] \\ &= (1 - s_k)\left[f_i^2\left(v, \rho_{pl}^{H\setminus\{k\}}\right) - f_j^2\left(v, \rho_{pl}^{H\setminus\{k\}}\right)\right] \\ &\quad + s_k\left[f_i^2\left(v_{N_{ij}^i}, (r_k)_{-ij}\right) - f_j^2\left(v_{N_{ij}^j}, (r_k)_{-ij}\right)\right] \\ &= f_i^2(u_T, \rho) - f_j^2(u_T, \rho). \end{aligned}$$

If $r_k(ij) = 0$ then there is a path from i to j, repeating the above reasoning we get the same equality, thus

$$f_j^2(u_T, \rho) - f_j^1(u_T, \rho) = f_i^2(u_T, \rho) - f_i^1(u_T, \rho) \quad \forall j \in B.$$

Hence,

$$\sum_{j \in B} f_j^2(u_T, \rho) - f_j^1(u_T, \rho) = |B|[f_i^2(u_T, \rho) - f_i^1(u_T, \rho)] = 0.$$

We have finally $f_i^1(u_T, \rho) = f_i^2(u_T, \rho)$. □

Next proposition enumerates several properties of the pl-Myerson–Owen values. The proof follows from Propositions 5.6 and 5.7.

Proposition 5.10 *The pl-Myerson–Owen value satisfies the following properties for a proximity relation ρ with $pl(\rho) = \{(r_k, s_k)\}_{k=1}^h$ and game v.*

(1) For all $i \in N$ it holds

$$\omega_i^{pl}(v, \rho) = \sum_{k=1}^{r} v_{p_k}^{N/r_k}(\{i\}) - \omega_i^{pl}(v^{svg}, \rho)$$

where $i \in B_{p_k} \in N/r_k$ for each k.

(2) If v is convex and $i \in N$ then $\omega_i^{pl}(v, \rho) \geq v(\{i\})$.

5.6 Similarity Relations

The pl-Myerson–Owen value can be seen as fuzzy versions of the Myerson–Owen value (Definition 5.6) for games with proximity relation. Similarity relations is the subfamily of proximity relations associated to the a priori unions structures of Owen [13] because the bilateral relations among the players are transitive.

Definition 5.14 A similarity relation is a fuzzy bilateral relation ρ over N satisfying

(1) Reflexivity: $\rho(i, i) = \rho(ii) = 1$ for all $i \in N$,
(2) Symmetry: $\rho(i, j) = \rho(j, i) = \rho(ij)$, for all $i, j \in N$and
(3) Transitivity: $\rho(i, j) \geq \rho(i, k) \wedge \rho(k, j)$ for all $i, j, k \in N$.

The set of similarity relations is denoted by FP^N

Obviously it is possible to analyze other models using different T-norms.

We consider again, as in Sect. 4.6, only extensions pl which are *transitive* (Definition 4.29), namely if ρ is transitive and $pl(\rho) = \{(r_k, s_k)\}_{k=1}^h$ then r_k is transitive too for all k. Proposition 4.27 showed that the cg-extension is transitive for communication structures and then also for proximity relations. We test in the next proposition that the pp-extension is also transitive.

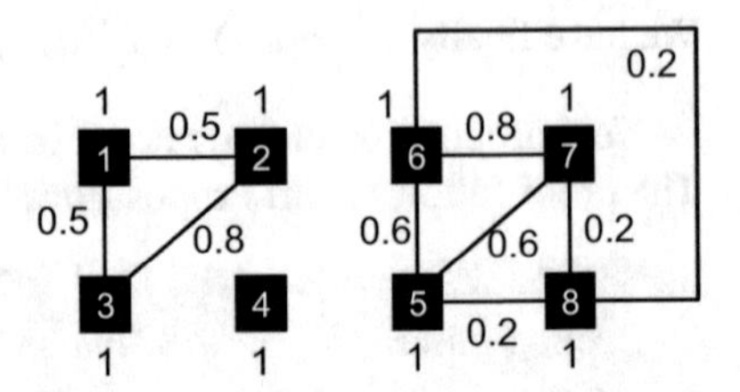

Fig. 5.9 Similarity relation

Proposition 5.11 *The prox-proportional extension for proximity relations is transitive.*

Proof Let ρ be a similarity relation and $pp(\rho) = \{(r_k, s_k)\}_{k=1}^{h}$. We must to prove that for all $k = 1, \ldots, h$ the graph r_k is transitive. Consider r_1. Suppose three different players $i, i', i'' \in N$ such that $r_1(ii') = r_1(i'i'') = 1$ (otherwise the result is trivial). Remember that $s_1 = \vee\{\rho(ij) : ij \in L(\rho)\}$ and $r_1(ij) = 1$ with $i \neq j$ if and only if $\rho(ij) = s_1$. As ρ is transitive

$$s_1 \geq \rho(ii'') \geq \rho(ii') \wedge \rho(i'i'') = s_1.$$

Therefore we obtain $\rho(ii'') = s_1$ and $r_1(ii'') = 1$. Now we test that, in the next step of the algorithm, the new fuzzy graph $\rho_1 = \rho - s_1 r_1$ is also a transitive fuzzy bilateral structure (although it is not a proximity relation). Observe that

$$\rho_1(ij) = \rho(ij) \wedge (1 - s_1),$$

if $i \neq j$ and $\rho_1(ij) > 0$, because the links in $L(r_1)$ are deleted and the others only reduce the level if their vertices obtain now less level than them. Using again the transitivity condition of ρ, as $\rho(ii'') \geq \rho(ii') \wedge \rho(i'i'')$ we get

$$\rho(ii'') \wedge (1 - s_1) \geq \rho(ii') \wedge \rho(i'i'') \wedge (1 - s_1).$$

Finally the reasoning can be repeated with $\rho = \rho_1$ and go on. □

Example 5.16 We represent in Fig. 5.9 the similarity relation with matrix

$$\rho = \begin{bmatrix} 1 & 0.5 & 0.5 & 0 & 0 & 0 & 0 & 0 \\ 0.5 & 1 & 0.8 & 0 & 0 & 0 & 0 & 0 \\ 0.5 & 0.8 & 1 & 0 & 0 & 0 & 0 & 0 \\ 0 & 0 & 0 & 1 & 0 & 0 & 0 & 0 \\ 0 & 0 & 0 & 0 & 1 & 0.6 & 0.6 & 0.2 \\ 0 & 0 & 0 & 0 & 0.6 & 1 & 0.8 & 0.2 \\ 0 & 0 & 0 & 0 & 0.6 & 0.8 & 1 & 0.2 \\ 0 & 0 & 0 & 0 & 0.2 & 0.2 & 0.2 & 1 \end{bmatrix}$$

Let pl be a transitive reflexive extension. The *pl-Owen value* is a value for games with similarity relation defined as the pl-Myerson–Owen value of the game with the similarity relation as a proximity one. We can obtain an axiomatization for the pl-Owen value from the axiomatization of the pl-Myerson–Owen value, but not all the axioms are feasible into the family of similarity relations. Next lemma will be useful for this purpose.

Lemma 5.3 *Let pl be a transitive reflexive extension. If $\rho \in FP^N$ with $pl(\rho) = \{(r_k, s_k)\}_{k=1}^h$ and $H = \{1, \ldots, h\}$ then $\rho_{pl}^K \in FP^N$ for all $K \subseteq H$.*

Proof Let $K \subseteq H$. We know that ρ_{pl}^K is a proximity relation, thus we only need to test transitivity. Let i, i', i'' three different players. As pl is reflexive and transitive then r_k is also reflexive, symmetric and transitive for all $k \in H$, namely r_k is an a priori union system. Hence $r_k(ii'') \geq r_k(ii') \wedge r_k(i'i'')$. So,

$$\rho_{pl}^K(ii'') = \sum_{k \in K} \frac{s_k}{s_K} r_k(ii'') \geq \sum_{k \in K} \frac{s_k}{s_K} [r_k(ii') \wedge r_k(i'i'')]$$

$$= \left[\sum_{k \in K} \frac{s_k}{s_K} r_k(ii')\right] \wedge \left[\sum_{k \in K} \frac{s_k}{s_K} r_k(i'i'')\right] = \rho_{pl}^K(ii') \wedge \rho_{pl}^K(i'i'').$$

□

Observe that the modified pl-fairness is not feasible because if we reduce the level of a pair of players we can break up the transitivity. In exchange, we introduce this other axiom used for the Owen value. Let pl be a transitive and reflexive extension. Let ρ be a similarity relation with $pl(\rho) = \{(r_k, s_k)\}_{k=1}^h$ and $H = \{1, \ldots, h\}$. For two different players $i, j \in N$ we denote

$$K_{ij}^+ = \{k \in H : \exists B \in N/r_k \text{ with } i, j \in B\},$$

and $K_{ij}^- = H \setminus K_{ij}^+$.[6] Suppose f a value for games with a similarity relation.

***pl*-Symmetry in a group** Let pl be a transitive extension and ρ be a similarity relation over N. If i, j are symmetric for the game v then

$$f_i(v, \rho) - f_j(v, \rho) = s_{K_{ij}^-} \left[f_i\left(v, \rho_{pl}^{K_{ij}^-}\right) - f_j\left(v, \rho_{pl}^{K_{ij}^-}\right)\right].$$

[6] Set $K_{ij} \subseteq K_{ij}^+$ because of being in the same component does not imply a direct link.

Theorem 5.7 *Let pl be an inner, transitive and reflexive extension. The pl-Owen value is the only value for games with a similarity relation which satisfies efficiency, pl-null group, pl-symmetry in the quotient, pl-symmetry in a group and linearity.*

Proof Theorem 5.6 and Lemma 5.3 ensure that the pl-Owen value satisfies efficiency, pl-null group, pl-symmetry in the quotient and linearity. Hence we only have to check that the pl-Owen value verifies pl-symmetry in a group. Let $i, j \in N$ be two symmetry players in a game v. We have

$$\omega_i(v, \rho) = s_{K_{ij}^+}\omega_i\left(v, \rho_{pl}^{K_{ij}^+}\right) + s_{K_{ij}^-}\omega_i\left(v, \rho_{pl}^{K_{ij}^-}\right),$$

and also

$$\omega_j(v, \rho) = s_{K_{ij}^+}\omega_j\left(v, \rho_{pl}^{K_{ij}^+}\right) + s_{K_{ij}^-}\omega_j\left(v, \rho_{pl}^{K_{ij}^-}\right).$$

The Owen value satisfies symmetry in a union, therefore for all $k \in K_{ij}^+$ we get $\omega_i(v, r_k) = \omega_j(v, r_k)$. So,

$$\omega_i\left(v, \rho_{pl}^{K_{ij}^+}\right) = \omega_j\left(v, \rho_{pl}^{K_{ij}^+}\right).$$

The uniqueness part is similar to Theorem 5.6 using pl-symmetry in a group instead of modified pl-fairness. □

References

1. Albizuri, M.J.: Axiomatizations of the Owen value without efficiency. Math. Soc. Sci. **55**, 78–89 (2008)
2. Aumann, R.J., Dreze, J.H.: Cooperative games with coalition structures. Int. J. Game Theory **3**(4), 217–237 (1974)
3. Barberá, S., Gerber, A.: On coalition formation: durable coalition structure. Math. Soc. Sci. **45**(2), 185–203 (2003)
4. Calvo, E., Gutiérrez, E.: The Shapley-Solidarity value for games with coalition structure. Int. Game Theory Rev. **15**(1) (2013)
5. Calvo, E., Lasaga, J., van den Nouweland, A.: Values for games with probabilistic graphs. Math. Soc. Sci. **37**, 70–95 (1999)
6. Casajus, A.: Beyond basic structures in game theory. Ph.D. thesis. University of Leipzig, Germany (2007)
7. Fernández, J.R., Gallego, I., Jiménez-Losada, A., Ordóñez, M.: Cooperation among agents with a proximity relation. Eur. J. Oper. Res. **250**(2), 555–565 (2016)
8. Hamiache, G.: The Owen value values friendship. Int. J. Game Theory **29**(4), 517–532 (2001)
9. Hart, S., Kurz, M.: Endogenous formation of coalitions. Econometrica **51**, 1047–1064 (1983)

10. Jiménez-losada, A., Fernández, J.R., Ordóñez, M.: Myerson values for games with fuzzy communication structure. Fuzzy Sets Syst. **213**, 74–90 (2013)
11. Kamijo, Y.: A two-step Shapley value in a cooperative game with a coalition structure. Int. Game Theory Rev. **11**(2), 207–214 (2009)
12. Myerson, R.B.: Graphs and cooperation in games. Math. Oper. Res. **2**(3), 225–229 (1977)
13. Owen, G.: Values of games with a priori unions. Mathematical Economics and Game Theory. Lecture Notes in Economics and Mathematical Systems, vol. 141, pp. 76–88 (1977)
14. Owen, G.: Game Theory, 4th edn. Emerald Group Publishing, Bingley (2013)
15. Sánchez-Soriano, J., Pulido, M.A.: On the core, the weber set and convexity in games with a priori unions. Eur. J. Oper. Res. **193**, 468–475 (2009)
16. van den Brink, R., Khmelnitskaya, A., van der Laan, G.: An Owen-type value for games with two-level communication structure. Ann. Oper. Res. **243**(1), 179–198 (2016)
17. Vázquez-Brage, M., García-Jurado, I., Carreras, F.: The Owen value applied to games with graph-restricted communication. Games Econ. Behav. **12**(1), 42–53 (1996)

Chapter 6
Fuzzy Permission

6.1 Introduction

In this chapter we present another interesting model of games with a bilateral relation among the agents different to the communication model [6] (see Chap. 4) and the a priori unions one [7] (see Chap. 5). Hierarchical structures appears in a lot of human systems: firms, economic organizations, social networks, economic networks, protocolos... We consider several models in the literature with a same line which can be studied from bilateral relations. Gilles et al. [4] in 1992 introduced the conjunctive approach of permission structure. Permission structures analyze hierarchical organizations with a certain interpretation, the influence of an agent on the action of another one is based on permission: the last player needs the participation of the first one to be considered as an active element. Huettner and Wiese [5] showed another different vision of the hierarchical systems, the coercive structures. Now the influence is thought as coercivity, the decision of certain agents implies the action of others.

We focus the chapter in the conjunctive approach [4] for permission systems. The conjunctive approach in the permission way supposes that each player needs the participation of all her superiors (understanding superiors not only those predecessors with direct relation over the agent, also those influencing the before processors and go on). We study in the chapter a broader option following van den Brink and Dietz [11] and Gallardo et al. [1]. First we consider only direct relations and later, by a transitivity extension, we arises the before conjunctive approach. This option has also related a particular disjunctive version of the coercive model, where the presence of an agent implies the action of all her successors. In 1997 van den Brink [9] analyzed the disjunctive approach. In this approach players only need permission from one of the direct predecessors, but this one needs one from her direct predecessors and go on.

The fuzzy version of the conjunctive approach was studied by Gallardo et al. [1] in 2014 following a particular fuzziness based on the Choquet integral. Later, in 2015, Gallardo et al. [2] introduced authorization structures. These systems permit to study hierarchical situations in a more general sense although they are not based on

A. Jiménez-Losada, *Models for Cooperative Games with Fuzzy Relations among the Agents*, Studies in Fuzziness and Soft Computing 355,
DOI 10.1007/978-3-319-56472-2_6

bilateral relations. In that context the reader can find the fuzzy version of the others models in the permission line. A deep work about authorization structure is given in Gallardo [3].

6.2 Permission Structures: The Conjunctive Approach

Hierarchical organizations are usual systems in firms, economic institutions and networks. We understand, following van den Brink [8], a hierarchical structure as an organization where "there are agents who have a direct influence on the actions taken by other agent in the organization". In this book we will use a concept of hierarchy in a broad sense but into our context.[1] Let N be our finite set of players.

Gilles et al. [4] introduced a *permission structure* over the set of players N as a mapping $su : N \to 2^N$ which is asymmetric, namely if $i \in su\,(j)$ then $j \notin su\,(i)$ (note that this condition implies $i \notin su\,(i)$). Players in $su\,(i)$ are named *successors* of i and the *predecessors* of i are $su^{-1}(i) = \{j \in N : i \in su\,(j)\}$. Let su be a permission structure. A player $j \in N$ is *subordinate* of another one i if there exists $\{i_q\}_{q=1}^{p}$ with $i_1 = i$, $i_p = j$ and $i_q \in su\,(i_{q-1})$ for all $q = 2, \ldots, p$. The set of subordinates of i is $\hat{su}\,(i)$ and the *superiors* of i are $\hat{su}^{-1}(i) = \{j \in N : i \in \hat{su}\,(j)\}$. The *conjunctive approach*[2] supposes that a player i needs the permission of all her predecessors to play. If a coalition S is formed in a game, only those players with all their superiors into the coalition can actively play, so they only obtain the profit for the subcoalition,

$$sov(S) = \{i \in S : \hat{su}^{-1}(i) \subseteq S\}. \tag{6.1}$$

Following the way of the general model in this book, Gilles et al. [4] introduced the information in the characteristic function of the game using the worths of the sovereign part, so the *conjunctive permission game* for each coalition S is

$$v^{sov}(S) = v(sov(S)). \tag{6.2}$$

The above structure supposes that all the players are active in the play time, we explain a slight different concept in the context of bilateral relations where players can obtain profits without being active. The modification is based on two points. The first one, we suppose that although an agent can be necessary to give permission she may not be active. In [4, 11] the authors introduced the idea of unproductive superiors but as a property from the game, now we consider it into the structure following Gallardo

[1] The concept of hierarchy in van den Brink [8] for instance involves transitivity, reflexivity and also only one top in the structure.

[2] van den Brink [9] introduced another approach, the disjunctive one, for hierarchical permission structures. The fuzzy version of the model has been studied in Gallardo et al. [2] in the context of authorization structures (a broader field for hierarchical structures which are not necessarily bilateral relations).

Fig. 6.1 Permission structure

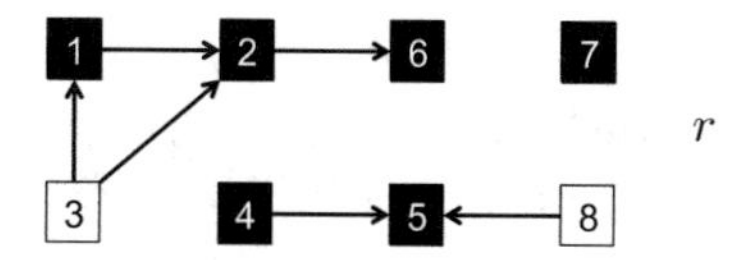

et al. [2]. So, a predecessor can study if it is advisable to participate or not as active player. The second one, following van den Brink and Dietz [11] and Gallardo et al. [1], we propose a first step of dependency, directly from the predecessors without mandatory transitivity. We define a permission structure here in this way.

Formally, a permission structure can be any bilateral relation, but we will impose the anti-symmetric condition that is equivalent of working with acycle relations (van den Brink [9] named these kind of permission structures as hierarchical).

Definition 6.1 A permission structure is a bilateral relation r over N satisfying antisymmetry, namely if $r(i, j) = 1$, with $i \neq j$, then $r(j, i) = 0$. The family of permission structures is denoted as A^N.

In this case: if $r(i, i) = 1$ then player i is active and if $r(i, j) = 1$ with $i \neq j$ then j needs the authorization of player i so as to enable her to work effectively.

Example 6.1 In Fig. 6.1 we show a permission structure over eight agents, $N = \{1, 2, 3, 4, 5, 6, 7, 8\}$, with matrix

$$r = \begin{bmatrix} 1\,1\,0\,0\,0\,0\,0\,0 \\ 0\,1\,0\,0\,0\,1\,0\,0 \\ 1\,1\,0\,0\,0\,0\,0\,0 \\ 0\,0\,0\,1\,1\,0\,0\,0 \\ 0\,0\,0\,0\,1\,0\,0\,0 \\ 0\,0\,0\,0\,0\,1\,0\,0 \\ 0\,0\,0\,0\,0\,0\,1\,0 \\ 0\,0\,0\,0\,1\,0\,0\,0 \end{bmatrix}.$$

The structure represents a hierarchical situation where players 3 and 8 are not active but they are needed for the action of several other players. Players 4 and 7 can play without the permission of another player.

Van den Brink and Dietz [11] introduced a local conjunctive option of dependence, modifying the concept of sovereign part. We adapt the definition to our context. Now, its formulation is as follows.

Definition 6.2 Let S be a coalition. The (conjunctive) sovereign part of S in a permission structure $r \in A^N$ is

$$\sigma^r(S) = \{i \in S \cap N^r : j \in S \, \forall r(j,i) = 1\}.$$

Example 6.2 Following Example 6.1, we determine the sovereign part of several coalitions. We have $\sigma^r(\{1,2\}) = \emptyset$ but $\sigma^r(\{1,2,3\}) = \{1,2\}$, namely player 1 is not active but her presence permits to play players 1 and 2. The sovereign part $\sigma^r(\{4,5,6,7\}) = \{4,7\}$ because 5 and 6 need all their direct predecessors in the directed graph, but there is none of 6 and only one of 5 (player 4). Now if player 8 joints to the before coalition we get $\sigma_r^c(\{4,5,6,7,8\}) = \{4,5,7\}$. Observe that

$$\sigma^r(N) = \{1,2,4,5,6,7\}.$$

The sovereign part satisfies the following properties.

Proposition 6.1 *The sovereign part in a permission structure r satisfies the following properties:*

(1) $\sigma^r(\emptyset) = \emptyset$ *and* $\sigma^r(N) = dom\,(r)$.
(2) *If* $S \subseteq T$ *then* $\sigma^r(S) \subseteq \sigma^r(T)$.
(3) $\sigma^r(S) \cup \sigma^r(T) \subseteq \sigma^r(S \cup T)$.
(4) $\sigma^r(S) \cap \sigma^r(T) = \sigma^r(S \cap T)$.

Proof (1) It is trivial by definition of the sovereign part.
(2), (3) Suppose $i \in \sigma^r(S)$. If $r(j,i) = 1$ then $j \in S \subseteq T$, and therefore we obtain $i \in \sigma^r(T)$ because also $i \in T$. Now, as $S, T \subseteq S \cup T$ then $\sigma^r(S) \cup \sigma^r(T) \subseteq \sigma^r(S \cup T)$.
(4) As $S \cap T \subseteq S$ and $S \cap T \subseteq T$ then $\sigma^r(S \cap T) \subseteq \sigma^r(S) \cap \sigma^r(T)$. Let $i \in \sigma^r(S) \cap \sigma^r(T)$ then we have for all $j \in N$ with $r(j,i) = 1$ that $j \in S \cap T$. Hence $i \in \sigma^r(S \cap T)$. □

Observe that the sovereign part is different than (6.1), proposed by Gilles et al. [4], because here the action of a player is feasible only with the permission of her direct superiors and not the others, but also is slight different than that in van den Brink and Dietz [11] because not all the players need to be active (they coincide when $N^r = N$). We relate our structures with those given by Gilles et al. [4].

Proposition 6.2 *Each structure su is identified with an order relation, a reflexive and transitive permission structure r.*

Proof In a permission structure following [4] all players are active, thus any relation defining the situation must be reflexive. Now, for $i \neq j$ we take $r_{su}(i, j) = 1$ if and only if $j \in \hat{su}(i)$. So $\sigma^{r_{su}}(S) = sov(S)$ for all coalition S. Obviously r_{su} is transitive because if $r_{su}(i, j) = 1$ and $r_{su}(j, k) = 1$ then there is a path from i to j and another one from j to k. Adding both paths we get a path from i to j. □

We will consider next reduction and extension in order to get better conditions for our permission structure.

Definition 6.3 Let r be a permission structure.

(1) The quasi-reflexive interior of r is another permission structure r° verifying $r^{\circ}(i, j) = 0$ if $r(j, j) = 0$ and $r^{\circ}(i, j) = r(i, j)$ otherwise.
(2) The transitive closure of r is a new permission structure $\hat{r}$ such that $\hat{r}(i, j) = 1$ if and only if there is a path in r from i to j, namely there exists $(j_k)_{k=0}^{p} \subseteq N$ with $j_0 = i$, $j_p = j$ and $r(i_{k-1}, j_k) = 1$ for all $k = 1, \ldots, p$.

Observe that quasi-reflexivity (see Sect. 3.3) in a permission structure implies that we only use the permission relations when they affect to active players, otherwise they are irrelevant. We show now that it is enough to analyze quasi-reflexivity permission structures for our conjunctive option. Let $r \in A^N$ be a permission structure. It is trivial to prove that the quasi-reflexive interior of r satisfies

$$\sigma^{r^{\circ}}(S) = \sigma^{r}(S). \tag{6.3}$$

So, given a permission structure we will usually consider its sovereign part as the sovereign part of its quasi-reflexive interior. The transitive closure permits to change from the direct model to the usual permission model. Next example shows these operations over a permission structure. Generally $\hat{r^{\circ}} \neq \hat{r}^{\circ}$, because of the first one loses important information from the original structure r.

Example 6.3 Figure 6.2 represents a hierarchical structure r in a firm by a permission structure with matrix

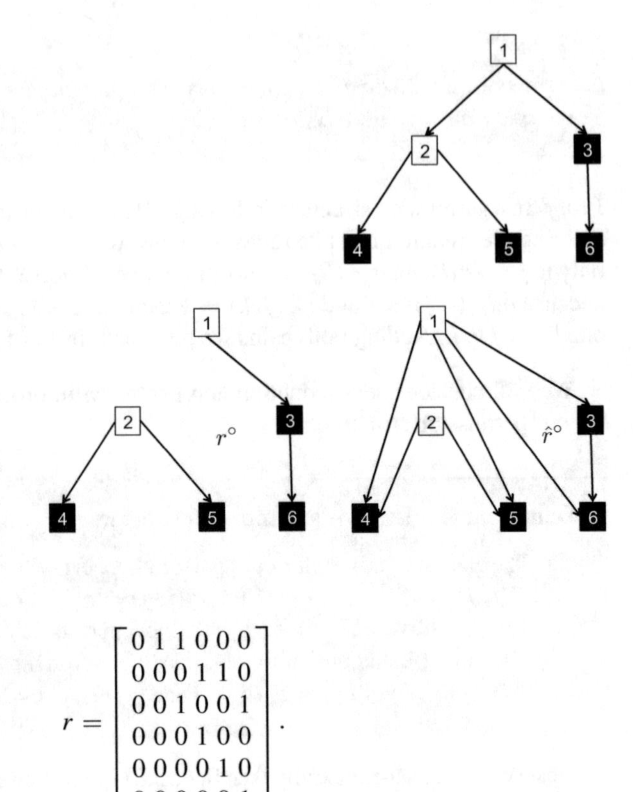

Fig. 6.2 Hierarchical structure

Fig. 6.3 Hierarchical structure

$$r = \begin{bmatrix} 0\,1\,1\,0\,0\,0 \\ 0\,0\,0\,1\,1\,0 \\ 0\,0\,1\,0\,0\,1 \\ 0\,0\,0\,1\,0\,0 \\ 0\,0\,0\,0\,1\,0 \\ 0\,0\,0\,0\,0\,1 \end{bmatrix}.$$

In this classical hierarchical organizations each player has at most one predecessor, and there is only one big boss on the top. The direct model suppose that each element in the hierarchy needs the permission of her immediately predecessor to play but not all her superiors. For instance, player 6 needs permission of player 3 to act, but not of the big boss 1. Hence link 12 is not useful in the direct model. In this case player 1 cannot obtain profits from the action of players 4, 5 or 6. Thus we can use only permission structure r° on the left in Fig. 6.3. But if we use the transitive closure then we suppose that each player needs the permission of all her superiors, so $\hat{r}^\circ$ on the right in Fig. 6.3 represents this new situation. Now player 5 needs the participation of players 1,2. We take in the second option also the quasi-reflexive interior but from the transitive closure, in order to reduce the relations. So, we delete again link 12.

$$r^\circ = \begin{bmatrix} 0\,0\,1\,0\,0\,0 \\ 0\,0\,0\,1\,1\,0 \\ 0\,0\,1\,0\,0\,1 \\ 0\,0\,0\,1\,0\,0 \\ 0\,0\,0\,0\,1\,0 \\ 0\,0\,0\,0\,0\,1 \end{bmatrix} \quad \hat{r}^\circ = \begin{bmatrix} 0\,0\,1\,1\,1\,1 \\ 0\,0\,0\,1\,1\,0 \\ 0\,0\,1\,0\,0\,1 \\ 0\,0\,0\,1\,0\,0 \\ 0\,0\,0\,0\,1\,0 \\ 0\,0\,0\,0\,0\,1 \end{bmatrix}.$$

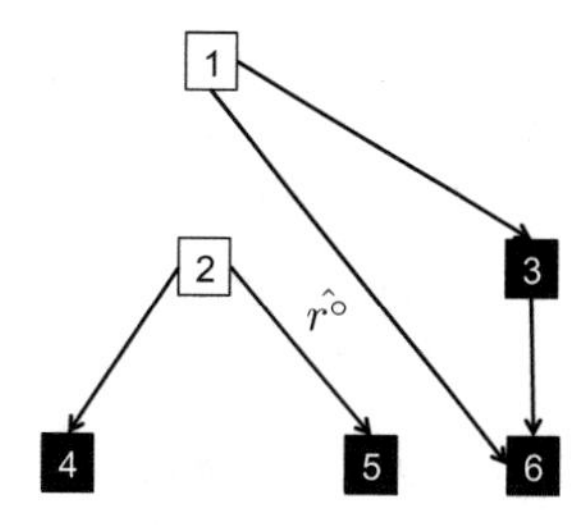

Fig. 6.4 Different ordering closure and interior

Observe that we cannot change the ordering of closure and interior. Figure 6.4 represents $\hat{r^\circ}$. We have lost information with respect to the another option, players 4 and 5 do not depend on player 1 in spite of the transitivity.

Given a cooperative game and a permission structure we construct a new game using the worths of the sovereign parts of the coalitions.

Definition 6.4 Let $v \in \mathcal{G}^N$ and $r \in A^N$. The (conjunctive) permission game is defined for all coalition S as

$$v^r(S) = v(\sigma^r(S)) = v(\sigma^{r^\circ}(S)).$$

We use here v^r as permission game because there in not incongruence.

Example 6.4 Consider r the permission structure in Fig. 6.5 for four players. If we take the anonymous game $v(S) = |S|$, the conjunctive permission game for the non-empty coalitions is in Table 6.1. For instance, player 2 needs permission from both players, 1 and 3, and player 3 is not active, thus

$$v^r(\{2, 3, 4\}) = v(\sigma^r(\{2, 3, 4\}) = v(\{4\}) = 1.$$

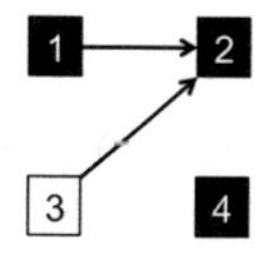

Fig. 6.5 The permission structure in Example 6.4

Table 6.1 Game v^r

S	$\{1\}$	$\{2\}$	$\{3\}$	$\{4\}$	$\{1,2\}$	$\{1,3\}$	$\{1,4\}$	$\{2,3\}$
$v^r(S)$	1	0	0	1	1	1	2	0
S	$\{2,4\}$	$\{3,4\}$	$\{1,2,3\}$	$\{1,2,4\}$	$\{1,3,4\}$	$\{2,3,4\}$	N	
$v^r(S)$	1	1	2	2	2	1	3	

The profit obtained by a coalition taking into account the permission structure is actually the profit that its sovereign part can earn. Next properties are satisfied by the direct conjunctive game.

Proposition 6.3 *Let $v \in \mathcal{G}^N$ be a game and $r \in A^N$ be a permission structure.*

(1) If v is monotonic then v^r is monotonic.
(2) If v is monotonic and superadditive then so is v^r
(3) If v is monotonic and convex then v^r is convex.
(4) If v is simple then so is v^r.
(5) $(av + bw)^r = av^r + bw^r$ for all $a, b \in \mathbb{R}$. Hence $(-v)^r = -(v^r)$.

Proof (1) Let $v \in \mathcal{G}_m^N$. From Proposition 6.1 $\sigma^r(S) \subseteq \sigma^r(T)$ when $S \subseteq T$. Hence

$$v^r(S) = v(\sigma^r(S)) \leq v(\sigma^r(T)) = v^r(T).$$

(2) Suppose $v \in \mathcal{G}_{sa}^N \cap \mathcal{G}_m^N$. Let S, T be coalitions with $S \cap T = \emptyset$. Proposition 6.1 implies $\sigma^r(S) \cup \sigma^r(T) \subseteq \sigma^r(S \cup T)$. Moreover, as $\sigma^r(S) \cap \sigma^r(T) = \emptyset$. Thus, monotonicity and superadditivity of v imply

$$\begin{aligned} v^r(S \cup T) &= v(\sigma^r(S \cup T)) \geq v(\sigma^r(S) \cup \sigma^r(T)) \\ &\geq v(\sigma^r(S)) + v(\sigma^r(T)) = v^r(S) + v^r(T). \end{aligned}$$

(3) Let $v \in \mathcal{G}_c^N$. The proof is similar because $\sigma^r(S \cap T) = \sigma^r(S) \cap \sigma^r(T)$. We get

$$\begin{aligned} v^r(S \cup T) + v^r(S \cap T) &= v(\sigma^r(S \cup T)) + v(\sigma^r(S \cap T)) \\ &\geq v(\sigma^r(S)) + v(\sigma^r(T)) = v^r(S) + v^r(T). \end{aligned}$$

(4) The first step said that v^r is monotone if so is v. But the worth of a coalition in the direct conjunctive game is the worth of one of its sub-coalitions then if v is a $\{0, 1\}$-game then so is v^r.
(5) A simple exercise for the reader. □

Observe that we have needed monotonicity for getting that superaditivity and convexity are inherited. This is either because it can be $\sigma^r(S) \cup \sigma^r(T) \subsetneq \sigma^r(S \cup T)$. Unlike the other models in the above chapters, additivity is not inheritable.

Example 6.5 Suppose the permission structure with matrix

$$r = \begin{bmatrix} 1 & 1 \\ 0 & 1 \end{bmatrix},$$

representing two players with 2 depending on 1. If $v = (v_1, v_2)$ is an additive game then $v^r(\{1\}) = v_1$, $v^r(\{2\}) = 0$ but $v^r(N) = v_1 + v_2$. Except if $v_2 = 0$ game v^r is not additive. If we take $v_2 = -v_1$ with $v_1 > 0$ then v superadditive but it is not monotone. The permission game is not superadditive, $v^r(\{1\}) = v_1$, $v^r(\{2\}) = 0$ and $v^r(N) = 0$.

Hence, if two games are strategically equivalent their direct conjunctive games are not always strategically equivalent. Also the saving game has problems. If $v \in \mathcal{G}^N$ and $r \in A^N$ then

$$(v^{svg})^r(S) = v^{svg}(\sigma^r(S)) = \sum_{i\in\sigma^r(S)} v(\{i\}) - v(\sigma^r(S)) = \sum_{i\in\sigma^r(S)} v(\{i\}) - v^r(S),$$

but

$$(v^r)^{svg}(S) = \sum_{i\in S} v^r(\{i\}) - v^r(S) = \sum_{i\in ind_r\cap S} v(\{i\}) - v^r(S),$$

where $ind_r = \{i \in N^r : r(j,i) = 0 \,\forall j \neq i\}$ is the set of independent active players in r. Obviously $ind_r \cap S \subseteq \sigma^r(S)$ but the equality is not always true. Thus generally $(v^r)^{svg} \neq (v^{svg})^r$. If for a game v we define the additive game $r^v \in \mathbb{R}^N$ for each $i \in N$ as

$$r_i^v = \begin{cases} 0, & \text{if } i \in ind_r \\ v(\{i\}), & \text{otherwise} \end{cases}$$

then we get the following property from the above equalities.

Proposition 6.4 *Let $v \in \mathcal{G}^N$ and $r \in A^N$. It holds*

$$(v^{svg})^r = (r^v)^r + (v^r)^{svg}.$$

Proof From the above equalities,

$$(v^{svg})^r(S) - (v^r)^{svg}(S) = \sum_{i\in\sigma^r(S)\setminus ind_r} v(\{i\}) = r^v(\sigma^r(S)) = (r^v)^r(S).$$

□

Furthermore, last proposition says that $(v^r)^{svg}$ is the permission game of certain game strategically equivalent to v^{svg} because of

$$(v^r)^{svg} = (v^{svg} - r^v)^r,$$

by Proposition 6.3.

The dual game of a permission game is

$$(v^{dual})^r(S) = v^{dual}(\sigma^r(S)) = v(N) - v(N \setminus \sigma^r(S))$$

but in the other hand

$$(v^r)^{dual}(S) = v^r(N) - v^r(N \setminus S) = v(N^r) - v(\sigma^r(N \setminus S)).$$

We can use the T-dual game (3.7) with $T = N^r$ but so we only solve the problem of non active players,

$$(v^{N^r dual})^r(S) = v^{N^r dual}(\sigma^r(S)) = v(N^r) - v(N^r \setminus \sigma^r(S)), \tag{6.4}$$

and $N^r \setminus \sigma^r(S) \neq \sigma^r(N \setminus S)$. In fact, if we use again Example 6.1, we obtained in Example 6.2 that $\sigma^r(\{1, 2, 3\} = \{1, 2\}$ and then $N^r \setminus \sigma^r(\{1, 2, 3\}) = \{4, 5, 6, 7\}$. But we also got $\sigma^r(\{4, 5, 6, 7, 8\}) = \{4, 5, 7\}$.

Huettner and Wiese [5] introduced disjunctive coercive structures.[3] Now the permission structure is understood as a coercive relation, in the sense that if a player decides to participate (actively or not) in a coalition then all her successors must be active in the coalition. In our context the coercive game is given as follows.

Definition 6.5 The (disjunctive) coercive game is defined for a game $v \in \mathcal{G}^N$ and a permission structure $r \in A^N$ as

$$\bar{v}^r(S) = v(\bar{\sigma}^r(S)),$$

where $\bar{\sigma}^r(S) = \{i \in N^r : \exists j \in S \text{ with } r(j, i) = 1\}$ is named the coercive part.

Observe that we have in this case $S \cap N^r \subseteq \bar{\sigma}^r(S)$ for any coalition because $r(i, i) = 1$ if $i \in S \cap N^r$. We name this version disjunctive because a player is obliged to participate only with the presence of one of her predecessors. The reader can also see that $\bar{\sigma}^r = \bar{\sigma}^{r^\circ}$.

Example 6.6 In this example we look at Fig. 6.5 as a coercive structure. Obviously we see the same but we change our interpretation. Now if player 3 is in the coalition then the activity of players 1, 2 is guaranteed. So

$$\bar{\sigma}^r(\{3\}) = \{1, 2\}.$$

[3] They introduced the model as the permission one in Gilles et al. [4], we present the model in a local way, as the permission one in [11]. It is also posible to define a coercive option in the conjunctive sense.

Table 6.2 Game $\bar{v}^r$

S	$\{1\}$	$\{2\}$	$\{3\}$	$\{4\}$	$\{1,2\}$	$\{1,3\}$	$\{1,4\}$	$\{2,3\}$
$\bar{v}^r(S)$	2	1	1	1	2	2	3	2
S	$\{2,4\}$	$\{3,4\}$	$\{1,2,3\}$	$\{1,2,4\}$	$\{1,3,4\}$	$\{2,3,4\}$	N	
$\bar{v}^r(S)$	2	3	2	3	3	3	3	

But in this case agents do not need the predecessors to play, so $\bar{\sigma}^r(\{2\}) = \{2\}$. Table 6.2 shows the coercive game of game $v(S) = |S|$.

There is a strong relation between both models, coercive and permission. Next proposition explains this goal in our context.

Proposition 6.5 *Let r be a permission structure. For all game $v \in \mathcal{G}^N$ it holds*

$$(\bar{v}^r)^{dual} = (v^{N^r dual})^r.$$

Proof We use the worth of the T-dual game in (6.4) with $T = N^r$. For the great coalition we get $v^r(N) = \bar{v}^r(N) = v(N^r)$. Let $S \subseteq N$ be a coalition. We prove the claim

$$N^r \setminus \sigma^r(S) = \bar{\sigma}^r(N \setminus S).$$

As $\sigma^r(S) \subseteq S$ then $N^r \cap (N \setminus S) \subseteq N^r \setminus \sigma^r(S)$. Also,

$$N^r \cap (N \setminus S) \subseteq N \setminus S \subseteq \bar{\sigma}^r(N \setminus S).$$

Observe that players in both sets must be in N^r. Suppose then $i \in N^r \cap S$. Player $i \notin \sigma^r(S)$ if and only if there is $j \in N \setminus S$ with $r(j,i) = 1$ if and only if $i \in \bar{\sigma}^r(N \setminus S)$. The claim is true. Now we have

$$\begin{aligned}(v^{N^r dual})^r(S) &= v(N^r) - v(N^r \setminus \sigma^r(S)) \\ &= \bar{v}^r(N) - v(\bar{\sigma}^r(N \setminus S)) = (\bar{v}^r)^{dual}(S).\end{aligned}$$

□

Hence, the coercive model has actually a strong relation with the permission one.

6.3 Values for Games with Permission Structure

In this section we study several values inspired on the conjunctive approach and following Gilles et al. [4]. Each pair $(v, r) \in \mathcal{G}^N \times A^N$ is called a *game over N with a permission structure*. A value function for games with permission structure is

$$f : \mathcal{G}^N \times A^N \to \mathbb{R}^N.$$

We define a first Shapley value for this model using the Shapley value of the direct conjunctive sovereign part, following [11].

Definition 6.6 The local (conjunctive) permission value is a value for games over N with hierarchical structure defined for each $v \in \mathcal{G}^N$ and $r \in A^N$ as

$$\delta(v, r) = \phi(v^r) = \phi(v^{r^\circ}).$$

Example 6.7 We consider game $v(S) = |S|^2$ with the permission structure r in Fig. 6.6 which is quasi-reflexive. Table 6.3 shows the sovereign part of each coalition and the worth in the permission game. Player 1 is not active in the structure but her permission is needed for some players. So, her payoff uses only the non-null marginal contributions in the permission game

$$\delta_1(v, r) = \phi_1(v^r) = \frac{1}{12}[v^r(\{1, 2\} - v^r(\{2\})] + \frac{1}{12}[v^r(\{1, 2, 3\}) - v^r(\{2, 3\})]$$
$$+ \frac{1}{12}[v^r(\{1, 2, 4\}) - v^r(\{2, 4\})] + \frac{1}{4}[v^r(N) - v^r(\{2, 3, 4\})] = \frac{32}{12} = \frac{8}{3}.$$

The local permission value is

$$\delta(v, r) = \left(\frac{16}{6}, \frac{19}{6}, \frac{9}{6}, \frac{10}{6}\right).$$

Gilles et al. [4] axiomatized his value using five axioms. Another axiomatization was given in [10]. We follow Gallardo et al. [1] here taking similar axioms. Let $f : \mathcal{G}^N \times A^N \to \mathbb{R}^N$ a value function for games over N with permission structure.

Only active players can participate obtaining profits, thus efficiency is restricted to the domain of the relation.

Restricted efficiency. For all $v \in \mathcal{G}^N$ and $r \in A^N$, $f(v, r)(N) = v(N^r)$.

The payoff of an agent not only comes from her own activity also from the activity of her successors. Hence, to get a null player we need the same condition for all

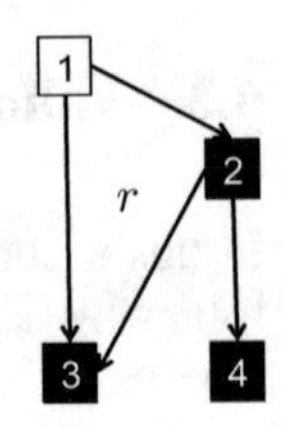

Fig. 6.6 Hierarchical structure in Example 6.7

Table 6.3 Game v^r and sovereign part in Example 6.7

S	$\{1\}$	$\{2\}$	$\{3\}$	$\{4\}$	$\{1,2\}$	$\{1,3\}$	$\{1,4\}$	$\{2,3\}$
$\sigma^r(S)$	$\emptyset$	$\emptyset$	$\emptyset$	$\emptyset$	$\{2\}$	$\emptyset$	$\emptyset$	$\emptyset$
$v^r(S)$	0	0	0	0	1	0	0	0
S	$\{2,4\}$	$\{3,4\}$	$\{1,2,3\}$	$\{1,2,4\}$	$\{1,3,4\}$	$\{2,3,4\}$	N	
$\sigma^r(S)$	$\{4\}$	$\emptyset$	$\{2,3\}$	$\{2,4\}$	$\emptyset$	$\{4\}$	$\{2,3,4\}$	
$v^r(S)$	1	0	4	4	0	1	9	

her successors. In our context an *inessential player*[4] for a game v and a permission structure r is a player $i \in N$ satisfying that j is a null player in v when $r(i, j) = 1$

Inessential player.[5] Let $r \in A^N$ and $v \in \mathcal{G}^N$. If $i \in N$ is an inessential player it holds $f_i(v, r) = 0$.

Next classical axiom (see Sect. 1.3) is true independently of the hierarchical structure.

Linearity For all $v, w \in \mathcal{G}^N$, $a, b \in \mathbb{R}$ and $r \in A^N$ it holds

$$f(av + bw, r) = af(v, r) + bf(w, r).$$

Players with veto power over necessary players are also necessaries.

Veto power over necessary players.[6] Let $v \in \mathcal{G}_m^N$ be a monotone game and $r \in A^N$. If $i \in N$ is a necessary player for v then for any $k \in N$ with $r(k, i) = 1$ it holds $f_k(v, r) \geq f_j(v, r)$ for all $j \in N$.

These axioms allow us to determine the local permission value in our context.

Theorem 6.1 *The local permission value is the only value for games over N with permission structure satisfying restricted efficiency, inessential player, veto power over necessary player, linearity.*

Proof We test that the local permission value verifies the four axioms. Suppose r a permission structure. As $\sigma^r = \sigma^{r^\circ}$ we can consider without losing generality that r is quasi-reflexive.

---

[4]The concept of inessential player here differs of the definition in [1, 11], because player i may not be active.

[5]van den Brink [8] called this axiom weakly inessential player because he used another stronger version for the disjunctive approach.

[6]This axiom is named strong necessary player in [11] and slight different of this one in [1].

RESTRICTED EFFICIENCY. Let $v \in \mathcal{G}^N$, as Shapley value is an efficient value (Proposition 1.9) we have

$$\sum_{i \in N} \delta_i(v, r) = \sum_{i \in N} \phi_i(v^r) = v^r(N) = v(N^r).$$

INESSENTIAL PLAYER. Let $i \in N$ such that j is a null player in a game v when $r(i, j) = 1$. We will prove the claim i is a null player in v^r. If there is not $j \in N$ with $r(i, j) = 1$ then $\sigma^r(S \cup \{i\})) = \sigma^r(S)$ for all $S \subseteq N \setminus \{i\}$ and we get the claim. Suppose then $\sigma^r(S \cup \{i\}) \neq \sigma^r(S)$. From Proposition 6.1 it holds that $\sigma^r(S) \subset \sigma^r(S \cup \{i\})$, so we denote

$$\sigma^r(S \cup \{i\}) \setminus \sigma^r(S) = \{i_1, \ldots, i_m\}.$$

Since Definition 6.3 each i_k with $k = 1, \ldots, m$ verifies $r(i, i_k) = 1$ because of $i_k \notin \sigma^r(S)$ but $i_k \in \sigma^r(S \cup \{i\})$. Hence they are null players. Using the concept of null player,

$$v(\sigma^r(S \cup \{i\})) = v(\sigma^r(S \cup \{i\}) \setminus \{i_1\}) = v(\sigma^r(S \cup \{i\}) \setminus \{i_1, i_2\}) = \cdots = v(\sigma^r(S)).$$

Furthermore

$$v^r(S \cup \{i\}) = v(\sigma^r(S \cup \{i\})) = v(\sigma^r(S)) = v^r(S).$$

Shapley value verifies null player (Proposition 1.11), thus by the claim $\delta_i^l(v, r) = 0$.
VETO POWER OVER NECESSARY PLAYERS. Let i be a necessary player in $v \in \mathcal{G}_m^N$. Suppose $r(k, i) = 1$. We see that k is a necessary player in v^r. Let S be a coalition with $k \notin S$. Obviously $i \notin \sigma^r(S)$ and then

$$v^r(S) = v(\sigma^r(S)) = 0.$$

Shapley value satisfies necessary player (Proposition 1.11), therefore for all j

$$\delta_k(v, r) = \phi_k(v^r) \geq \phi_j(v^r) = \delta_j(v, r).$$

LINEARITY. The result follows from Proposition 6.2 because of $(av + bw)^r = av^r + bw^r$.

Consider now f a value for games with permission structure satisfying the five axioms. From linearity we only have to obtain the uniqueness for the unanimity games. Let $T \subseteq N$ be a non-empty coalition. Remember that, for the unanimity game u_T, players out of T are null players and players within T are necessary players. We set

$$R_T = \{i \in N : \exists j \in T \text{ with } r(i, j) = 1\}.$$

Let $i \notin R_T$. In that case there is not $j \in T$ with $r(i, j) = 1$, namely if $r(i, j) = 1$ then $j \notin T$. Hence j is a null player for u_T. We get that player i is inessential. The inessential player axiom says that $f_i(u_T, r) = 0$. As players in T are necessary players, the veto power over necessary players axiom implies that all the players in R_T must have the same payoff, K. Now restricted efficiency says

$$\sum_{i \in N} f_i(u_T, r) = |R_T| K = u_T(N^r).$$

So, we obtain

$$K = \begin{cases} \dfrac{1}{|R_T|}, & T \subseteq N^r \\ 0, & \text{otherwise.} \end{cases}$$

□

We have needed four axioms to characterize the local permission value, we must analyze the independence of the axioms.

Remark 6.1 We find values different from the local permission value verifying all the axioms except one of them.

- Consider value f^1 defined for each $v \in \mathcal{G}^N$ and $r \in A^N$ as

$$f_i^1(v, r) = \sum_{\{j \in N^r : r(i,j)=1\}} \frac{v(\{j\})}{|\{h \in N : r(h, j) = 1\}|}$$

  for all player $i \in N$ (0, if there is not j with $r(i, j) = 1$). This value satisfies all the axioms except restricted efficiency. Linearity is trivial. If i is an inessential player then every player j with $r(i, j) = 1$ is a null player for game v, therefore $v(\{j\}) = 0$ and $f_i^1(v, r) = 0$. Let i a necessary player for v. We have $v(\{j\}) = 0$ for all $j \in N \setminus \{i\}$. If $v \in \mathcal{G}_m^N$ then $v(\{i\}) \geq 0$. Observe that

$$f_k^1(v, r) = \begin{cases} \dfrac{v(\{i\})}{|\{h \in N : r(h, i) = 1\}|}, & \text{if } r(k, i) = 1 \\ 0, & \text{otherwise.} \end{cases}$$

  Obviously $f^1 \neq \delta$.
- The value defined as $f_i^2(v, r) = \phi(v, N^r)$ satisfies all the axioms except veto power over the necessary players. We take into account the axiomatization of the extended Shapley value, Theorem 3.1. Observe that inessential player is true because of the extended Shapley value satisfies restricted null player and carrier. The extended Shapley value also verifies restricted efficiency and linearity. But $f^2 \neq \delta$.

- Similar to the egalitarian value (see Remark 1.4) we defined

$$f_i^3(v, \mathscr{B}) = \frac{v(N^r)}{n}.$$

This value satisfies all the axioms except inessential player axiom.

- Following Remark 1.4 again we take

$$Ine^r(v) = \{i \in N : j \text{ null player in } v \text{ if } r(i, j) = 1\}.$$

It defines

$$f_i^4(v) = \begin{cases} \dfrac{v(N^r)}{|N \setminus Ine(v)|}, & \text{if } i \notin Ine(v) \\ 0, & \text{if } i \in Ine(v). \end{cases}$$

This value satisfies all the axioms except additivity. As f^3 this value satisfies restricted efficient and veto power over the necessary players property. But also the inessential player axiom by construction. The problem is that the denominator in the formula depends of the game ($Ine^r(v)$ changes with the game).

Remark 6.2 The above axioms imply particularly a logical property of the local permission value from its definition, the payoffs of the players do not depend of the superfluous links. We can determine values such that the payoff of a player in the structure is not the same than in the quasi-reflexive interior. Consider for instance a pair $(v, r) \in \mathscr{G}^N \times A^N$ the following new permission structure $r_v \in A^N$. Let $i \notin N^r$ and $j \in N^r \setminus null(v)$ (remember $null(v)$ is the set of null players in v) then $r_v(i, j) = 1$ if they are connected in r by a path $\{i_k\}_{k=1}^m$ ($i_1 = i, i_m = j$) with $i_k \notin N^r$ for all $k = 1, \ldots, m-1$, and $r_v(i, j) = r(i, j)$ otherwise. Let $f(v, r) = \phi(v^{r_v})$. If we take for instance

$$r = \begin{bmatrix} 0\,1\,0 \\ 0\,0\,1 \\ 0\,0\,1 \end{bmatrix} \quad r_{u_{\{3\}}} = \begin{bmatrix} 0\,1\,1 \\ 0\,0\,1 \\ 0\,0\,1 \end{bmatrix} \quad (r^\circ)_{u_{\{3\}}} = \begin{bmatrix} 0\,0\,0 \\ 0\,0\,1 \\ 0\,0\,1 \end{bmatrix}$$

then $f(u_{\{3\}}, r) = (1/3, 1/3, 1/3)$ and $f(u_{\{3\}}, r^\circ) = (0, 1/2, 1/2)$.

Next we prove some properties of the local permission value.

Proposition 6.6 *The local permission value satisfies the following properties for a permission structure r.*

(1) Let $v \in \mathscr{G}_c^N \cap \mathscr{G}_m^N$. If $S \subseteq N$ then $\delta(v, r)(S) \geq v(\sigma_r^c(S))$.
(2) If $v \in \mathscr{G}^N$ then

$$\delta_i(v^{svg}, r) = v^r(\{i\}) + \delta_i(r^v, r) - \delta_i(v, r).$$

Proof (1) Proposition 6.3 implies that v^r is convex. Hence, Proposition 1.14 says

$$\delta(v,r)(S) = \phi(v^r)(S) \geq v^r(S) = v(\sigma^r(S)).$$

(2) We use Proposition 6.4, the linearity of the Shapley value (Proposition 1.8) and Proposition 1.15 to get the equality. So, for every $i \in N$

$$\begin{aligned}\delta_i(v^{svg}, r) &= \phi_i((v^{svg})^r) = \phi_i((r^v)^r) + \phi_i((v^r)^{svg}) \\ &= \phi_i((r^v)^r) + v^r(\{i\}) - \phi_i(v^r) = \delta_i(r^v, r) + v^r(\{i\}) - \delta_i(v, r).\end{aligned}$$

□

The proof of the next proposition is removed from the proof of the uniqueness in Theorem 6.1. Remember that e^S represents the canonical vector for coalition S. We obtain the payoffs of the players in a unanimity game working in a permission systems r.

Proposition 6.7 *Let $T \subseteq N$ be a non empty coalition. Let $r \in A^N$ and $R^r_T = \{i \in N : \exists j \in T \text{ with } r(i,j) = 1\}$. For each $i \in N$ it holds*

$$\delta(u_T, r) = \begin{cases} \dfrac{1}{|R^r_T|} e^{R^r_T}, & \text{if } T \subseteq N^r \\ 0, & \text{otherwise.} \end{cases}$$

We can describe the local permission value from the dividends using the above proposition.

Theorem 6.2 *For each $v \in \mathcal{G}^N$ and a permission structure $r \in A^N$ the local permission value of a player $i \in N$ is*

$$\delta_i(v,r) = \sum_{\{T \subseteq N^r : i \in R^r_T\}} \frac{\Delta^v_T}{|R^r_T|},$$

where $R^r_T = \{i \in N : \exists j \in T \text{ with } r(i,j) = 1\}$.

Proof From Proposition 1.1 and Proposition 6.5 we get

$$\delta_i(v,r) = \sum_{i \in T \subseteq N} \Delta^v_T \delta_i(u_T, r) = \sum_{\{T \subseteq N^r : i \in R^r_T\}} \Delta^v_T \frac{1}{|R^r_T|}.$$

□

Table 6.4 Dividends of the game and sets R^r_T in Example 6.8

T	{2}	{3}	{4}	{2, 3}	{2, 4}	{3, 4}	{2, 3, 4}
R^r_T	{1, 2}	{1, 2, 3}	{2, 4}	{1, 2, 3}	{1, 2, 4}	N	N
Δ^v_T	1	1	1	2	2	2	0

Example 6.8 We test the above formula over the game $v(S) = |S|^2$ in Example 6.7. The permission structure r is in Fig. 6.6. Table 6.4 shows the dividends (calculated using formula (1.2)) of the game and the corresponding sets R^r_T. Observe that we only use coalitions $T \subseteq N^r$. Now we determine the payoff of Player 1 as

$$\delta_1(v, r) = \frac{\Delta^v_{\{2\}}}{2} + \frac{\Delta^v_{\{3\}}}{3} + \frac{\Delta^v_{\{2,3\}}}{3} + \frac{\Delta^v_{\{2,4\}}}{3} + \frac{\Delta^v_{\{3,4\}}}{4} = \frac{8}{3}.$$

We get again

$$\delta(v, r) = \left(\frac{16}{6}, \frac{19}{6}, \frac{9}{6}, \frac{10}{6}\right).$$

The second Shapley value for games with a permission structure uses the Shapley value of the sovereign part in the transitive closure, following [1].

Definition 6.7 The (conjunctive) permission value is a value for games over N with permission structure defined for each $v \in \mathcal{G}^N$ and $r \in A^N$ as

$$\hat{\delta}(v, r) = \delta(v, \hat{r}) = \phi(v^{\hat{r}}).$$

This value coincides with the permission value in Gilles et al. [4] when r is reflexive.

Example 6.9 We consider again game $v(S) = |S|^2$ in Example 6.7 with the permission structure r in Fig. 6.6. Figure 6.7 shows the transitive closure. We use again formula in Theorem 6.2, and then we can take the dividends of the game calculated before. We only need to change to sets $R^{\hat{r}}_T$. The payoff of player 1 for the permission value is

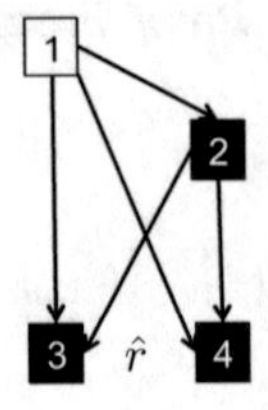

Fig. 6.7 Transitive closure in Example 6.9

Table 6.5 Dividends of the game and sets $R_T^{\hat{r}}$ in Example 6.9

T	{2}	{3}	{4}	{2, 3}	{2, 4}	{3, 4}	{2, 3, 4}
$R_T^{\hat{r}}$	{1, 2}	{1, 2, 3}	{1, 2, 4}	{1, 2, 3}	{1, 2, 4}	N	N
Δ_T^v	1	1	1	2	2	2	0

$$\hat{\delta}_1(v, r) = \frac{\Delta^v_{\{2\}}}{2} + \frac{\Delta^v_{\{3\}}}{3} + \frac{\Delta^v_{\{4\}}}{3} + \frac{\Delta^v_{\{2,3\}}}{3} + \frac{\Delta^v_{\{2,4\}}}{3} + \frac{\Delta^v_{\{3,4\}}}{4} = 3.$$

The permission value is, using Table 6.5,

$$\hat{\delta}(v, r) = (3, 3, 1.5, 1.5).$$

Next axioms determine the permission value. Only active players can participate obtaining profits, thus efficiency is restricted to the domain of the relation. For all $v \in \mathcal{G}^N$ and $r \in A^N$, $f(v, r)(N) = v(N^r)$.
Transitive inessential player. Let $r \in A^N$ and $v \in \mathcal{G}^N$. If $i \in N$ is a player satisfying that j is an inessential player in $\hat{r}$ for v, it holds $f_i(v, r) = 0$.

Transitive veto power over necessary players. Let $v \in \mathcal{G}_m^N$ be a monotone game and $r \in A^N$. If $i \in N$ is a necessary player for v then $f_k(v, r) \geq f_j(v, r)$ for all $j \in N$ and $\hat{r}(k, i) = 1$.

The proof of next theorem is like that one of Theorem 6.1.

Theorem 6.3 *The permission value is the only value for games over N with permission structure satisfying restricted efficiency, transitive inessential player, transitive veto power over necessary player and linearity.*

Also Proposition 6.5 implies the following properties of the permission value. Observe that for all $r \in A^N$ we have $ind_{\hat{r}} = ind_r$, and then $\hat{r}^v = r^v$.

Proposition 6.8 *The permission value satisfies the following properties for a permission structure r.*

(1) Let $v \in \mathcal{G}_c^N \cap \mathcal{G}_m^N$. If $S \subseteq N$ then $\hat{\delta}(v, r)(S) \geq v(\sigma^{\hat{r}}(S))$.
(2) If $v \in \mathcal{G}^N$ then

$$\hat{\delta}_i(v^{svg}, r) = \hat{\delta}_i(r^v, r) + v^r(\{i\}) - \hat{\delta}_i(v, r).$$

Finally we study a coercive value using the coercive game given in Definition 6.5. It was introduced by Huettner and Wiese [5].

Definition 6.8 The local coercive value is a value for games with permission structure defined for each game v and permission structure r as

$$\bar{\delta}(v, r) = \phi(\bar{v}^r).$$

Next proposition permits to calculate the coercive value from the permission value using the T-dual game (3.7).

Theorem 6.4 *Let $v \in \mathcal{G}^N$ and $r \in A^N$. The local coercive value satisfies*

$$\bar{\delta}(v, r) = \delta(v^{N^r dual}, r).$$

Proof Proposition 6.5 established a relation between the duals of the local permission game and the local coercive game. So, from Proposition 1.15

$$\begin{aligned}\bar{\delta}(v, r) &= \phi(\bar{v}^r) = \phi\left((\bar{v}^r)^{dual}\right) = \phi\left((v^{N^r dual})^r\right) \\ &= \delta(v^{N^r dual}, r).\end{aligned}$$

□

Example 6.10 Now we use the permission structure in Fig. 6.6 as a coercive structure. We consider again game $v(S) = |S|^2$. Table 6.6 shows the active part of each coalition in the coercive sense and the worth in the coercive game. Player 1 is not active in the structure but her coercive activity imposes the activity of some players. So, her payoff uses only the non-null marginal contributions in the coercive game

$$\begin{aligned}\bar{\delta}_1(v, r) = \phi_1(\bar{v}^r) &= \frac{1}{4}\bar{v}^r(\{1\}) + \frac{1}{12}[\bar{v}^r(\{1, 3\}) - \bar{v}^r(\{3\})] \\ &+ \frac{1}{12}[v^r(\{1, 4\}) - v^r(\{4\})] + \frac{1}{12}[v^r(\{1, 3, 4\}) - v^r(\{3, 4\})] = \frac{28}{12} = \frac{7}{3}.\end{aligned}$$

The local coercive value is

$$\bar{\delta}(v, r) = \left(\frac{14}{6}, \frac{29}{6}, \frac{3}{6}, \frac{8}{6}\right).$$

We calcule now the value by dividends using formulas in Theorems 6.2 and 6.4. We need to do the T_r-dual of v,

$$v^{N^r dual}(S) = 9 - |\{2, 3, 4\} \setminus S|^2.$$

Table 6.6 Game $\bar{v}^r$ and coercive part in Example 6.10

S	{1}	{2}	{3}	{4}	{1, 2}	{1, 3}	{1, 4}	{2, 3}
$\bar{\sigma}^r(S)$	{2, 3}	{2, 3, 4}	{3}	{4}	{2, 3, 4}	{2, 3}	{2, 3, 4}	{2, 3, 4}
$\bar{v}^r(S)$	4	9	1	1	9	4	9	9
S	{2, 4}	{3, 4}	{1, 2, 3}	{1, 2, 4}	{1, 3, 4}	{2, 3, 4}	N	
$\bar{\sigma}^r(S)$	{2, 3, 4}	{3, 4}	{2, 3, 4}	{2, 3, 4}	{2, 3, 4}	{2, 3, 4}	{2, 3, 4}	
$\bar{v}^r(S)$	9	4	9	9	9	9	9	

Table 6.7 Dividends of the T-dual game and sets R_T^r in Example 6.10

T	{2}	{3}	{4}	{2, 3}	{2, 4}	{3, 4}	{2, 3, 4}
R_T^r	{1, 2}	{1, 2, 3}	{2, 4}	{1, 2, 3}	{1, 2, 4}	N	N
$\Delta_T^{v^{T_r-dual}}$	5	5	5	−2	−2	−2	0

Table 6.7 determines the T-dual dividends and its local permission value. We can test that $\bar{\delta}(v, r) = \delta(v^{N^r dual}, r)$.

Huettner and Wiese [5] introduced a new axiom to distinguish the coercive model from the permission value. We present here a slight different axiom. Let v be a game. A player $i \in N$ is named *sufficient* in v if

$$v(\{i\}) = \bigvee_{T \subseteq N} v(T). \tag{6.5}$$

Coercion power over sufficient players. Let $v \in \mathcal{G}_m^N$ be a monotone game and $r \in A^N$. If $i \in N$ is a sufficient player for v and $k \in N$ with $r(k, i) = 1$ then $f_k(v, r) \geq f_j(v, r)$ for all $j \in N$.

This axiom replaces veto power over necessary players property in the axiomatization for the new value.

Theorem 6.5 *The local coercive value is the only value for games over N with permission structure satisfying restricted efficiency, inessential player, coercion power over sufficient player and linearity.*

Proof We test that the local coercive value verifies the four axioms. Suppose r a permission structure (quasi-reflexive without losing generality).

EFFICIENCY. Let $v \in \mathcal{G}^N$, efficiency of the Shapley value (Proposition 1.9) implies

$$\sum_{i \in N} \bar{\delta}_i(v, r) = \sum_{i \in N} \phi_i(\bar{v}^r) = \bar{v}^r(N) = v(N^r).$$

INESSENTIAL PLAYER. If i is a null player for v then she is also a null player for $v^{N^r dual}$. In fact, if $S \subseteq N \setminus \{i\}$ then

$$\begin{aligned} v^{N^r dual}(S \cup \{i\}) &= v(N^r) - v(N^r \setminus (S \cup \{i\})) \\ &= v(N^r) - v(N^r \setminus S) = v^{N^r dual}(S). \end{aligned}$$

Suppose now i satisfies the conditions of the axiom, Theorem 6.1 implies that

$$\bar{\delta}_i(v, r) = \delta_i(v^{N^r dual}, r) = 0.$$

COERCION POWER OVER SUFFICIENT PLAYERS. First we prove that if v is monotone then $v^{N^r dual}$ is also monotone. Let $T \subseteq S$. Obviously $N^r \setminus S \subseteq N^r \setminus T$. We get

$$\begin{aligned} v^{N^r dual}(T) &= v(N^r) - v(N^r \setminus T) \\ &\leq v(N^r) - v(N^r \setminus S) = v^{N^r dual}(S). \end{aligned}$$

Let $i \in N^r$ be a sufficient player in $v \in \mathcal{G}_m^N$. We prove that i is a necessary player in $v^{T_r - dual}$. Let $S \subseteq N \setminus \{i\}$

$$\begin{aligned} 0 \leq v^{N^r dual}(S) &= v(N^r) - v(N^r \setminus S) \\ &\leq v(\{i\}) - v(N^r \setminus S) \leq 0. \end{aligned}$$

Hence $v^{N^r dual}(S) = 0$. We take k with $r(k, i) = 1$ (therefore $i \in N^r$) and i sufficient. Theorem 6.1 implies

$$\bar{\delta}_k(v, r) = \delta_k(v^{N^r dual}, r) \geq \delta_j(v^{N^r dual}, r) = \bar{\delta}_j(r, v),$$

for all $j \in N$.

LINEARITY. The result follows also from Theorem 6.1.

For each non-empty coalition T we define the *anyone game* w_T as

$$w_T(S) = \begin{cases} 1, & S \cap T \neq \emptyset \\ 0, & \text{otherwise.} \end{cases} \tag{6.6}$$

These games constitute a base of the games as the unanimity games. The cardinality of anyone games is the same as the cardinality of unanimity games, so it is enough if we see that they are independent. Anyone games are actually the dual of the unanimity games,

$$(u_T)^{dual}(S) = u_T(N) - u_T(N \setminus S) = 1 - u_T(N \setminus S) = \begin{cases} 0, & \text{if } T \subseteq N \setminus S \\ 1, & \text{otherwise} \end{cases} = w_T(S),$$

because of $T \subseteq N \setminus S$ if and only if $T \cap S = \emptyset$. Furthermore it is easy to test that $(av + bw)^{dual} = av^{dual} + bw^{dual}$. Thus, if we consider

$$\sum_{\{T \subseteq N : T \neq \emptyset\}} a_T w_T = 0 = \sum_{\{T \subseteq N : T \neq \emptyset\}} a_T (w_T)^{dual} = \sum_{\{T \subseteq N : T \neq \emptyset\}} a_T u_T$$

we deduce that $a_T = 0$ for all T. Consider f a value for games with permission structure satisfying the five axioms. From linearity we only have to obtain the uniqueness for the anyone games. Let $T \subseteq N$ be a non-empty coalition. Players out of T are null. We denote as

$$R_T = \{k \in N : r(k, i) = 1 \text{ for some } i \in T\}.$$

Players out of R_T are inessential in r for w_T thus their payoffs are zero. Players in T are sufficient in the anyone game. Coercion power over sufficient players implies that all the players in R_T must have the same payoff, K. Now efficiency says

$$\sum_{i \in N} f_i(u_T, r) = |R_T| K = w_T(N^r).$$

So, we obtain

$$K = \begin{cases} \dfrac{1}{|R_T|}, & T \subseteq N^r \\ 0, & \text{otherwise.} \end{cases}$$

□

6.4 Games with Fuzzy Permission Structure

The concept of fuzzy permission structure was introduced in Gallardo et al. [1]. They considered any reflexive fuzzy bilateral relation, but here we suppose weakly anti-transitivity in the sense of Sect. 3.3, following the crisp case. Also we consider reflexivity non necessary. The reader can think about other models only changing the T-norm in the concept of anti-symmetry.

Definition 6.9 A fuzzy permission structure is a fuzzy bilateral relation r which is weakly anti-transitive, namely $\rho(i, j) + \rho(j, i) \leq 1$ for all $i, j \in N$ with $i \neq j$. The family of fuzzy permission structures over N is FA^N.

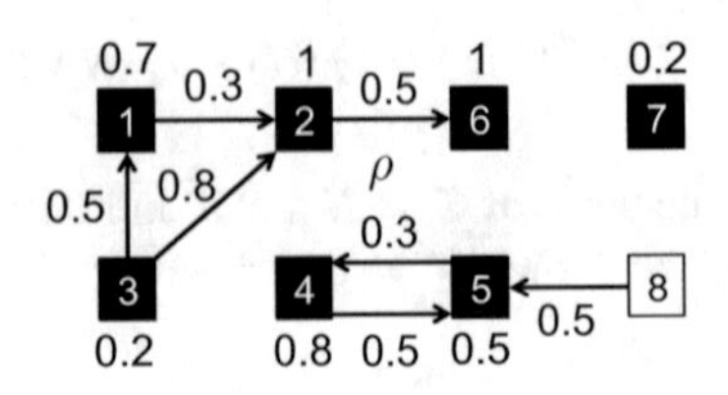

Fig. 6.8 Fuzzy permission structure

In this case, if $\rho(i, i)$ is the level of i as active agent, $\rho(i, j)$ with $i \neq j$ means the level of player j that needs the authorization of player i.

Example 6.11 Figure 6.8 shows a fuzzy permission structure over eight agents, $N = \{1, 2, 3, 4, 5, 6, 7, 8\}$ similar to the crisp one in Fig. 6.1, with matrix

$$\rho = \begin{bmatrix} 0.7 & 0.3 & 0 & 0 & 0 & 0 & 0 & 0 \\ 0 & 1 & 0 & 0 & 0 & 0.5 & 0 & 0 \\ 0.5 & 0.8 & 0.2 & 0 & 0 & 0 & 0 & 0 \\ 0 & 0 & 0 & 0.8 & 0.5 & 0 & 0 & 0 \\ 0 & 0 & 0 & 0.3 & 0.5 & 0 & 0 & 0 \\ 0 & 0 & 0 & 0 & 0 & 1 & 0 & 0 \\ 0 & 0 & 0 & 0 & 0 & 0 & 0.2 & 0 \\ 0 & 0 & 0 & 0 & 0.5 & 0 & 0 & 0 \end{bmatrix}.$$

Player 5 depends on players 4 and 8 as in Fig. 6.1 but she only acts at level 0.5. Player 2 needs permission of player 1 and player 3 but at different levels. Player 6 needs player 2 at level 0.5 and then she has level 0.5 of autonomy. Player 4 depends on player 5 but player 5 depends also on player 4, for disjoint reasons.

Gallardo [1] introduced the local fuzzy conjunctive option of dependence. In our context we change it in the following sense. Remember from Sect. 3.3 that the fuzzy domain τ^ρ for a fuzzy relation ρ is given by $\tau^\rho(i) = \rho(i, i)$ for each player i (3.4).

Definition 6.10 Let S be a coalition. The fuzzy (conjunctive) sovereign part of S in a fuzzy permission structure $\rho \in FA^N$ is $\sigma^\rho(S) \in [0, 1]^N$ with

$$\sigma^\rho(S)(i) = \left[\tau^\rho(i) - \bigvee_{j \in N \setminus S} \rho(j, i)\right] \vee 0,$$

for each player $i \in N$.

The fuzzy sovereign part is actually a fuzzy coalition (Definition 2.1), namely given a coalition the fuzzy permission structure determines the level of activity of each player.

Example 6.12 Following Example 6.11, we determine the fuzzy sovereign part of several coalitions. We obtain $\sigma^{\rho}(\{1, 2\}) = (0.2, 0.2, 0, 0, 0, 0, 0, 0)$ but

$$\sigma^{\rho}(\{1, 2, 3\}) = (0.7, 1, 0.2, 0, 0, 0, 0, 0),$$

namely player 3 is not very active but her presence permits to play players 1 and 2 at maximum level. Now we get $\sigma^{\rho}(\{2\}) = (0, 0.2, 0, 0, 0, 0, 0, 0)$. The fuzzy sovereign part

$$\sigma^{\rho}(\{4, 5, 6, 7\}) = (0, 0, 0, 0.8, 0, 0.6, 0.2, 0)$$

but $\sigma^{\rho}(\{4, 6, 7\}) = (0, 0, 0, 0.5, 0, 0.5, 0.2, 0)$. Observe that

$$\sigma^{\rho}(N) = (0.7, 1, 0.2, 0.8, 0.5, 1, 0.2, 0) = \tau^{\rho}.$$

As in the crisp case, the fuzzy relation can be reduced in those superfluous levels of links. We look for the fuzzy weakly reflexive interior of the fuzzy permission structure. The concept follows the definition of weakly reflexive fuzzy binary relation (see Sect. 3.3).

Definition 6.11 Let $\rho \in FA^N$ be a fuzzy permission structure. The weakly reflexive interior of ρ is another permission structure ρ° verifying for all $i, j \in N$

$$\rho^{\circ}(i, j) = \rho(i, j) \wedge \rho(j, j).$$

Observe that weakly reflexivity in a fuzzy permission structure implies again that we only use the links while they affect to active players. We show now that it is enough to analyze weakly reflexivity fuzzy permission structures for our conjunctive option. Let $\rho \in FA^N$ be a fuzzy permission structure. We test the claim

$$\sigma^{\rho^{\circ}}(S) = \sigma^{\rho}(S). \tag{6.7}$$

In fact, for each $i \in N, \sigma^{\rho^{\circ}}(S)(i) \neq \sigma^{\rho}(S)(i)$ if and only if $\bigvee_{j \in N \setminus S} \rho(j, i) < \rho(i, i)$.

Example 6.13 Figure 6.9 represents a fuzzy permission structure ρ and its weakly reflexive interior. Observe that player 2 depends on players 1, although the level of dependency is 0.5 the level of activity of player 2 is only 0.2. Hence the level of permission of player 1 over agent 2 can be reduced.

We summarize some properties of the fuzzy sovereign part.

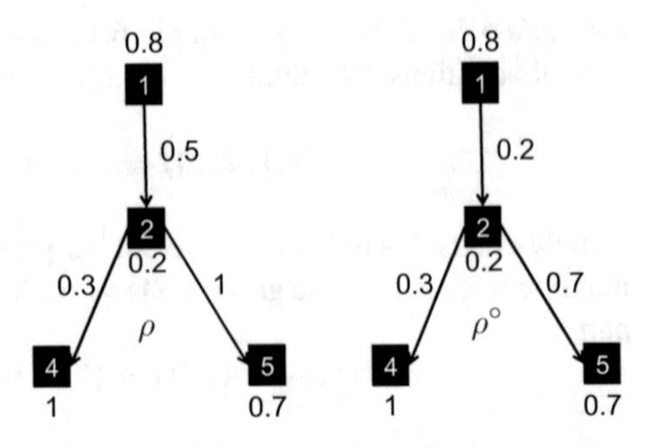

Fig. 6.9 weakly reflexive interior

Proposition 6.9 *The fuzzy sovereign part in a fuzzy permission structure* ρ *satisfies the following properties:*

(1) $\sigma^{\rho}(S) \leq \tau^{\rho} \times e^{S}$*, moreover* $\sigma^{\rho}(\emptyset) = 0$ *and* $\sigma^{\rho}(N) = \tau^{\rho}$*.*
(2) If $\rho = r \in A^{N}$ *then* $\sigma^{\rho}(S) = e^{\sigma^{r}(S)}$ *for all* $S \subseteq N$*.*
(3) If $S \subseteq T$ *then* $\sigma^{\rho}(S) \leq \sigma^{\rho}(T)$*.*
(4) $\sigma^{\rho}(S) \vee \sigma^{\rho}(T) \leq \sigma^{\rho}(S \cup T)$*.*
(5) $\sigma^{\rho}(S) \wedge \sigma^{\rho}(T) = \sigma^{\rho}(S \cap T)$*.*

Proof (1) It is trivial by definition of fuzzy sovereign part.
(2) Let $\rho = r \in A^{N}$. If $i \notin N^{r}$ then $\sigma^{\rho}(S)(i) = 0$. If $i \notin S$ but $i \in N^{r}$ then $\sigma^{\rho}(S)(i) = 0$ because there exists $i \in N \setminus S$ with $\rho(i,i) = 1$. Suppose then $i \in S \cap N^{r}$, the fuzzy sovereign part verifies $\bigvee_{j \in N \setminus S} \rho(j,i) \in \{0,1\}$. If $\bigvee_{j \in N \setminus S} \rho(j,i) = 0$ then $\sigma^{\rho}(S)(i) = 1$ and $r(j,i) = 0$ for all $j \notin S$, so $i \in \sigma^{r}(S)$. If $\bigvee_{j \in N \setminus S} \rho(j,i) = 1$ then $\sigma^{\rho}(S)(i) = 0$ and there is $j \notin S$ with $r(j,i) = 0$, so $i \notin \sigma^{r}(S)$.
(3) If $S \subseteq T$ then $N \setminus T \subseteq N \setminus S$, therefore, for all $\in N$

$$\bigvee_{j \in N \setminus T} \rho(j,i) \leq \bigvee_{j \in N \setminus S} \rho(j,i).$$

We get then

$$\rho(i,i) - \bigvee_{j \in N \setminus S} \rho(j,i) \leq \rho(i,i) - \bigvee_{j \in N \setminus T} \rho(j,i).$$

So, $\sigma^{\rho}(S)(i) \leq \sigma^{\rho}(T)(i)$.
(4) As $S, T \subseteq S \cup T$ then, following the above step, $\sigma^{\rho}(S), \sigma^{\rho}(T) \leq \sigma^{\rho}(S \cup T)$. So,

$$\sigma^{\rho}(S) \vee \sigma^{\rho}(T) \leq \sigma^{\rho}(S \cup T).$$

(5) Obviously $\sigma^{\rho}(S) \wedge \sigma^{\rho}(T) \geq \sigma^{\rho}(S \cap T)$. Let $i \in N$. We prove the another inequality. There exists $k \notin S$ (or $k \notin T$) such that $\sigma^{\rho}(S \cap T)(i) = \rho(i,i) - \rho(k,i)$, namely

$$\rho(k,i) = \bigvee_{j \in N \setminus (S \cap T)} \rho(j,i) = \bigvee_{j \in N \setminus S} \rho(j,i).$$

So, $\sigma^{\rho}(S \cap T)(i) = \sigma^{\rho}(S)(i) \geq \sigma^{\rho}(S)(i) \wedge \sigma^{\rho}(T)(i)$. □

The fuzzy sovereign part determines a fuzzy coalition into each coalition. Now we can use a partition by levels (Definition 2.7) of this fuzzy coalition to calculate the worth of the coalition taking into account the information. Given a cooperative game and a fuzzy permission structure we construct a new game using the worths of the sovereign parts of the coalitions.

Definition 6.12 Let $v \in \mathcal{G}^N$ and $\rho \in FA^N$. Let pl be an extension for fuzzy coalitions. The pl-permission game is defined for all coalition S as

$$v^{\rho}_{pl}(S) = v^{pl}(\sigma^{\rho}(S)) = \sum_{k=1}^{m} s_k v(S_k),$$

where $pl(\sigma^{\rho}(S)) = \{(S_k, s_k)\}_{k=1}^{m}$.

Gallardo et al. [1] analyzed this value using the Choquet extension and only for reflexive fuzzy relations.

Example 6.14 Consider ρ the fuzzy permission structure in Fig. 6.10 for four players. Suppose game $v(S) = |S|^2$. We take the Choquet extension (see Definition 2.12). The ch-permission game for the non-empty coalitions is in Table 6.8, namely for any coalition S

$$v^{\rho}_{ch}(S) = \int_c \sigma^{\rho}(S)\, dv.$$

For instance, the worth of the great coalition is the Choquet integral of $\sigma^{\rho}(N) = (1, 1, 0.2, 0.7)$

$$v^{\rho}_{ch}(N) = \int_c \sigma^{\rho}(N)\, dv = 0.2v(N) + 0.5v(\{1,2,4\}) + 0.3v(\{1,2\}) = 8.9.$$

If we use the proportional extension (Definition 2.11) the game changes as we show in Table 6.9. Now, the worth of the great coalition is

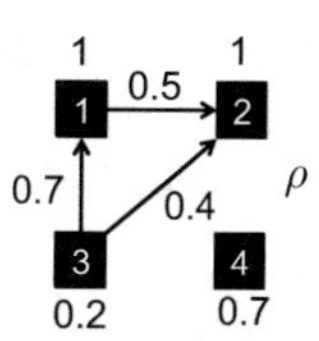

Fig. 6.10 The fuzzy permission structure in Example 6.14

Table 6.8 Game v^{ρ}_{ch}

S	{1}	{2}	{3}	{4}	{1, 2}
$\sigma^{\rho}(S)$	(0.3, 0, 0, 0)	(0, 0.5, 0, 0)	(0, 0, 0.2, 0)	(0, 0, 0, 0.7)	(0.3, 0.6, 0, 0)
$v^{\rho}_{ch}(S)$	0.3	0.5	0.2	0.7	1.5
S	{1, 3}	{1, 4}	{2, 3}	{2, 4}	{3, 4}
$\sigma^{\rho}(S)$	(1, 0, 0.2, 0)	(0.3, 0, 0, 0.7)	(0, 0.5, 0, 2, 0)	(0, 0.5, 0, 0.7)	(0, 0, 0.2, 0.7)
$v^{\rho}_{ch}(S)$	1.6	1.6	1.1	2.2	1.3
S	{1, 2, 3}	{1, 2, 4}	{1, 3, 4}	{2, 3, 4}	N
$\sigma^{\rho}(S)$	(1, 1, 0.2, 0)	(0.3, 0.6, 0, 0.7)	(1, 0, 0.2, 0.7)	(0, 0.5, 0.2, 0.7)	(1, 1, 0.2, 0.7)
$v^{\rho}_{ch}(S)$	5	4	4.1	3.2	8.9

Table 6.9 Game v^{ρ}_{pr}

S	{1}	{2}	{3}	{4}	{1, 2}
$\sigma^{\rho}(S)$	(0.3, 0, 0, 0)	(0, 0.5, 0, 0)	(0, 0, 0.2, 0)	(0, 0, 0, 0.7)	(0.3, 0.6, 0, 0)
$v^{\rho}_{pr}(S)$	0.3	0.5	0.2	0.7	0.9
S	{1, 3}	{1, 4}	{2, 3}	{2, 4}	{3, 4}
$\sigma^{\rho}(S)$	(1, 0, 0.2, 0)	(0.3, 0, 0, 0.7)	(0, 0.5, 0, 2, 0)	(0, 0.5, 0, 0.7)	(0, 0, 0.2, 0.7)
$v^{\rho}_{pr}(S)$	1.2	1	0.7	1.2	0.9
S	{1, 2, 3}	{1, 2, 4}	{1, 3, 4}	{2, 3, 4}	N
$\sigma^{\rho}(S)$	(1, 1, 0.2, 0)	(0.3, 0.6, 0, 0.7)	(1, 0, 0.2, 0.7)	(0, 0.5, 0.2, 0.7)	(1, 1, 0.2, 0.7)
$v^{\rho}_{pr}(S)$	4.2	1.6	1.9	1.4	4.9

$$v^{\rho}_{pr}(N) = 1v(\{1, 2\}) + 0.2v(\{3\}) + 0.7v(\{4\}) = 4.9.$$

Remark 6.3 There exist another way to analyze fuzzy permission, following the strategy in the above chapters: partition by levels of the fuzzy graph. Taking into account the philosophy of the book, we prefer to present a different technique.[7] First we should consider for this way anti-symmetric fuzzy relations and not the weak condition because of bringing problems to the model. If ρ is a permission structure and $pl(\rho) = \{(r_k, s_k)\}_{k=1}^{m}$ a partition by levels of the fuzzy graph, then we can use the game

$$v^{\rho}_{pl}(S) = \sum_{k=1}^{m} s_k v^{r_k}(S), \ \forall S \subseteq N$$

where v^{r_k} is the permission game (Definition 6.4) for each permission structure r_k. Look at the next example. We consider the fuzzy permission structure in Fig. 6.10. Table 6.10 shows the Choquet by graphs partition of the directed fuzzy graph

[7]Furthermore this option is not yet available in the literature.

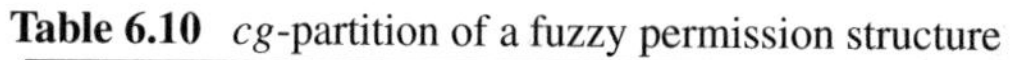

Table 6.10 cg-partition of a fuzzy permission structure

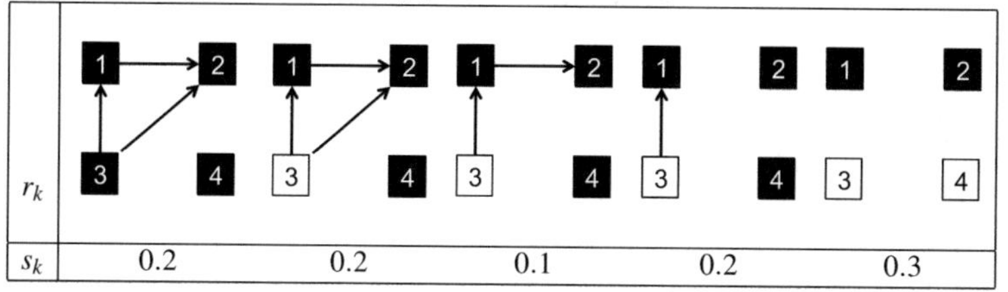

r_k					
s_k	0.2	0.2	0.1	0.2	0.3

following Definition 4.17. So, taking the same game in Example 6.14, $v(S) = |S|^2$, the worth for instance of coalition $\{2, 3\}$ is

$$\begin{aligned} v^{\rho}_{cg}(\{2,3\}) &= 0.2v^{r_1}(\{2,3\}) + 0.2v^{r_2}(\{2,3\}) + 0.1v^{r_3}(\{2,3\}) \\ &\quad +0.2v^{r_4}(\{2,3\}) + 0.3v^{r_5}(\{2,3\}) \\ &= 0.2 + 0 + 0 + 0.2 + 0.3 = 0.7. \end{aligned}$$

Hence $v^{\rho}_{cg} \neq v^{\rho}_{ch}$.

Following the crisp case, next properties are true for the pl-permission game using any extension for fuzzy coalitions.

Proposition 6.10 *Let v be a game and ρ be a fuzzy permission structure. For every extension pl over coalitions it holds:*

(1) If $\rho = r \in A^N$ then $v^{\rho}_{pl} = v^r$.

(2) For all $w \in \mathcal{G}^N$ and $a, b \in \mathbb{R}$,

$$(av + bw)^{\rho}_{pl} = av^{\rho}_{pl} + bw^{\rho}_{pl}.$$

Particularly, $(-v)^{\rho}_{pl} = -(v^{\rho}_{pl})$.

(3) For each coalition S,

$$(v^{svg})^{\rho}_{pl}(S) = (v^{\rho}_{pl})^{svg}(S) + \sum_{i \in S} \left[\sigma^{\rho}_i(S) - \sigma^{\rho}_i(\{i\})\right] v(\{i\}).$$

Proof (1) As pl is an extension (Definition 2.9) then we have $pl(e^T) = \{(T, 1)\}$ for all $T \subseteq N^r$. Proposition 6.9 implies that $\sigma^{\rho}(S) = e^{\sigma^r(S)}$. Hence $pl(\sigma^{\rho}(S)) = \{(\sigma^r(S), 1)\}$ and

$$v^{\rho}_{pl}(S) = 1v(\sigma^r(S)) = v^r(S).$$

(2) It is easy to test for the reader.

(3) Let $S \subseteq N$ and $pl(\sigma^\rho(S)) = \{(S_k, s_k)\}_{k=1}^m$. Remember that $\sigma^\rho(S)(i) = 0$ if $i \notin S$. We obtain from (2.6),

$$\begin{aligned}(v^{svg})^\rho_{pl}(S) &= \sum_{k=1}^m s_k v^{svg}(S_k) = \sum_{k=1}^m s_k \left[\sum_{i \in S_k} v(\{i\}) - v(S_k)\right] \\ &= \left[\sum_{k=1}^m s_k \sum_{i \in S_k} v(\{i\})\right] - \sum_{k=1}^m s_k v(S_k) \\ &= \left[\sum_{i \in N}\left(\sum_{\{k: i \in S_k\}} s_k\right) v(\{i\})\right] - v^\rho_{pl}(S) = \left[\sum_{i \in S} \sigma^\rho(S)(i) v(\{i\})\right] - v^\rho_{pl}(S).\end{aligned}$$

On the other hand, as $v^\rho_{pl}(\{i\}) = \sigma^\rho(\{i\})(i)v(\{i\})$ for all player i,

$$(v^\rho_{pl})^{svg}(S) = \left[\sum_{i \in S} v^\rho_{pl}(\{i\})\right] - v^\rho_{pl}(S) = \left[\sum_{i \in S} \sigma^\rho(\{i\})(i)v(\{i\})\right] - v^\rho_{pl}(S)$$

The subtraction of both equalities gets

$$(v^{svg})^\rho_{pl}(S) - (v^\rho_{pl})^{svg}(S) = \sum_{i \in S} \left[\sigma^\rho(S)(i) - \sigma^\rho(\{i\})(i)\right] v(\{i\}).$$

□

It is not possible to analyze for all the extensions pl the inherited properties from the original game. Following Sect. 2.4, for the proportional extension the only inherited property is the superadditivity, but for the Choquet extension all the properties are inherited. However now we need monotonicity since Proposition 6.9. We can say from Propositions 6.9 and 2.8 for the Choquet extension the following properties.

Proposition 6.11 *Let v be a game and ρ be a fuzzy permission structure.*

(1) If v is monotonic then v^ρ_{ch} is monotonic.
(2) If v is monotonic and superadditive then so is v^ρ_{ch}
(3) If v is monotonic and convex then v^ρ_{ch} is convex.

Following Huettner and Wiese [5] the coercive version of the model is introduced.

Definition 6.13 Given an extension pl for fuzzy coalitions. The pl-coercive game is defined for a game $v \in \mathcal{G}^N$ and a fuzzy permission structure $\rho \in FA^N$ as

$$\bar{v}^{\rho}_{pl}(S) = v^{pl}(\bar{\sigma}^{\rho}(S)),$$

where the fuzzy coercion part for each $i \in N$ is

$$\bar{\sigma}^{\rho}(S)(i) = \tau^{\rho}(i) \wedge \bigvee_{j \in S} \rho(j, i).$$

Observe that we have in this case $\rho(i, i) \leq \bar{\sigma}^{r}(S)(i)$ if $i \in S$. We can also test that $\bar{\sigma}^{\rho} = \bar{\sigma}^{\rho^{\circ}}$. In fact, given a coalition S, if we take i with some $j \in S$ with $\rho(j, i) > \rho(i, i)$ then $\rho^{\circ}(j, i) = \rho(i, i)$. Hence $\bar{\sigma}^{\rho}(S)(i) = \rho(i, i) = \bar{\sigma}^{\rho^{\circ}}(S)(i)$.

Example 6.15 In this example we look at Fig. 6.10 as a coercive structure. Obvious we say see the same but we change our interpretation. Now the coercive coalition of $\{3\}$ is

$$\bar{\sigma}^{\rho}(\{3\}) = (0.7, 0.4, 0.2, 0).$$

So, the worth using the Choquet extension for the game in Example 6.14 is

$$\bar{v}^{\rho}_{ch}(\{3\}) = \int_{c} \bar{\sigma}^{\rho}(\{3\})\, dv = 2.9.$$

Table 6.11 shows the ch-coercive game of that example.

There is also relation between both models, the fuzzy permission one and the fuzzy coercive one, in a dual way as in the crisp case (Proposition 6.5).

Table 6.11 Game $\bar{v}^{\rho}_{ch}$

S	$\{1\}$	$\{2\}$	$\{3\}$	$\{4\}$	$\{1, 2\}$
$\bar{\sigma}^{\rho}(S)$	$(1, 0.5, 0, 0)$	$(0, 1, 0, 0)$	$(0.7, 0.4, 0.2, 0)$	$(0, 0, 0, 0.7)$	$(1, 1, 0, 0)$
$\bar{v}^{\rho}_{ch}(S)$	2.5	1	2.9	0.7	4
S	$\{1, 3\}$	$\{1, 4\}$	$\{2, 3\}$	$\{2, 4\}$	$\{3, 4\}$
$\bar{\sigma}^{\rho}(S)$	$(1, 0.5, 0.2, 0)$	$(1, 0.5, 0, 0.7)$	$(0.7, 1, 0.2, 0)$	$(0, 1, 0, 0.7)$	$(0.7, 0.4, 0.2, 0.7)$
$\bar{v}^{\rho}_{ch}(S)$	3.5	5.6	4.1	3.1	6.5
S	$\{1, 2, 3\}$	$\{1, 2, 4\}$	$\{1, 3, 4\}$	$\{2, 3, 4\}$	N
$\bar{\sigma}^{\rho}(S)$	$(1, 1, 0.2, 0)$	$(1, 1, 0, 0.7)$	$(1, 0.5, 0.2, 0.7)$	$(0.7, 1, 0.2, 0.7)$	$(1, 1, 0.2, 0.7)$
$\bar{v}^{\rho}_{ch}(S)$	5	7.5	7	8	8.9

Proposition 6.12 *Let pl be an extension for fuzzy coalitions and let $\rho \in FA^N$ be a fuzzy permission structure. For all $v \in \mathcal{G}^N$ it holds*

$$(\bar{v}^{\rho}_{pl})^{dual}(S) = v^{pl}(\tau^{\rho}) - v^{pl}(\tau^{\rho} - \sigma^{\rho}(S)),$$

with $S \subseteq N$.

Proof For the great coalition we get $\bar{\sigma}^{\rho}(N) = \tau^{\rho}$ because for each player i,

$$\bar{\sigma}^{\rho}(N)(i) = \rho(i,i) \wedge \bigvee_{j \in N} \rho(j,i) = \rho(i,i) = \tau^{\rho}(i).$$

So, $v^{pl}(\bar{\sigma}^{\rho}(N)) = v^{pl}(\tau^{\rho})$. We prove the claim

$$\tau^{\rho} - \sigma^{\rho}(S) = \bar{\sigma}^{r}(N \setminus S).$$

We can suppose ρ weakly reflexive. Let $i \in N$. We obtain

$$\begin{aligned}\tau^{\rho}(i) - \sigma^{\rho}(S)(i) &= \rho(i,i) - \left[\rho(i,i) - \bigvee_{j \in N \setminus S} \rho(j,i)\right] \\ &= \bigvee_{j \in N \setminus S} \rho(j,i) = \bar{\sigma}^{\rho}(N \setminus S)(i).\end{aligned}$$

Using the definition of the dual game (1.5), we get for each coalition S

$$\begin{aligned}(\bar{v}^{\rho}_{pl})^{dual}(S) &= \bar{v}^{\rho}_{pl}(N) - \bar{v}^{\rho}_{pl}(N \setminus S) = v^{pl}(\bar{\sigma}^{\rho}(N)) - v^{pl}(\bar{\sigma}^{\rho}(N \setminus S)) \\ &= v^{pl}(\tau^{\rho}) - v^{pl}(\tau^{\rho} - \sigma^{\rho}(S)).\end{aligned}$$

□

6.5 Values for Games with Fuzzy Permission Structure

We study several values inspired on the fuzzy conjunctive approach. Each pair $(v, \rho) \in \mathcal{G}^N \times FA^N$ is called a *game over N with a fuzzy permission structure.* A value for games with fuzzy permission structure is

$$f : \mathcal{G}^N \times FA^N \to \mathbb{R}^N.$$

We extend first the local permission value (Definition 6.6) to the fuzzy context.

Definition 6.14 The local pl-permission value is a value for games over N with fuzzy permission structure defined for each $v \in \mathcal{G}^N$ and $\rho \in FA^N$ as

$$\delta^{pl}(v, \rho) = \phi(v^{\rho}_{pl}).$$

Example 6.16 We consider again game $v(S) = |S|^2$ with the permission structure ρ in Fig. 6.10. Table 6.8 determined the fuzzy sovereign parts of the coalitions and the ch-permission game. The local ch-permission value is

$$\delta^{ch}(v, r) = \phi(v^{\rho}_{ch}) = (2.48, 2.27, 2.07, 2.08)\,.$$

But also we can define the local pr-permission value using the proportional extension. Now we take Table 6.9 obtaining

$$\delta^{pr}(v, r) = \phi(v^{\rho}_{pr}) = (1.5, 1.35, 1.35, 0.7)\,.$$

In this model we follow Gallardo et al. [1] using an axiomatization only for the Choquet model and reflexive fuzzy permission structures. So, we can see how the models can be studied using particular properties of the Choquet integral (see Sect. 2.1).

We look for axioms to determine the local ch-permission value for games with a broader family of permission structures.

Definition 6.15 A fuzzy permission structure $\rho \in FA^N$ is z-reflexive with $z \in (0, 1]$ if for all $\rho(i, j) > 0$ it holds $\rho(i, j) \leq \rho(j, j) = z$. The family of fuzzy permission structures which are z-reflexive for some z is denoted as FA^N_0.

If ρ is a z-reflexive permission structure then ρ is weakly reflexive and the fuzzy domain satisfies $\tau^{\rho} = ze^{N^{\rho}}$ being N^{ρ} the domain of ρ. Figure 6.11 represents a 0.7-reflexive permission structure. We use $\rho \in FA^N_0$ with constant z to say z-reflexive permission structure.

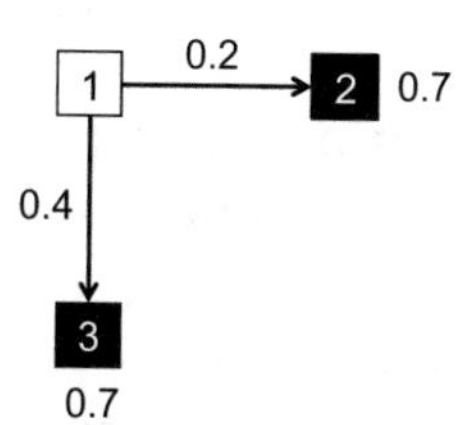

Fig. 6.11 0.7-reflexive permission structure

Efficiency would depend now on the chosen partition by levels of the sovereign part of the great coalition, the fuzzy domain, but as the structure is z-reflexive coincides with the crisp domain.

Restricted efficiency. For all $v \in \mathcal{G}^N$ and $\rho \in FA_0^N$ with constant z,

$$f(v, \rho)(N) = zv(N^\rho).$$

We say that $i \in N$ is *fuzzy inessential* for $\rho \in FA_0^N$ and $v \in \mathcal{G}^N$ if j is a null player in v when $\rho(i, j) > 0$.

Fuzzy inessential player. Let $\rho \in FA_0^N$ and $v \in \mathcal{G}^N$. If $i \in N$ is a fuzzy inessential player, it holds $f_i(v, \rho) = 0$.

Linearity. For all $v, w \in \mathcal{G}^N$, $a, b \in \mathbb{R}$ and $r \in FA_0^N$ it holds

$$f(av + bw, r) = af(v, r) + bf(w, r).$$

Players with veto power over necessary players are also necessaries, but in this case veto power implies total control of the activity of the another agent.

Fuzzy veto power over necessary players.[8] Let $v \in \mathcal{G}_m^N$ be a monotone game and $\rho \in FA_0^N$ with constant z. If $i \in N$ is a necessary player for v and $k \in N$ with $\rho(k, i) = z$ then $f_k(v, r) \geq f_j(v, r)$ for all $j \in N$.

For the fuzzy case we need another axiom linked to a usual characteristic of the Choquet integral, comonotonicity (2.3). If $\rho, \rho' \in FA_0^N$ with $\tau^\rho = \tau^{\rho'} = ze^{N^\rho}$ then for all $t \in [0, 1]$ we have $t\rho + (1 - t)\rho' \in FA_0^N$. If $i, j \in N$ then

$$\begin{aligned} &t\rho(i, j) + (1 - t)\rho'(i, j) + t\rho(j, i) + (1 - t)\rho'(j, i) \\ &= t[\rho(i, j) + \rho(j, i)] + (1 - t)[\rho'(i, j) + \rho'(j, i)] \leq 1. \end{aligned}$$

Also when we suppose $\rho(i, j) > 0$ (or $\rho'(i, j) > 0$)

$$t\rho(i, j) + (1 - t)\rho'(i, j) \leq t\rho(j, j) + (1 - t)\rho'(j, j) = z,$$

because if $\rho(j, j) = z$ then $\rho'(j, j) = z$.

Comonotonicity.[9] Let $v \in \mathcal{G}^N$ and $t \in [0, 1]$. For all $\rho, \rho' \in FA_0^N$ comonotone (as fuzzy sets, see Sect. 3.3) with $\tau^\rho = \tau^{\rho'}$ it holds

$$f(v, t\rho + (1 - t)\rho') = tf(v, \rho) + (1 - t)f(v, \rho').$$

Gallardo et al. [1] proved that these axioms (similar to them) permit to determine the local permission value.

[8] This axiom is not satisfied for all extension pl.

[9] This axiom is also not satisfied for any extension pl.

Theorem 6.6 *The local ch-permission value is the only value for games over N with z-reflexive fuzzy permission structure satisfying restricted efficiency, inessential player, veto power over necessary player, linearity and comonotonicity.*

Proof We test that the local ch-permission value verifies the axioms.
RESTRICTED EFFICIENCY. As ρ is z-reflexive then $\sigma^{\rho}(N) = \tau^{\rho} = ze^{N^{\rho}}$. Let $v \in \mathcal{G}^N$, efficiency of the Shapley value (Proposition 1.9), Proposition 6.9 and property (C7) of the Choquet integral imply

$$\sum_{i \in N} \delta_i^{ch}(v, \rho) = \sum_{i \in N} \phi_i(v_{ch}^{\rho}) = v_{ch}^{\rho}(N) = v^{ch}(\sigma^{\rho}(N)) = \int_c ze^{N^{\rho}}\, dv = zv(N^{\rho}).$$

INESSENTIAL PLAYER. We denote the different worths in $\{z - \rho(j,i)\}_{(j,i) \in N \times N}$ as the set $D = \{0 = t_0 < \cdots < t_m\}$. This set satisfies

$$im_0(\sigma^{\rho}(S)) \subseteq D$$

for all coalition S. In fact, if $t \in im(\sigma^{\rho}(S))$ then there exist $i \in S$ and $j \in N \setminus S$ with $t = z - \rho(j,i)$. Property (C10) of the Choquet integral in Sect. 2.1 implies that

$$\int_c \sigma^{\rho}(S)\, dv = \sum_{p=1}^{m} (t_p - t_{p-1}) v\left(\left[\sigma^{\rho}(S)\right]_{t_p}\right).$$

Let $i \in N$ such that j is a null player in a game v when $\rho(i,j) > 0$. We will prove that i is a null player in v_{ch}^{ρ}. We take S with $i \in S$. Fixed t_p, with $p = 1, \ldots, m$, if player $j \in [\sigma^{\rho}(S)]_{t_p} \setminus [\sigma^{\rho}(S \setminus \{i\})]_{t_p}$ then:

- $z - \bigvee_{k \in N \setminus S} \rho(k,j) \geq t_p$ thus $z - t_p \geq 0$,
- and $z - \rho(i,j) < t_p$ therefore $\rho(i,j) > z - t_p \geq 0$.

As in the crisp case (Theorem 6.1), we get $v\left([\sigma^{\rho}(S)]_{t_p}\right) = v\left([\sigma^{\rho}(S \setminus \{i\})]_{t_p}\right)$ and finally

$$v_{ch}^{\rho}(S) = v_{ch}^{\rho}(S \setminus \{i\}).$$

Shapley value satisfies null player, hence $\delta_i^{ch}(v, \rho) = 0$.
VETO POWER OVER NECESSARY PLAYERS. Let i a necessary player and $j \in N$ with $\rho(j,i) = z$. We will prove that j is a necessary player for v_{ch}^{ρ}. Suppose $S \subseteq N \setminus \{j\}$,

$$\sigma^{\rho}(S)(i) = z - \bigvee_{k \in N \setminus S} \rho(k,i) = 0.$$

For all t the corresponding cut verifies $i \notin [\sigma^{\rho}(S)]_t$, so $v([\sigma^{\rho}(S)]_t) = 0$. Property (C6) implies $v^{\rho}_{ch}(S) = 0$. Necessary player is satisfied by Shapley value (Proposition 1.11),

$$\delta_j^{ch}(v, \rho) = \phi_j(v^{\rho}_{ch}) \geq \phi_k(v^{\rho}_{ch}) = \delta^{\rho}_{ch}(v, \rho),$$

for all $k \in N$.
LINEARITY. The result follows from Proposition 6.10 and the linearity of the Shapley value.
COMONOTONICITY. Let $t \in [0, 1]$. Let also $\rho, \rho' \in FA_0^N$ comonotone with $\tau^{\rho} = \tau^{\rho'}$. Remember from (2.3) that for all $(i, j), (i', j') \in N \times N$ it holds

$$\left[\rho(i, j) - \rho(i', j')\right]\left[\rho'(i, j) - \rho'(i', j')\right] \geq 0.$$

Hence for any coalition S and $i \in N$ we obtain the same $k_i \notin S$ satisfying

$$\bigvee_{k \in N \setminus S} \rho(k, i) = \rho(k_i, i) \text{ and } \bigvee_{k \in N \setminus S} \rho'(k, i) = \rho'(k_i, i).$$

This last reason implies two things for a coalition S:

- Fuzzy coalitions $\sigma^{\rho}(S), \sigma^{\rho'}(S)$ are comonotone (and then also $t\sigma^{\rho}(S), (1 - t)\sigma^{\rho'}(S)$). Let $i, j \in N$ with $\sigma^{\rho}(S)(i) \geq \sigma^{\rho}(S)(j)$. We have $\rho(k_i, i) \leq \rho(k_j, j)$. Comonotonicity of ρ, ρ' implies $\rho'(k_i, i) \leq \rho'(k_j, j)$, and then we get the result because of $\tau^{\rho} = \tau^{\rho'}$.
- The fuzzy sovereign part verifies

$$\sigma^{t\rho+(1-t)\rho'} = t\sigma^{\rho} + (1 - t)\sigma^{\rho'},$$

because we obtain for all $S \subseteq N$

$$\begin{aligned} t \bigvee_{j \in N \setminus S} \rho(j, i) + (1 - t) \bigvee_{j \in N \setminus S} \rho'(j, i) &= t\rho(k_i, i) + (1 - t)\rho'(k_i, i) \\ &= \bigvee_{j \in N \setminus S} \left[t\rho(j, i) + (1 - t)\rho'(j, i)\right]. \end{aligned}$$

These two facts allow us to calculate the ch-permission game of the convex combination for each coalition S. We use properties (C4) and (C5) of the Choquet integral obtaining

$$\begin{aligned} v_{ch}^{t\rho+(1-t)\rho'}(S) &= \int_c \sigma^{t\rho+(1-t)\rho'}(S)\, dv = \int_c t\sigma^{\rho}(S) + (1 - t)\sigma^{\rho'}(S)\, dv \\ &= t \int_c \sigma^{\rho}(S)\, dv + (1 - t) \int_c \sigma^{\rho'}(S)\, dv = tv^{\rho}_{ch}(S) + (1 - t)v^{\rho'}_{ch}(S). \end{aligned}$$

Consider f a value for games with fuzzy reflexive permission structure satisfying the four axioms. Suppose $\rho \in FA_0^N$ with constant z. From linearity we only have to obtain the uniqueness for the unanimity games. Let $T \subseteq N$ be a non-empty coalition. We reason by induction on $|im(\rho)|$. If $|im(\rho)| = 1$ then $im(\rho) = \{z\}$ because ρ is z-reflexive. Hence, $\rho = zr$ with $r \in A^N$. We can repeat the uniqueness part of the proof in Theorem 6.1. Players out of T are null. We denote as

$$R_T = \{k \in N : \exists i \in T \text{ with } r(k, i) = 1\}.$$

Players out of R_T are fuzzy inessential in ρ for u_T thus their payoffs are zero. Players in T are necessary in the unanimity game. Fuzzy veto power over necessary players implies that all the players in R_T must have the same payoff, K. Now efficiency says

$$\sum_{i \in N} f_i(u_T, r) = |R_T| K = u_T(N).$$

So, we obtain $K = 1/|R_T|$. Suppose true the uniqueness when $|im(\rho)| < d$ with $d > 1$. Finally we apply comonotonicity when we have $\rho \in FA_0^N$ with $|im(\rho)| = d$. If $im(\rho) = \{\lambda_1 < \cdots < \lambda_d = z\}$. Consider the fuzzy permission structures,

$$\rho_1(i, j) = \begin{cases} z, & \text{if } \rho(i, j) = z \\ 0, & \text{otherwise} \end{cases} \qquad \rho_2(i, j) = \begin{cases} z, & \text{if } \rho(i, j) = z \\ \dfrac{z\rho(i, j)}{\lambda_{d-1}}, & \text{otherwise.} \end{cases}$$

Obviously,

$$\rho = \left(1 - \frac{\lambda_{d-1}}{z}\right)\rho_1 + \frac{\lambda_{d-1}}{z}\rho_2,$$

and $\tau^{\rho_1} = \tau^{\rho_2}$. The fuzzy relations ρ_1, ρ_2 are comonotone because if we take $(i, j), (i', j')$ with $\rho_1(i, j) = z$ and $\rho_1(i', j') = 0$ then

$$\rho_2(i', j') = z\frac{\rho(i, j)}{\lambda_{d-1}} \leq z = \rho_2(i, j).$$

Furthermore $|im(\rho_1)| = 1$ and $|im(\rho_2)| = d - 1$. □

Remark 6.4 Certain conditions are necessary for working comonotonicity in the above theorem. The problem is that we cannot guarantee the comonotonicity of the sovereign parts from the comonotonicity of the structures. Look at the next example. Consider the structures in Fig. 6.12. Obviously ρ and ρ' are comonotone but for instance $\sigma^\rho(\{2, 3\}) = (0, 0.7, 0)$ and $\sigma^{\rho'}(\{2, 3\}) = (0, 0.117, 0.35)$, thus they are not comonotone.

Next we prove some properties of the local permission value.

Fig. 6.12 Comonotonicity needs reflexivity

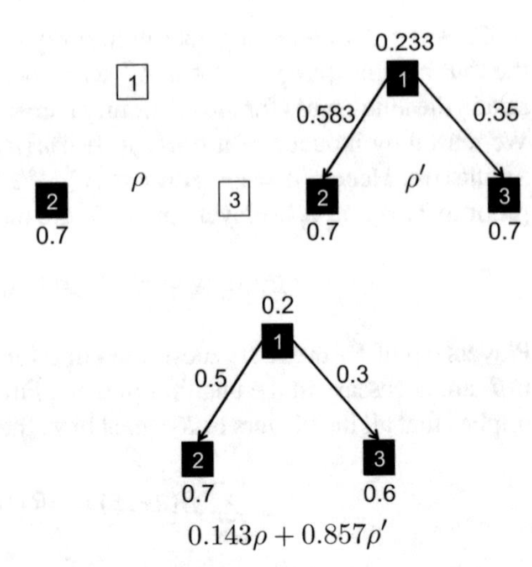

Proposition 6.13 *The local pl-permission value satisfies the following properties for a permission structure ρ.*

(1) Let $v \in \mathcal{G}_c^N \cap \mathcal{G}_m^N$. If $S \subseteq N$ then $\delta^{pl}(v, \rho)(S) \geq v(\sigma_{pl}^{\rho}(S))$.
(2) For all game v it holds $\delta^{pl}(v, \rho)(N) = v^{pl}(\tau^{\rho})$.

Proof (1) Proposition 6.3 implies that v^r is convex. Hence, Proposition 1.14 says

$$\delta(v, r)(S) = \phi(v^r)(S) \geq v^r(S) = v(\sigma^r(S)).$$

(2) We have $\sigma^{\rho}(N) = \tau^{\rho}$. Let $v \in \mathcal{G}^N$, efficiency of the Shapley value (Proposition 1.9) and Proposition 6.9 imply

$$\sum_{i \in N} \delta_i^{pl}(v, \rho) = \sum_{i \in N} \phi_i(v_{pl}^{\rho}) = v_{pl}^{\rho}(N) = v^{pl}(\tau^{\rho}).$$

□

Gallardo et al. [1] introduced also a fuzzy transitive closure for fuzzy permission structures in order to define a fuzzy permission value. This concept can change if we use another T-norms.

Definition 6.16 Let $\rho \in FA^N$. The fuzzy transitive closure of ρ is a fuzzy permission structure $\hat{\rho}$ defined, for all $i, j \in N$, with $i \neq j$ by

$$\hat{\rho}(i, j) = \bigvee_{\{i_p\}_{p=0}^{q} \in P_{ij}} \bigwedge_{p=1}^{q} \rho\left(i_{p-1}, i_p\right)$$

where $P_{ij} = \left\{\{i_p\}_{p=0}^{q} \subseteq N : i_0 = i,\ i_q = j\right\}$ and $\hat{\rho}(i, i) = \rho(i, i)$ for any player $i \in N$.

The fuzzy permission structure $\widehat{\rho}$ is obviously transitive. Moreover, if ρ is transitive then $\widehat{\rho} = \rho$. If $\rho = r \in A^N$ then $\hat{\rho} = \hat{r}$, namely it is an extension of the transitive closure (Definition 6.3) in the crisp case. Also generally $\hat{\rho^{\circ}} \neq \hat{\rho}^{\circ}$.

Example 6.17 Suppose the fuzzy permission structure ρ in Fig. 6.13. The fuzzy transitive closure of ρ is in the same figure. In this case transitivity not only implies new links, for instance $\hat{\rho}(1, 3) = 0.3$, but also the level of a feasible link can increase, for instance $\rho(1, 4) = 0.2$ but $\hat{\rho}(1, 4) = 0.5$.

The second Shapley value for games with a fuzzy permission structure will use the Shapley value of the fuzzy sovereign part in the transitive closure depending on the chosen extension.

Definition 6.17 Let pl be an extension for fuzzy coalitions. The pl-permission value is a value for games over N with fuzzy permission structure defined for each $v \in \mathcal{G}^N$ and $\rho \in FA^N$ as

$$\hat{\delta}^{pl}(v, \rho) = \delta^{pl}(v, \hat{\rho}).$$

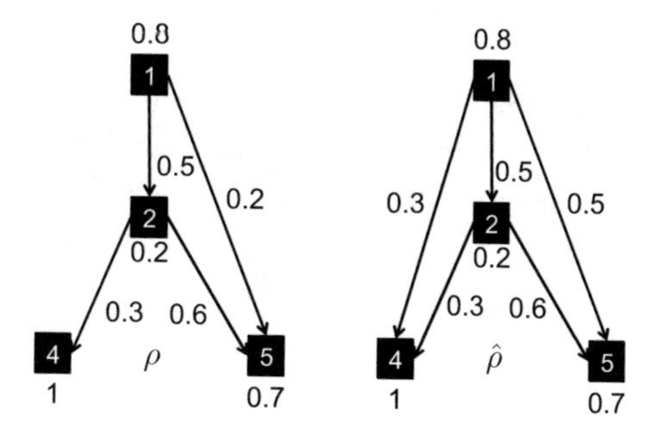

Fig. 6.13 Fuzzy transitive closure for fuzzy permission structure

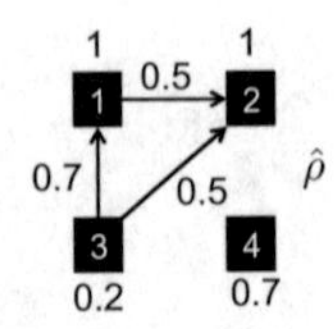

Fig. 6.14 The transitive closure of the fuzzy permission structure in Example 6.14

Table 6.12 Game $\hat{v}^{\rho}_{ch}$

S	{1}	{2}	{3}	{4}	{1, 2}
$\sigma^{\hat{\rho}}(S)$	(0.3, 0, 0, 0)	(0, 0.5, 0, 0)	(0, 0, 0.2, 0)	(0, 0, 0, 0.7)	(0.3, 0.5, 0, 0)
$v^{\rho}_{ch}(S)$	0.3	0.5	0.2	0.7	1.4
S	{1, 3}	{1, 4}	{2, 3}	{2, 4}	{3, 4}
$\sigma^{\rho}(S)$	(1, 0, 0.2, 0)	(0.3, 0, 0, 0.7)	(0, 0.5, 0, 2, 0)	(0, 0.5, 0, 0.7)	(0, 0, 0.2, 0.7)
$v^{\hat{\rho}}_{ch}(S)$	1.6	1.6	1.1	2.2	1.3
S	{1, 2, 3}	{1, 2, 4}	{1, 3, 4}	{2, 3, 4}	N
$\sigma^{\hat{\rho}}(S)$	(1, 1, 0.2, 0)	(0.3, 0.5, 0, 0.7)	(1, 0, 0.2, 0.7)	(0, 0.5, 0.2, 0.7)	(1, 1, 0.2, 0.7)
$v^{\hat{\rho}}_{ch}(S)$	5	3.7	4.1	3.2	8.9

Example 6.18 Consider the transitive closure $\hat{\rho}$ of the fuzzy permission structure in Fig. 6.10 for four players (see Fig. 6.14) an again the game $v(S) = |S|^2$. The *ch*-permission game for the non-empty coalitions is in Table 6.12. Observe that there is only an slight difference between the closure and the original structure, link 32 increases from 0.4 to 0.5. Tables 6.8 and 6.12 have two different worths for coalitions {1, 2} and {1, 2, 4}. The *ch*-permission value is

$$\hat{\delta}^{ch}(v, \rho) = (2.45, 2.23, 2.15, 2.06)$$

Next axioms determine the *ch*-permission value for games over FA^N_0.

Transitive fuzzy inessential player. Let $\rho \in FA^N_0$ and $v \in \mathcal{G}^N$. If $i \in N$ is a player satisfying that j is a null player in v when $\hat{\rho}(i, j) > 0$, it holds $f_i(v, \rho) = 0$.

Transitive fuzzy veto power over necessary players. Let $v \in \mathcal{G}^N_m$ be a monotone game and $\rho \in FA^N_0$ with constant z. If $i \in N$ is a necessary player for v then $f_k(v, r) \geq f_j(v, r)$ for all $j \in N$ and $\hat{\rho}(k, i) = z$ or $k = i$.

Theorem 6.7 *The ch-permission value is the only value for games over N with z-reflexive fuzzy permission structure satisfying restricted efficiency, transitive fuzzy inessential player, transitive fuzzy veto power over necessary player, linearity and comonotonicity.*

Proof The proof is like that one of Theorem 6.6. We only need to prove that closure works well with comonotonicity. We take $\rho, \rho' \in FA_0^N$ with $\tau^\rho = \tau^{\rho'}$ and comonotone. Let $t \in [0, 1]$. If $\rho'' = t\rho + (1-t)\rho'$ then we show that $\hat{\rho''} = t\hat{\rho} + (1-t)\hat{\rho'}$, for $t \in [0, 1]$. Indeed, given $i, j \in N$, it holds that

$$\begin{aligned}\hat{\rho''}(i,j) &= \bigvee_{\{i_p\}_{p=0}^q \in P_{ij}} \bigwedge_{p=1}^{q} \left(t\rho + (1-t)\rho'\right)\left(i_{p-1}, i_p\right) \\ &= t \bigvee_{\{i_p\}_{p=0}^q \in P_{ij}} \bigwedge_{p=1}^{q} \rho\left(i_{p-1}, i_p\right) + (1-t) \bigvee_{\{i_p\}_{p=0}^q \in P_{ij}} \bigwedge_{p=1}^{q} \rho'\left(i_{p-1}, i_p\right) \\ &= t\hat{\rho}(i,j) + (1-t)\hat{\rho'}(i,j)\end{aligned}$$

These equalities are true because ρ, ρ' are comonotone and then the minimum is obtained for the same $\rho\left(i_{p-1}, i_p\right)$ and the maximum is obtained by the same sequence $\left\{i_p\right\}_{p=0}^q$. Moreover, it is easy to check that $\hat{\rho}$ and $\hat{\rho'}$ are also comonotone. □

Finally we study a fuzzy coercive value using the coercive game given in Definition 6.13.

Definition 6.18 Let pl be an extension for fuzzy coalitions. The local pl-coercive value is a value for games with fuzzy permission structure defined for each game v and fuzzy permission structure ρ as

$$\bar{\delta}^{pl}(v, \rho) = \phi(\bar{v}_{pl}^{\rho}).$$

Example 6.19 We look again at Fig. 6.10 as a coercive structure. Table 6.13 determine the ch-coercive game, thus the local ch-coercive value is

$$\bar{\delta}^{ch}(v, \rho) = (2.041, 1.56, 2.51, 2.79)$$

We follow Huettner and Wiese [5] to axiomatize the new value for games with reflexive fuzzy permission structure.

Fuzzy coercion power over sufficient players. Let $v \in \mathcal{G}_m^N$ be a monotone game and $\rho \in FA_0^N$ with constant z. If $i \in N$ is a sufficient player for v then $f_k(v, r) \geq f_j(v, r)$ for all $j \in N$ and $\rho(k, i) = z$.

The proof of the following theorem is similar to that in Theorem 6.6.

Table 6.13 Game $\bar{v}^{\rho}_{ch}$

S	$\{1\}$	$\{2\}$	$\{3\}$	$\{4\}$	$\{1,2\}$
$\bar{\sigma}^{\rho}(S)$	$(1, 0.5, 0, 0)$	$(0, 1, 0, 0)$	$(0.7, 0.4, 0.2, 0)$	$(0, 0, 0, 0.7)$	$(1, 1, 0, 0)$
$\bar{v}^{\rho}_{ch}(S)$	2.5	1	2.9	0.7	4
S	$\{1,3\}$	$\{1,4\}$	$\{2,3\}$	$\{2,4\}$	$\{3,4\}$
$\bar{\sigma}^{\rho}(S)$	$(1, 0.5, 0.2, 0)$	$(1, 0.5, 0, 0.7)$	$(0.7, 1, 0.2, 0)$	$(0, 1, 0, 0.7)$	$(0.7, 0.4, 0.2, 0.7)$
$\bar{v}^{\rho}_{ch}(S)$	3.5	5.6	4.1	3.1	6.5
S	$\{1,2,3\}$	$\{1,2,4\}$	$\{1,3,4\}$	$\{2,3,4\}$	N
$\bar{\sigma}^{\rho}(S)$	$(1, 1, 0.2, 0)$	$(1, 1, 0, 0.7)$	$(1, 0.5, 0.2, 0.7)$	$(0.7, 1, 0.2, 0.7)$	$(1, 1, 0.2, 0.7)$
$\bar{v}^{\rho}_{ch}(S)$	5	7.5	7	8	8.9

Theorem 6.8 *The local ch-coercive value is the only value for games over N with z-reflexive fuzzy permission structure satisfying efficiency, fuzzy inessential player, fuzzy coercion power over sufficient player, linearity and comonotonicity.*

Proof We test that the local coercive value verifies the four axioms. Suppose ρ a reflexive fuzzy permission structure.
EFFICIENCY. Let $v \in \mathcal{G}^N$, as the fuzzy permission structure is z-reflexive $\tau^{\rho} = ze^{N^{\rho}}$ again. By efficiency of the Shapley value (Proposition 1.9) we get by property (C7) of the Choquet integral,

$$\sum_{i \in N} \bar{\delta}^{ch}_i(v, \rho) = \sum_{i \in N} \phi_i(v^{\rho}_{ch}) = \int ze^{N^{\rho}}\, dv = zv(N^{\rho}).$$

FUZZY INESSENTIAL PLAYER. We have

$$\{\rho(j,i)\}_{(i,j) \in N \times N} \cup \{0\} = \{0 = t_0 < \cdots < t_m = z\} \supseteq im_0(\bar{\sigma}^{\rho}(S))$$

for all coalition S. In fact, if $t \in im(\bar{\sigma}^{\rho}(S))$ then there exist $j \in S$ with $t = \rho(j,i)$. Property (C10) of the Choquet integral implies again that

$$\int_c \bar{\sigma}^{\rho}(S)\, dv = \sum_{p=1}^{m} (t_p - t_{p-1}) v\left([\bar{\sigma}^{\rho}(S)]_{t_p}\right).$$

Let $i \in N$ such that j is a null player in a game v when $\rho(i,j) > 0$. We will prove that i is a null player in $\bar{v}^{\rho}_{ch}$. We take S with $i \in S$. Fix t_p, with $p = 1, \ldots, m$. Player $j \in [\bar{\sigma}^{\rho}(S)]_{t_p} \setminus [\bar{\sigma}^{\rho}(S \setminus \{i\})]_{t_p}$ if and only if $\rho(i,j) \geq t_p > 0$ from Definition 6.13.

Hence j is a null player in v. As in the crisp case (Theorem 6.1), we get

$$v\left(\left[\bar{\sigma}^{\rho}(S)\right]_{t_p}\right) = v\left(\left[\bar{\sigma}^{\rho}(S \setminus \{i\})\right]_{t_p}\right)$$

and finally $\bar{v}^{\rho}_{ch}(S) = \bar{v}^{\rho}_{ch}(S \setminus \{i\})$. Shapley value satisfies null player, therefore $\bar{\delta}^{ch}_i(v, \rho) = 0$.

FUZZY COERCION POWER OVER SUFFICIENT PLAYERS. We define game $w \in \mathcal{G}^N$ by

$$w(S) = zv(N^{\rho}) - \int_c \left[ze^{N^{\rho}} - \sigma^{\rho}(S)\right] dv$$

for each coalition S. We prove that if $v \in \mathcal{G}^N_m$ then $w \in \mathcal{G}^N_m$. Let $T \subseteq S$. Proposition 6.9 implies $\sigma^{\rho}(T) \subseteq \sigma^{\rho}(S)$ and then $ze^{N^{\rho}} - \sigma^{\rho}(S) \leq ze^{N^{\rho}} - \sigma^{\rho}(S)$. As v is monotone we can use property (C8) of the Choquet integral (Sect. 2.1) to obtain

$$\int_c [ze^{N^{\rho}} - \sigma^{\rho}(S)] \leq \int_c [ze^{N^{\rho}} - \sigma^{\rho}(T)].$$

We get

$$w(T) = zv(N^{\rho}) - \int_c [ze^{N^{\rho}} - \sigma^{\rho}(S)]\, dv \leq zv(N^{\rho}) - \int_c [ze^{N^{\rho}} - \sigma^{\rho}(T)] = w(S).$$

Let $k \in N$ be a fuzzy sufficient player in v and i with $\rho(i, k) = z$. We prove now that i is a necessary player in w. If we consider $S \subseteq N \setminus \{i\}$, we have that $ze^{\{k\}} \leq ze^{N^{\rho}} - \sigma^{\rho}(S)$ because $\sigma^{\rho}_k(S) = 0$ by Definition 6.10. So, using (C7) and (C8),

$$\begin{aligned} 0 \leq w(S) &= zv(N^{\rho}) - \int_c [ze^{N^{\rho}} - \sigma^{\rho}(S)] \\ &\leq zv(\{k\}) - \int_c [ze^{N^{\rho}} - \sigma^{\rho}(S)] = \int_c ze^{\{k\}}\, dv - \int_c [ze^{N^{\rho}} - \sigma^{\rho}(S)] \leq 0. \end{aligned}$$

Hence $w(S) = 0$. Observe that Proposition 6.12 says $\left(\bar{v}^{\rho}_{ch}\right)^{dual} = w$. Propositions 1.15 and 1.11 imply

$$\bar{\delta}^{ch}_i(v, \rho) = \phi_i(\bar{v}^{\rho}_{ch}) = \phi_i\left(\left(\bar{v}^{\rho}_{ch}\right)^{dual}\right) = \phi_i(w) \geq \phi_j(w) = \bar{\delta}^{ch}_j(v, \rho)$$

for all $j \in N$.

LINEARITY. The result is easy and similar to Theorem 6.6.

COMONOTONICITY. Let $t \in [0, 1]$. Let also $\rho, \rho' \in FA^N_0$ comonotone with $\tau^{\rho} = \tau^{\rho'}$. For any coalition S and $i \in N$ we obtain the same $k_i \notin S$ satisfying

$$\bigvee_{k \in N \setminus S} \rho(k, i) = \rho(k_i, i) \text{ and } \bigvee_{k \in N \setminus S} \rho'(k, i) = \rho'(k_i, i).$$

Thus fuzzy coalitions $\bar{\sigma}^{\rho}(S), \bar{\sigma}^{\rho'}(S)$ are comonotone (and then also $t\sigma^{\rho}(S)$, $(1-t)\sigma^{\rho'}(S)$). Furthermore it holds

$$\bar{\sigma}^{t\rho+(1-t)\rho'} = t\bar{\sigma}^{\rho} + (1-t)\bar{\sigma}^{\rho'},$$

because we obtain for all $S \subseteq N$

$$\begin{aligned} t \bigvee_{j\in N\setminus S} \rho(j,i) + (1-t) \bigvee_{j\in N\setminus S} \rho'(j,i) &= t\rho(k_i,i) + (1-t)\rho'(k_i,i) \\ &= \bigvee_{j\in N\setminus S} \left[t\rho(j,i) + (1-t)\rho'(j,i)\right]. \end{aligned}$$

We use properties (C4) and (C5) of the Choquet integral obtaining

$$\begin{aligned} \bar{v}_{ch}^{t\rho+(1-t)\rho'}(S) = \int_c \bar{\sigma}^{t\rho+(1-t)\rho'}(S)\,dv &= \int_c t\bar{\sigma}^{\rho}(S) + (1-t)\bar{\sigma}^{\rho'}(S)\,dv \\ = t\int_c \bar{\sigma}^{\rho}(S)\,dv + (1-t)\int_c \bar{\sigma}^{\rho'}(S)\,dv &= t\bar{v}_{ch}^{\rho}(S) + (1-t)\bar{v}_{ch}^{\rho'}(S). \end{aligned}$$

Consider f a value for games with fuzzy z-reflexive permission structure satisfying the axioms. Suppose $\rho \in FA_0^N$. From linearity we only have to obtain the uniqueness for the family of anyone games (6.6) w_T with T non-empty set. Let $T \subseteq N$ be a non-empty coalition. We follow by induction on $|im(\rho)|$. If $|im(\rho)| = 1$ then $im(\rho) = \{z\}$ because ρ is z-reflexive. Hence, $\rho \in A^N$. Players out of T are null in w_T. We denote as

$$R_T = \{k \in N : \rho(k,i) = z \text{ for some } i \in T\}$$

Players in T are sufficient in the anyone game. Coercion power over sufficient players implies that all the players in R_T must have the same payoff, K. We apply efficiency to get

$$\sum_{i\in N} f_i(w_T,\rho) = |R_T|K = w_T(N) = 1.$$

So, we obtain $K = 1/|R_T|$. Suppose true the uniqueness when $|im(\rho)| < d$ with $d > 1$. Let $\rho \in FA_0^N$ with $|im(\rho)| = d$. Consider as in Theorem 6.7 the fuzzy permission structures ρ_1, ρ_2 which are comonotone, and $|im(\rho_1)|, |im(\rho_2)| < d$. □

Another interesting extension studied in Chap. 2 is the multilinear one. But remember that this extension has a probabilistic sense, and then we should change the definition of fuzzy sovereign part. If we suppose $\rho \in FA_0^N$ then we interpret $\rho(i,j)$ as the probability of player j depends on player i. So, for all coalition S we can define

$$\eta_i^{\rho}(S) = \prod_{j\notin S}(1-\rho(j,i)),$$

namely, the probability of player i does not depend on players out of S. There exists one option where both concept of sovereign parts coincide, taking ρ as a forest of rooted tree.

Definition 6.19 A fuzzy permission structure ρ is named hierarchical if it is reflexive and for each $i \in N$ there is at most one $j \in N \setminus \{i\}$ with $\rho(j,i) > 0$. The family of hierarchical fuzzy permission structures is denoted as FA_1^N.

Let $\rho \in FA_1^N$ and $S \subset N$. If $i \in S$ such there is not $j \in N \setminus \{i\}$ with $\rho(j,i) > 0$ then $\eta_i^\rho(S) = 1 = \sigma_i^\rho(S)$. Otherwise there is only one $j_i \in N \setminus \{i\}$ with $\rho(j_i, i) > 0$. If $j_i \in S$ then $\eta_i^\rho(S) = 1 = \sigma_i^\rho(S)$ but if $j_i \notin S$ then $\eta_i^\rho(S) = 1 - \rho(j_i, i) = \sigma_i^\rho(S)$. Hence

$$\eta^\rho(S) = \sigma^\rho(S).$$

Let $\rho \in FA_1^N$. For all game v the ml-permission structure, following Definition 6.12, is

$$v_{ml}^\rho(S) = v^{ml}(\sigma^\rho(S)) = \sum_{T \subseteq S} \left[\prod_{i \in T} \sigma_i^\rho(S) \prod_{i \in S \setminus T} \sigma_i^\rho(S) \right] v(T).$$

Observe that we can reduce the sum in the above multilinear expression to only coalitions contain in S because otherwise the probability is zero.

Example 6.20 Figure 6.15 shows a hierarchical fuzzy permission structure with three players. We construct the ml-permission game using the multilinear extension (Definition 2.10) of game $v(S) = |S|^2$. For instance, if we take coalition $\{2, 3\}$ the fuzzy sovereign part is $\sigma^\rho(\{2,3\}) = (0, 0.8, 0.6)$, and the worth in the multilinear extension,

$$\begin{aligned} v_{ml}^\rho(\{2,3\}) &= v^{ml}(0, 0.8, 0.6) \\ &= 0.2 \cdot 0.4 v(\emptyset) + 0.8 \cdot 0.4 v(\{2\}) + 0.2 \cdot 0.6 v(\{3\}) + 0.8 \cdot 0.6 v(\{2,3\}) \\ &= 0.32 + 0.12 + 0.48 \cdot 4 = 2.36. \end{aligned}$$

Observe that $v_{ch}^\rho(\{2,3\}) = 0.6v(\{2,3\}) + 0.2v(\{2\}) = 2.8$. Table 6.14 determines all the worths of v_{ml}^ρ. Now we determine the local ml-permission value,

$$\delta^{ml}(v, \rho) = \phi(v_{ml}^\rho) = (3.64, 2.73, 2.63).$$

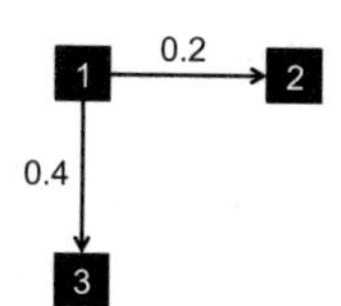

Fig. 6.15 Hierarchical fuzzy permission structure

Table 6.14 Game v^{ρ}_{ml}

S	$\{1\}$	$\{2\}$	$\{3\}$	$\{1,2\}$
$\sigma^{\rho}(S)$	$(1,0,0)$	$(0,0.8,0)$	$(0,0,0.6)$	$(1,1,0)$
$v^{\rho}_{ml}(S)$	1	0.8	0.6	4
S	$\{1,3\}$	$\{2,3\}$	$\{1,2,3\}$	
$\sigma^{\rho}(S)$	$(1,0,1)$	$(0,0.8,0.6)$	$(1,1,1)$	
$v^{\rho}_{ml}(S)$	4	2.36	9	

Next we get an axiomatization of the local ml-permission value. We introduce the following axiom for a value f for games with hierarchical fuzzy permission structure replacing comonotonicity.

Conditional reduction. Let $\rho \in FA^N_1$. If $\rho(i,j) > 0$ then for all game v,

$$f(v,\rho) = \rho(i,j) f(v,\rho^{ij}_1) + (1-\rho(i,j)) f(v,\rho^{ij}_0).$$

where $\rho^{ij}_1 = \rho$ except $\rho^{ij}_1(i,j) = 1$ and $\rho^{ij}_0 = \rho$ except $\rho^{ij}_0(i,j) = 0$.

Theorem 6.9 *The local ml-permission value is the only value for games with hierarchical fuzzy permission value satisfying efficiency, fuzzy inessential player, fuzzy veto power over necessary players, linearity and conditional reduction.*

Proof First we prove that our value satisfies all the axioms. Let $\rho \in FA^N_1$.
EFFICIENCY. As ρ is efficient the $\sigma^{\rho}(N) = e^N$. We obtain

$$\sum_{i\in N} \delta^{ml}_i(v,\rho) = \sum_{i\in N} \phi_i(v^{\rho}_{ml}) = v^{\rho}_{ml}(N) = v^{ml}(e^N) = v(N).$$

FUZZY INESSENTIAL PLAYER. Suppose $i \in N$ such that i and all player j with $\rho(i,j) > 0$ are null players for a game v. We test that i is a null player for v^{ρ}_{ml}. Let S be a coalition with $i \in S$. Let j with $\sigma^{\rho}_j(S) > \sigma^{\rho}_j(S \setminus \{i\})$. There must be $\rho(i,j) > 0$, $\sigma^{\rho}_j(S) = 1$ and $\sigma^{\rho}_j(S \setminus \{i\}) = 1 - \rho(i,j)$. So, j is a null player in v. We denote as

$$M = \{j \in S \setminus \{i\} : \sigma^{\rho}_j(S) > \sigma^{\rho}_j(S \setminus \{i\})\}.$$

If $R \subseteq S$ with $i \in R$ then we take $T = R \setminus \{i\}$ obtaining

$$\prod_{k\in R} \sigma^{\rho}_k(S) \prod_{k\in S\setminus R} (1-\sigma^{\rho}_k(S)) v(R) = \sigma^{\rho}_i(S) \prod_{k\in T} \sigma^{\rho}_k(S) \prod_{k\in (S\setminus\{i\})\setminus T} (1-\sigma^{\rho}_k(S)) v(T).$$

Also for any $T \subseteq S$ with $i \notin T$ we get

$$\prod_{k\in T}\sigma_k^\rho(S)\prod_{k\in S\setminus T}(1-\sigma_k^\rho(S))v(T)=(1-\sigma_i^\rho(S))\prod_{k\in T}\sigma_k^\rho(S)\prod_{k\in(S\setminus\{i\})\setminus T}(1-\sigma_k^\rho(S))v(T).$$

Adding the two above equalities

$$v_{ml}^\rho(S)=\sum_{T\subseteq S\setminus\{i\}}\prod_{k\in T}\sigma_k^\rho(S)\prod_{k\in(S\setminus\{i\})\setminus T}(1-\sigma_k^\rho(S))v(T).$$

Now we use that if $j \in M$ then $\sigma_j^\rho(S) = 1$, hence

$$v_{ml}^\rho(S)=\sum_{M\subseteq T\subseteq S\setminus\{i\}}\prod_{k\in T\setminus M}\sigma_k^\rho(S)\prod_{k\in(S\setminus\{i\})\setminus T}(1-\sigma_k^\rho(S))v(T\setminus M).$$

Now take coalition $S \setminus \{i\}$. Let $T \subseteq S \setminus \{i\}$ with $j \in M \setminus T$. We can give $T \cup \{j\}$ verifying

$$\begin{aligned}&\prod_{k\in T}\sigma_k^\rho(S\setminus\{i\})\prod_{k\in(S\setminus\{i\})\setminus T}(1-\sigma_k^\rho(S\setminus\{i\}))v(T)\\ &=(1-\sigma_j^\rho(S\setminus\{i\}))\prod_{k\in T}\sigma_k^\rho(S\setminus\{i\})\prod_{k\in(S\setminus\{i\})\setminus(T\cup\{j\})}(1-\sigma_k^\rho(\setminus\{i\}))v(T\cup\{j\}).\end{aligned}$$

and

$$\begin{aligned}&\prod_{k\in T\cup\{j\}}\sigma_k^\rho(S\setminus\{i\})\prod_{k\in(S\setminus\{i\})\setminus(T\cup\{j\})}(1-\sigma_k^\rho(S\setminus\{i\}))v(T\cup\{j\})\\ &=\sigma_j^\rho(S\setminus\{i\})\prod_{k\in T}\sigma_k^\rho(S\setminus\{i\})\prod_{k\in(S\setminus\{i\})\setminus(T\cup\{j\})}(1-\sigma_k^\rho(\setminus\{i\}))v(T\cup\{j\}).\end{aligned}$$

If we repeat the reasoning with all the players in $M \setminus T$ and we add all the two obtained expressions we have

$$v_{ml}^\rho(S\setminus\{i\})=\sum_{M\subseteq T\subseteq S\setminus\{i\}}\prod_{k\in T\setminus M}\sigma_k^\rho(S)\prod_{k\in(S\setminus\{i\})\setminus T}(1-\sigma_k^\rho(S))v(T\setminus M).$$

FUZZY VETO POWER OVER NECESSARY PLAYERS. Let j be a necessary player for $v \in \mathcal{G}_m^N$. If $i \in N$ satisfying $\rho(i, j) = 1$ then i is a necessary player for v_{ml}^ρ. In fact, if S is a coalition with $i \notin S$ then $\sigma_j^\rho(S) = 0$. So, as j is necessary for v

$$v_{ml}^\rho(S)=\sum_{j\in T\subseteq S}\prod_{k\in T}\sigma_k^\rho(S)\prod_{k\in S\setminus T}(1-\sigma_k^\rho(S))v(T)=0.$$

LINEARITY. The proof is similar to the others axiomatizations in the section.
CONDITIONAL REDUCTION. Suppose $\rho(i, j) > 0$ with $i \neq j$. Observe that for all coalition S it holds $v^{\rho}_{ml}(S) = v^{\rho^{ij}_1}_{ml}(S) = v^{\rho^{ij}_0}_{ml}(S)$ except if $j \in S$ and $i \notin S$. Let S verifying these conditions, we get $\sigma^{\rho}_j(S) = 1 - \rho(i, j)$ and

$$\begin{aligned} v^{\rho}_{ml}(S) &= \sum_{T \subseteq S} \prod_{k \in T} \sigma^{\rho}_k(S) \prod_{k \in S \setminus T} (1 - \sigma^{\rho}_k(S)) v(T) \\ &= (1 - \rho(i, j)) \sum_{j \in T \subseteq S} \prod_{k \in T \setminus \{j\}} \sigma^{\rho}_k(S) \prod_{k \in S \setminus T} (1 - \sigma^{\rho}_k(S)) v(T) \\ &\quad + \rho(i, j) \sum_{j \notin T \subseteq S} \prod_{k \in T} \sigma^{\rho}_k(S) \prod_{k \in S \setminus (T \cup \{j\})} (1 - \sigma^{\rho}_k(S)) v(T) \\ &= (1 - \rho(i, j)) v^{\rho^{ij}_1}_{ml}(S) + \rho(i, j) v^{\rho^{ij}_0}_{ml}(S). \end{aligned}$$

Last equality is true because we have $\rho^{ij}_1(i, j) = 1$ and $\rho^{ij}_1(i, j) = 0$.

The first part of the uniqueness is the same that in Theorem 6.6, reasoning by induction in this case on the number of links with $\rho(i, j) \in (0, 1)$. Conditional reduction gets the result. □

References

1. Gallardo, J.M., Jiménez, N., Jiménez-Losada, A., Lebrón, E.A.: Games with fuzzy permission structure: a conjunctive approach. Inf. Sci. **278**, 510–519 (2014)
2. Gallardo, J.M., Jiménez, N., Jiménez-Losada, A., Lebrón, E.A.: Games with fuzzy authorization structure. Fuzzy Sets Syst. **272**, 115–125 (2015)
3. Gallardo, J.M.: Values for games with authorization structure. Ph.D. thesis. University of Seville. Spain (2015)
4. Gilles, R.P., Owen, G., van den Brink, R.: Games with permission structures: the conjunctive approach. Int. J. Game Theory **20**(3), 277–293 (1992)
5. Huettner, F., Wiese, H.: The need for permission, the power to enforce, and duality in cooperative games with a hierarchy. Int. Game Theory Rev. **18**, 4 (2016)
6. Myerson, R.B.: Graphs and cooperation in games. Math. Oper. Res. **2**(3), 225–229 (1977)
7. Owen, G.: Values of games with a priori unions. In: Henn, R., Moeschlin, O. (eds.) Mathematical Economics and Game Theory. Lecture Notes in Economics and Mathematical Systems, vol. 141, pp. 76–88. Springer, Berlin (1982)
8. van den Brink, R.: Relational power in hierarchical organizations. Ph.D. thesis. Tilburg University. The Netherlands (1994)
9. van den Brink, R.: An axiomatization of the disjunctive permission value for games with a permission structure. Int. J. Game Theory **26**, 27–43 (1997)
10. van den Brink, R., Gilles, R.P.: Axiomatizations of the conjunctive permission value for games with permission structures. Games Econ. Behav. **12**, 113–126 (1996)
11. van den Brink, R., Dietz, C.: Games with a local permission structure: separation of authorization and value generation. Theory Decision **76**(3), 343–361 (2014)

Index

A. Jiménez-Losada, *Models for Cooperative Games with Fuzzy Relations among the Agents*, Studies in Fuzziness and Soft Computing 355,
DOI 10.1007/978-3-319-56472-2

Printed in the United States
By Bookmasters